The Planetary Emerg

ALSO BY KENT D. SHIFFERD

# The Planetary Emergency

*Environmental Collapse and
the Promise of Ecocivilization*

KENT D. SHIFFERD

McFarland & Company, Inc., Publishers
*Jefferson, North Carolina*

Library of Congress Cataloguing-in-Publication Data

Names: Shifferd, Kent D., 1940– author.
Title: The planetary emergency : environmental collapse and the promise
of ecocivilization / Kent D. Shifferd.
Description: Jefferson, North Carolina : McFarland & Company, Inc.,
Publishers, 2021 | Includes bibliographical references and index.
Identifiers: LCCN 2020055482 | ISBN 9781476683461
(paperback : acid free paper) ∞
ISBN 9781476641287 (ebook)
Subjects: LCSH: Environmental degradation. | Global warming.
Classification: LCC GE140 .S485 2021 | DDC 304.2/8—dc23
LC record available at https://lccn.loc.gov/2020055482

British Library cataloguing data are available

ISBN (print) 978-1-4766-8346-1
ISBN (ebook) 978-1-4766-4128-7

Front cover: Background image by Vladimir Salman;
(inset) Bosco Verticale towers in Milan, Italy (Ivan Kurmyshov)

Printed in the United States of America

*McFarland & Company, Inc., Publishers*
*Box 611, Jefferson, North Carolina 28640*
*www.mcfarlandpub.com*

For the women who have taught me
the most important things I know

Madge Drummond
Francis Drummond Shifferd
Patricia Allen Shifferd
Sania Drummond Shifferd
Sarah Katherine Shifferd

# Table of Contents

# Introduction

## The Revolution of Hypercivilization

"In effect, the human race has entered into a great wager.
We are, so to speak, betting the planet."[1]—Charles C. Mann

We are on the brink of a comprehensive environmental and social catastrophe, a planetary scale emergency that involves much more than the disastrous effects of unchecked global warming. Humanity faces a stark choice—breakdown, if we go on as we are now, or fundamental change and breakthrough to Ecocivilization.

## The Biosphere

Planet Earth is a biosphere, a zone of life three billion eight hundred million years in the making, a web of complicated organisms and complex ecosystems developed in response to their surroundings, resulting in the living planet on which civilization depends for its life support. It's an extremely rare thing in the universe, a miracle on a dead rock, and possibly the only one. It is more a process than a thing, in which myriad parts exchange materials and energy with each other along the subtle pathways that make up the fantastically complicated web of life. We can only live within this thin zone of water, air and soil that envelops the planet.

It is only a few miles deep. For terrestrial plants and animals the natural limits are up around six or seven thousand meters above sea level. And while some very primitive creatures live at a depth of six thousand meters below the surface of the sea, the overwhelming portion of sea life lives in shallow areas along the continental shelves. A cross-section of the globe would show that the entire biosphere occupies only the thinnest, outermost skin comprising less than one percent of the whole diameter, like the skin of an apple. It evolved over eons, going through stages that are unrepeatable. We are our history. "Things are as they are because they were as they were."[2] If we undo it we can't go back and start over. We are the product of the past and must conform to its patterns and laws. We are not free.

Earth did not start as a living planet. It was born approximately 4.5 billion years ago. Early earth was an environment entirely alien to life, a planet whose surface was "gloomy, glowing with the splashes and eruptions of the red-hot magma ocean, covered only by a thin crust of rock under a dark, glowering sky."[3] We can think of this span of time as being placed on a calendar of forty-five days. Each day then represents one hundred million years. About ten days, or a billion years, went by before life emerged. Another eighteen days, 1.8 billion years, went by before the oxygen revolution resulted in life as we know it. Mammals did not appear until the last day and we humans only in the last hour, and our vaunted industrial civilization only in the last minute.[4]

## *The Miracle: Life*

Life began simply at first, but the molecules of living matter that eventually developed were far more complex than any inert molecules. And then in a rush, geologically speaking, it tumbled along the evolutionary path in myriad forms, flowering eventually into a consciousness of itself and of the universe. What evolved in this stunning process of emergent creation was not simply a panoply of living creatures, but a whole interlinked system of life. The planet metamorphosed. It became not merely a planet with life on it, but a living planet, a biotic planet, a biosphere.

The patterns of evolutionary branching have been interrupted by catastrophes that eliminated whole sections of the tree of life, leaving only a few twigs from which the branching then started again. This has implications for our own species. "We are," as Loren Eiseley wrote, "one of the many appearances of the thing called Life; we are not its perfect image, for it has no image except Life, and Life is multitudinous and emergent in the stream of time."[5] And, as William Howels has pointed out, there would be no starting over if, "supposing in a moment of idiot progress, we killed ourselves off."[6]

The biosphere is vastly more complex than can be described even in whole volumes if, in fact, we understood it perfectly, which we do not. It is composed of countless feedback cycles and highly subtle interconnections, and being so made acts as an infinitely complex web through which radiate all manner of effects. Every ecosystem, every individual, every cell is a through-put point. All boundaries are permeable. What we put into the environment gets into us. Furthermore, it is light, chemically speaking. Of the more than one hundred elements in the periodic table, living beings are principally made up of about six: carbon, hydrogen, nitrogen, oxygen, sulphur and phosphorus. These are among the lightest elements. While it is true that living creatures need trace elements of some of the heavier materials, such as iron and calcium, the biosphere is mostly light. The much heavier elements play havoc with life. Lead, arsenic, mercury, cadmium, uranium and plutonium—all have a tendency to depress or destroy living systems. By and large, life was shielded from these substances during nearly all of the evolutionary history until the advent of Hypercivilization. Viewed from space, the other bodies of solar system are lifeless—the dead rock of the moon, overheated Venus, Jupiter locked in ice. It is different here. "Out of a single molten planet the hummingbirds and pterodactyls and gray whales were all woven."[7] Earth viewed from space is a blue-green miracle.

## *The Revolution of Hypercivilization*

We are now threatening all of that. In a brief geological moment, the last 300 years, we have created Hypercivilization, a powerfully destructive way of interacting with nature that is radically altering and simplifying planetary ecosystems at an ever-accelerating rate.

Hypercivilization is a supreme discontinuity. It differs from all previous civilizations by both quantitative and qualitative revolutions, producing overwhelmingly more of everything, and introducing novel kinds of substances into the biosphere for which life has no evolutionary preparation or defense. It is characterized by an unprecedented overreach in population, energy capture and dispersion, rapid urbanization, and a chemical revolution, all leading to the toxification of the biosphere, massive habitat loss, extinctions, desertification, environmental diseases, food shortages and climate change. In the last 300 years, our technical reach has leapt into the stars and descended into the heart of the atom and the gene. We have changed the conditions in which life evolved and upon which it is dependent. Neither we humans nor the Earth has ever been here before.

Hypercivilization is a greatly exaggerated, globalized, and intensified form of civilization, a radical departure from both the evolutionary and the cultural past. It began to emerge first in the Western mind with a revolution in beliefs and values around 1600 CE, and then materialized in a wave of new institutions, the technologies of industrialization and the population explosion. Today, billions of hands are literally tearing at the web of life. Hypercivilization was firmly entrenched in Western Europe by 1900 CE and in the twentieth century it spread like a tidal wave over the rest of the Earth, adding the new technologies of nuclear fission and biogenics. Its main impact on Earth's life support system is destructive. In Hypercivilization, the good life is defined as acquiring ever more material things by pursuing endless economic growth at all costs. It depends on extreme rates of extraction of minerals and fossil fuels. All of this is called "Progress," in spite of the fact that most negative impacts on humans and nature are externalized from its economic system, to be assessed against us and our children for generations. Chemical pollution, deforestation, drought, erosion, extinctions, crowding, a deteriorating climate, and consequent social ills such as industrialized warfare and extreme poverty became normative. Seen in historical perspective, Hypercivilization burst upon the earth and trashed it in a comparatively few moments of evolutionary time. But from our limited perspective in the present, it was a long time in coming.

I don't mean to discount the wonderful, life-giving and enriching aspects of the modern world. None of us would give up anesthesia or all the rest of modern medicine. The advance of literacy is miraculous, and the internet has made more knowledge available to more people more rapidly than ever before. The world is more democratic. But all this is beside the point if it all comes crashing down.

Our species, *Homo sapiens*, has become a threat to the planet. Consider the definition of an invasive species: "an alien species whose introduction does or is likely to cause economic or environmental harm or harm to human health."[8] An invasive species spreads rapidly and out-competes indigenous species. In many cases, its population then overshoots and collapses. Are we the most dangerous threat to ourselves and thousands of other inhabitants of the planet? It does not have to be this way. I am not a misanthrope. I think we have a rightful place on this planet and have great potential, but at this historical moment we are way off course. Chapters 1 through 5 demonstrate this, taking us on a quick tour of the planetary emergency. The evidence is overwhelming. Chapter 6 looks at the prospect of collapse. In Part II I explain how it came to this, tracing our history from pre-agricultural times, and then the early agricultural civilizations when the foundations were laid for Hypercivilization, up through its emergence in scientific materialism, capitalism, the European conquest of the planet, and the industrial revolutions. In Part III I lay out a path forward to Ecocivilization.

## Our Myopia

Why do we not all see this? Why do we go on preparing for twentieth-century careers? First, our personal time scale—our own lifetime—is too short, so everything seems normal. It worked for our grandparents and it works for us, unless you live in Haiti or the slums of Milwaukee. We are like that mythical frog placed in a pan of water that is ever so gradually heating until it kills him. Placed in the hot water at the start, it would jump out, but the slow development of Hypercivilization *in relation to a single human life* is like the slowly heating pan of water. Second, in spite of all the advances of our science, we have failed to fully understand the fundamental nature of the biosphere. This ignorance is leading us to disaster. Hypercivilization is literally a dead end.

We act as if we don't understand four things. First, when we make our individual and social choices we fail to appreciate that civilization exists *within* the natural world. It is a small circle within a larger circle from which it draws its life. Human society is a throughput operation. Everything comes from somewhere and goes somewhere.

When the smaller circle gets too large, it destroys the capacity of the larger circle to sustain it. Then it collapses. The natural capital upon which all economic capital rests—breathable air, water, soil, stable climate and biodiversity—are depleted or destroyed outright. Second, we don't stop to appreciate that nature operates according to immutable laws which, while

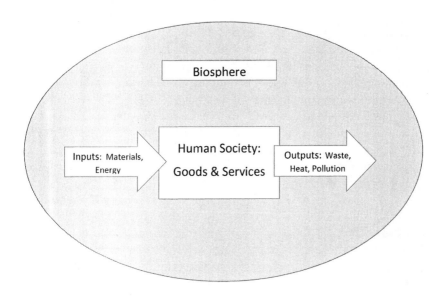

**Throughput. Every civilization draws materials and energy from the closed system of the biosphere, processes them into goods and services and ejects the wastes. What makes Hypercivilization unprecedented is the volume and rate of toxic and non-biodegradable wastes and that they are ejected without thought for the damage caused to the biosphere's life support systems.**

they can never be made to cease operating, can be ignored for a time but only with inevitable consequences. A man may jump from a tall building and ignore the law of gravity, but only for a moment. On a geological time scale, that moment is the era of Hypercivilization. Third, the problem is that we do not understand this geological time scale. Only in perceiving and acknowledging the immensity of the time scale can we perceive what a radical departure Hypercivilization is from all prior natural and human history. Think of it this way: Look at these immense time periods as a 24-hour clock. The Big Bang occurred at 12:01 a.m. Life on Earth did not appear until 8:00 p.m. At 10:30 the first vertebrates appeared, at 11:35 the dinosaurs, and at 11:59:50—ten seconds before now—humans! Hypercivilization occurs in the last .0001 second. Hypercivilization has come about in a mere eye-blink of Earth time.

A fourth problem is our failure to understand exponential growth, or doubling. In ten cycles of linear growth, as in the production of widgets, you produce two on the first day and one more each day after that, as in 2, 3, 4, 5, 6, 7, 8, 9, 10, 11. We get a total of 66 widgets at the end of ten days. But in ten cycles of doubling, as in 1, 2, 4, 6, 8, 16, 32, 64, 128, 256, 512, we get an astonishingly high number—1,023. When resource use grows at just seven percent per year, the doubling time is ten years, so in a century we would get ten doublings. Consider another analogy, a pond with water lilies growing in it. "The number of lilies double each day. If they were allowed to grow unchecked, they would completely cover the pond in 30 days, choking out all other forms of life in the water. For a long time, they seem small, so you don't worry about cutting them back until they cover half the pond. How much time will you have to avert disaster, once they cross your threshold for action?"[9] Just one more day. After that, it's too late. If resource extraction is doubling, and that has taken, say, 29 years to use up half the resource, the remaining half is gone after just one more year. Population, resource demand, waste production all grow exponentially in the state of Hypercivilization.

A few observers have noticed this. Susan Griffin has written that this culture has at last evolved "into an oddly ephemeral kind of giant, an electronic behemoth, busily feeding on the world,"[10] Brian Swimme wrote that we live in "a culture that congratulates itself as it commits geocide."[11] In Daniel Quinn's provocative book, *Ishmael,* the author has the sage say, "My subject is *captivity.* … You're the captive of a civilizational system that more or less compels you to go on destroying the world in order to live."[12] Rachel Carson's 1962 book, *Silent Spring*, had its day and then was largely forgotten. Paul Ehrlich's 1968 book, *The Population Bomb*, had its little day and then was forgotten. Donella and Dennis Meadows 1972 report to the Club of Rome, titled *The Limits to Growth*, was forgotten. E.F. Schumacher's 1973 classic, *Small Is Beautiful: Economics as If People Mattered*, was forgotten, as were others. This book puts together the whole picture, sees all of the crises—depletion, population, pollution, climate shift, extinctions, erosion, etc.—as the single crisis of Hypercivilization.

How shall we get this message across to all who think we are living in normal times? We can say that we are on a runaway train heading for the chasm where the bridge is out and there is no one in the engine. Or we can say it resembles one of its own huge oil tankers whose momentum is so great that it travels on for miles after the command has been given to reverse engines, but the here the analogy fails because no such command has been given, at least not from the bridge. If *Homo sapiens* does not become in fact wise, as our self-designated Latin name implies, there will be a new Dark Age made more severe by the desperate overcrowding of the planet.

In the past, there have been regional collapses—whole societies have driven themselves into extinction. The collapse I see ahead, if we do not radically alter course, will be global. We can now see the end of the road and it's ugly. This book is meant to be a warning. "Never send to know for whom the bell tolls; it tolls for thee."[13] But it is only by understanding where and how we went wrong as a species that we can find the right path out of this predicament and move on toward some kind of sustainable future. We need a new story. We need to learn both humility and a whole new, intelligent way of looking at "us," that is, at the single system in which humanity and nature are so obviously one.

Are we without hope? I hope not. I wrote this book to change our view of history, of where we have come from and where we are now and, above all, the choices we have about where we are going. If anything can save us it is the possibility that we will use the intelligence which nature has evolved in us. We will actually become *Homo sapiens*.

The concluding chapters provide many examples of the wonderful and encouraging work that is going on as we begin to shift away from Hypercivilization to Ecocivilization. We are already moving forward toward a different and far more sustainable way of life. In these chapters we examine our changing mind set about nature and our place in it, and new ways of doing education, energy, religion, communities, and international security.

# PART I

## BREAKDOWN
### *A Brief Tour Through the Global Environmental Crisis*

# 1

## The Perfect Storm

### A Warning from 15,000 Scientists

Hypercivilization created crises in the areas of economic growth, population overshoot, food production, urbanization, nature deficit disorder, over-consumption, chemical pollution, and extinctions, and this is only a partial list. In November of 2017 an alarming warning was published in *BioScience*, a leading scientific journal. It was titled "15,000 Scientists from 184 Countries Warn Humanity of Environmental Catastrophe,"[1] urging "global leaders to save the planet from environmental disaster. Signers included Jane Goodall, E.O. Wilson, and James Hansen. It was the scientists' 'second notice.'" The first, issued in 1992 and signed by only 1,700 scientists, began: "Human beings and the natural world are on a collision course." They noted the growing hole in the ozone layer, pollution, fresh water depletion, "overfishing, deforestation, plummeting wildlife populations, as well as unsustainable rises in greenhouse gas emissions, global temperatures and human population levels." Other than the ozone layer, fixed by the Montreal Protocol, things have gotten much worse, including "a 28.9 percent reduction of vertebrate wildlife, a 62.1 percent increase in $CO_2$ emissions, a 167.6 percent rise in global average annual temperature change and a 35.5 percent increase in the global population." They are adamant that time is running out and are urging leaders "take immediate action as a moral imperative to current and future generations of human and other life." They warn that "We are jeopardizing our future by not reining in our intense but geographically and demographically uneven material consumption and by not perceiving continued rapid population growth as a primary driver behind many ecological and even societal threats."

# A Mythology of Infinite Economic Growth on a Finite Planet

When we talk about the economy, we are talking about the productive interchange with nature—i.e., extracting materials and processing them into goods and unwanted by-products which are thrown "away" along with the goods when those wear out. As Senator Gaylord Nelson said, "The economy is a wholly owned subsidiary of the environment, not the other way around,"[2] a fact that most economists ignore. From the middle of the eighteenth century to the present, the world has experienced unprecedented economic growth, perhaps nowhere more dramatic than in the United States. "In 2015—236 years after independence—GDP per capita has increased more than 26-fold to $52,706. This means that the output per person in one year in the past was the same as the output per person in two weeks today."[3] And since the population has grown from 2.5 million to 326 million, we need to multiply that times 130 in order to gauge the impact on the environment. Because of such dramatic historic growth in extraction and throughput worldwide, people have come to believe that it is normal and unending.

Economic growth came to be seen as the solution for nearly all social, political and even spiritual problems, especially poverty. But for the developing countries to achieve the same levels of extraction, production and discard is impossible. Resources will run out, and in the attempt, we will severely damage the life support system of the biosphere long before that happens. Nevertheless, politicians make further growth their most powerful campaign promise and differ only over how to achieve it. In this there was no difference between command economies like the Soviet Union and the market economies of the West. Growth came to be the sole criterion of social good—more people with more things. Consumerism became like a religion. In 1955, economist Victor Lebow published a prophetic essay in the *Journal of Retailing*, arguing that "Our enormously productive economy demands that we make consumption our way of life, that we convert the buying and use of goods into rituals, that we seek our spiritual and our ego satisfaction in consumption. We need things consumed, burned up, worn out, replaced and discarded at an ever-increasing rate."[4] Unfortunately, they counted only the goods and discounted the negative side effects on society, including the correlation between continued growth and undemocratic levels of inequality,[5] as well as upon nature. Economists even developed an accounting system based on the dollar value of all transactions that included some of these negative impacts, but counted them as positives—for example, the cost of cleaning up pollution, or treating pollution-induced cancers—as part of the Gross Domestic Product. Any rise in that number was considered progress. They became so mystified by the concept that a decline in productivity or GDP was nevertheless given the oxymoron, "negative growth." Growth became an end in itself, an unquestioned good.

Infinite growth is unsustainable because civilization is a little world within the larger world of nature, upon which it depends for air, water, soil, stable climate, etc. It can only expand so far before it consumes all the resources or fouls the self-cleansing mechanisms of nature. And long before that, conflict over the remaining, scarce resources disorganizes civilization and threatens it with collapse. It is self-defeating behavior, and such a defeat will mean poverty, mass migrations, violent conflict, disease, and starvation. If continued, unthinking growth is a threat to the well-being of all, even the rich. Unsustainable growth is the common trend in all the manifestations of Hypercivilization, from food production to warfare. Fortunately, there are ways out of this bind, as we shall see in the last section of this book.

## Population Explosion

In the era of Hypercivilization the human population swarmed. It took over three million years for hominids to evolve and to reach the number

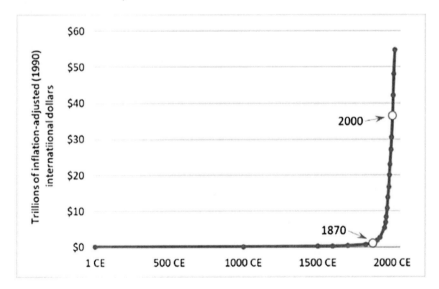

Global Gross World Product, 1 CE to 2015 CE. This graph and the following one (Human Population Timeline) is the famous j-curve that illustrates the sudden and explosive growth that has characterized Hypercivilization and its destructive impact on the natural world (courtesy Darrin Qualman).

one billion, a mark finally achieved around 1825. It had taken 123 years to increase from 1 billion to 2 billion. The next increments of a billion took 33 years, 14, years and 13 years, respectively. The transition from the fifth to the sixth has already occurred. The chart below shows the explosion of population in the era of Hypercivilization.[6]

As of this writing the world population is 7.6 billion, up from 1.7 billion in 1900. Millions more will have been added by the time you are reading this, and the UN estimates it will hit 11.2 billion by 2100.[7] Very small annual increases can add up to very short doubling times. If the annual

# WORLD POPULATION

## Past, Present and Future

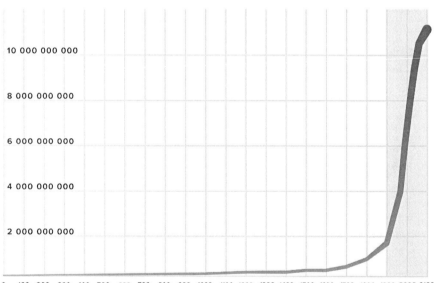

**Population Explosion: It took all of human history to reach a population of about a billion by 1825 when it took off, doubling by 1858, and again by 1872, and again by 1885, arriving at 6 billion, and as of April 2019 it had reached 7.7 billion. Coupled with rising per capita impact resulting from exponential economic growth, some experts believe we have already exceeded the long-term carrying capacity of the Earth (courtesy Ken Jorgustin).**

increase is only half of one percent (.5 percent), the doubling time is 135 years. When the rate of increase goes to one percent, it falls to seventy years and at two percent it falls to 35 years. A three percent rate of increase yields a doubling time of 23 years and a population and food production nightmare. The portent for the future is ominous. Nigeria, for example, had 189.5 million people in 2017. If the growth rate continues unchanged at its present 2.9 percent per year, in less than a century Nigeria will have to feed 1.8 billion people, a total impossibility.[8] Furthermore, annual rates of increase add up to surprisingly large percentage increases over the course of a century. At a one percent annual rate of increase, the absolute population will have increased 270 percent. A two percent annual increase yields a 724 percent overall increase and a three percent, a 1,922 percent increase.[9] Some scientists already consider the "current human population growth an 'eco-pathological process' that is out of control and injuring the earth."[10] The explosive growth of population in the era of Hypercivilization is a great discontinuity with all of our past, placing severe pressure on the environment, especially in the production of food, and on soil and water.

## Food Production in Hypercivilization

Hypercivilization brought immense and unprecedented changes in the ways of food production, preparation, and distribution. These included the industrialization of agriculture, which involved heavy machinery, massive increases in scale, corporate involvement in the actual farming and in the development and distribution of seeds, genetic manipulation, and the replacement of small-scale animal husbandry with giant, water- and air-polluting Concentrated Animal Feeding Operations. These required high inputs of antibiotics, which resulted in the development of resistant strains of disease organisms and also threatened human health. Another development was the replacement of natural manures with artificial inputs such as nitrogen fertilizer. A massive increase in the amount of wild habitat was given over to food production—especially through the destruction of tropical forests that were replaced with monocultures, resulting in a dramatic decline in biodiversity. Hedgerows and wood lots disappeared as farmers plowed from roadside to roadside, planting one or two crops, which insects and other pests saw as magnificent banquets. Hypercivilization then introduced an array of poisons into the environment with the development of pesticides, which then led to resistant strains of weeds and other "'pests' and were a hazard to human health. It was a growth industry as indicated by this graph of the dollar value of global pesticide production."[11]

Soil nutrients were mined. There was massive erosion of topsoil, silting

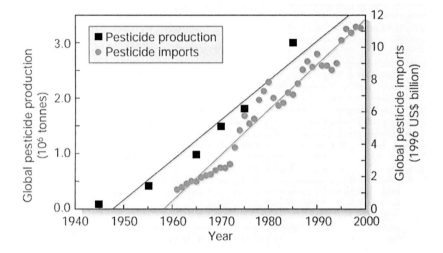

**Growth of Pesticides: Since 1940 pesticide use has doubled more than threefold. While it has leveled off in Europe and the U.S., it has exploded in China and the pesticides have become ever more toxic to wildlife (Tillman et al. [2001], https:// ourworldindata.org/pesticides).**

up rivers, lakes and ocean deltas, and a salinization of many over-irrigated lands, reducing or eliminating production altogether. In social terms there was a massive exodus from the land, divorcing people from an experiential knowledge of nature while destroying rural communities and contributing to the explosive growth of cities. In the U.S., Earl Butz, Secretary of Agriculture, 1971–76, said "Get big or get out" and urged farmers to plant "fence row to fence row."[12] Paul Gruchow, who lived through the exodus, described it:

> In the decade of my coming of age, millions of farm dwellers left the land and sought new lives in towns and cities, not because that was what they desired but because they had no alternative. This removal constituted one of the greatest mass migrations in history.… If you grew up on a farm in the last fifty years, as I did, and you were at all alert to what was happening there, you could not have missed the steady attrition of all kinds. You would have seen the empty farmhouses, the barns rotting and falling in on themselves.… You would have noticed the diminishing songbirds, the disappearing butterflies, the vanishing potholes, the uprooted fencerows, the balding hilltops.… You would have seen the empty churchyards. Only the cemeteries remained, odd temples of death jutting out of corn fields. You could still be buried in the countryside, but you could no longer be baptized there.[13]

Small, carefully husbanded, mixed farms were amalgamated into large, corporate entities mono-cropping by pouring on the artificial inputs. In third-world countries corporate export agriculture forced locals off the plains and up into the mountains, where they were forced to cut the forests, as in the case of the Selva Lacondona in Mexico. "Today the hills and canyons of the Lacondona are pocked with the bald remnants of abandoned pastures, some of them eroded to bedrock."[14]

While much larger amounts of food were produced, much of it was tainted with chemical additives and pesticide residues. Pesticides also began showing up in surface and ground waters and in human tissues and mother's milk. More and more food was transported over immense distances, as in the absurd example of asparagus being grown in Brazil, shipped to China for processing, and shipped to Europe for consumption, all of which led to increased burning of fossil fuels for transport, contributing, along with the tractors, gas powered grain drying, and harvesting machines, to an acceleration of global warming.

Another feature of Hypercivilization agriculture was an expropriation of food producing land from the poor as first-world corporations and nations bought up traditional farming lands in developing countries. Commercial fishing was also industrialized, which resulted in overharvesting and the collapse and near collapse of many fish stocks, to be replaced with genetically modified fish and shellfish in artificial farms or pens. Similar to land expropriation was an expropriation of vital fish protein from the poor in the developing world, as massive first-world trawlers invaded their traditional coastal fishing areas.

The development of genetically altered foods, whether by traditional breeding methods or laboratory manipulation, resulted in the so-called Green Revolution, which did dramatically increase output per acre. However, the new strains of crops required heavy inputs of fertilizer (a derivative of oil and gas), pesticides, and irrigation that lowered water tables. In some cases it resulted in social upheavals as the better-off farmers who were able to afford the inputs bought up the land of those who were not, and who had no choice but to move into the rural slums in search of paid work so they could buy food they used to grow. Increasing the food supply was a positive feedback to the population explosion, more eaters placing more impact on the biosphere. Another result was the replacement of crops that fed local people with crops grown for the global export market.

In short, the production of food in the era of Hypercivilization was unsustainable and highly damaging to the biosphere's ability to support civilization in the long term. Many low income countries found they were producing less food per capita in 1980 than they had in 1970.[15] Combined with other features of Hypercivilization—which we will shortly examine— it was, if unchecked, leading in the direction of collapse, leaving unskilled billions living in cities where they would not have recourse to food and other necessities should disaster occur.

## Urbanization and Nature Deficit Disorder

For most of history the overwhelming majority of people lived on the land. They knew the seasons, the soils, the limits and the possibilities first-hand. Cities that did exist were small. But part and parcel of Hypercivilization has been an explosion in the number and size of cities.

In the year 1900, less than one person in ten lived in a city—only about two hundred million total. By the 1980s, almost half of the now much larger population of the world lived in cities, about 2.5 billion people. In 1890 there were only nine cities in the world with populations over a million. By 1920 there were 27, and by 1980 there were 230 such cities, including 26 with populations over five million.[16] By 2005 there were 336 and that counted only the official city boundaries and did not include the surrounding urban agglomerations.[17] Already by 1945 Tokyo had grown to twice its 1920 area and its green belt had disappeared entirely.[18] In the strip of megalopolis that by century's end ran from Boston to Washington, D.C., one fourth of the U.S. population lived in cities where the original ecosystems had been totally altered and the habitats were almost entirely artefactual, from the chemistry of the water and air to the land forms and species which inhabited them. In industrialized countries, by the year 2000, 75 to 85 percent of the population lived in cities. This pattern spread into Latin America over the course of the twentieth century. In 1920 it was only fourteen percent urban, but by 1980 the percentage had risen to sixty-six.[19] By 2020 it is estimated to rise to 81 percent. Mexico City achieved a population of eighteen million. Sao Paulo grew to fourteen million, as did Calcutta in India.[20] By 1985, Teheran's population exceeded the entire population of Iran in 1914, with twice as many as in 1800.[21] In Africa, the least urbanized developing region, urban population was growing 5 percent yearly, and by 1987 175 million people lived in cities on that continent, having fled rural poverty and environmental degradation. In Alexandria, Egypt, four million were using a sewer system designed for one million, and parts of the city were "literally awash in raw sewage."[22] Karachi, in Pakistan, was a typical case in point. The population of this city grew from 400,000 in 1947 to nine million in the mid–1990s. There "are a million inhabitants of illegal squatter shanties"; uncollected garbage piles up, streets are flooded for days at a time, unemployment is at 25 percent, and the population is growing at 6 percent annually, homes are often without electricity or drinking water and large areas are controlled by drug barons.[23] Teheran is the fourth-most-polluted city on the planet, trailing only Delhi and Calcutta in India and Beijing in China.

Cities eat the surrounding farmland. They demand vast amounts of energy, especially as suburbanization has occurred. They often have large slums where lead poisoning and air pollution cause sickness. In the developing world they are woefully inadequate in supplying basic necessities like clean water, air and sanitation. The Ganges, the holy river where masses of pilgrims bathe, is the recipient of the raw sewage from 114 towns. In Colombia, the Bogota River downstream of the capital city had a fecal coliform bacterial count of 7.3 million. The safe drinking water limit is 100.[24]

Many who live in cities have little contact with the real world of nature, which produces their food and all their other goods. Young people suffer from nature deficit disorder. Goods appear in stores, are purchased, used, and then discarded, knowing not where nor with what impact, not knowing that the old cell phone ends up in an electronic dump in India, polluting soils and water and endangering the health of the disassemblers.

Millions have been divorced from a daily, direct and productive relationship with the land, effecting a cultural lobotomy such that the source of biological information necessary for humanity to avoid overshoot and collapse was cut off from the only species that had ever needed it. "In 1910, [U.S.] farm population accounted for a third of the total. By 1969 it was a mere twentieth."[25] By century's end, over 95 percent of Americans were gone from the land. In the Hypercivilization city, children in the street gangs in the slums of Sao Paulo and New York, and Yuppie kids in the arcades and shopping malls of Chicago, grew up entirely innocent of any experiential knowledge of the natural world, ignorant of nature as a realm of necessity where violating natural laws leads inevitably to disaster. By century's end, three generations of such people had been raised in the teeming cities of the world, millions who were without a clue that something catastrophic was going wrong with the biosphere.

Richard Louv notes in his book, *Last Child in the Woods*,[26] that there have been immense changes in childhood experience from the 1950s to the present. Now, children are not only severely restricted from free experience in the natural world but that even their educational experiences of nature has changed from direct experience to learning complex abstractions. In the U.S., legal and covenant restrictions on free play, combined with parental fear and an overwhelming imposition of TV, video and computers, are all preventing children from experiencing the natural world first hand. Children are becoming containerized: "They spend more and more time in car seats, high chairs, an even baby seats made expressly for watching TV. When small children do go outside they're often placed in containers—strollers—and pushed by walking or jogging parents."[27] But much of the time kids are indoors plastered to LED screens. "In the U.S., children 6–7 spend 30 hours a week looking at a TV or computer monitor."[28] He also notes that the great environmental leaders of the past all grew up playing freely out of doors, and wonders who will be the leaders of the future. American children, at least, are suffering from "nature deficit disorder."

Some environmentalists argue that cities are good for the environment, keeping people from cluttering up the land, but this argument has two flaws. First, under industrialized agriculture, the land is being abused and its long-term productivity diminished. Second, there is nothing wrong with humans on the land, in small villages and family farms, especially if they are doing mixed farming, permaculture, and other methods that restore the ecosystems, but in the cities, children are learning Lebow's religion of consumption.

## The Growth of Consumption

In the era of Hypercivilization the growth in the consumption of materials and the products into which they are made and the waste they leave behind has been mind-boggling. In the single decade of the 1960s, Americans alone bought 47 million electric hair dryers.[29] They consumed more fossil fuels and minerals in the 50 years after World War II than all the rest of humanity in the entirety of history.[30] In England, toward the end of the twentieth century, people were dumping 2.5 billion so-called disposable diapers into landfills each year. In 1890 Americans were consuming less than 20 kilograms of cement per person, but by 1990 it had risen to about 320 kilograms, and the population of these consumers had more than doubled. Paper consumption per person more than doubled from 1910 to 1990, and steel consumption per person tripled in the hundred years from 1890 to 1990.[31] Nearly all trends were up as the following world data show. Between 1950 and 2000:

- Registered vehicles went from 70 million to 723 million.
- Oil consumption from 3,800 million barrels per year to 27,635.
- Natural gas consumption from 6.5 trillion cubic feet per year to 94.5.
- Coal consumption from 1,400 million metric tons per year to 5,100.
- Electrical generation capacity in millions of kW, from 154 to 3,240.
- Corn was up from 131 million metric tons to 342, wheat from 143 to 356 million metric tons.
- Cotton from 5.4 million metric tons to 12 million.
- Wood pulp from 12 million metric tons to an astonishing 102 million.
- Iron from 134 to 408 million metric tons, steel from 185 to 651.
- Aluminum from 1.5 million metric tons to 12 million.[32]

The statistics roll on.

- In 1950, when there were only 2.6 billion people, we had 53 million cars, or one for every 50 persons. By the time we hit 6 billion people we had 500 million cars, or one for every dozen people. (Sierra Club)
- The world's consumption of plastic has grown from 5 million tons in the 1950s to 100 million tons today. (WasteOnline)
- Global coal consumption went from 3,752,183 short tons in 1980 to 7,994,703 in 2012.[33]

As a result, the waste stream exploded. Humans produce 2.6 trillion tons of garbage each year. Americans alone use 1,500 plastic drink bottles every second, and some 40 million are thrown away each year. They were discarding 2.7 billion batteries, and 350 million spray cans. The largest maker of multi-layer paper-foil-plastic drink boxes made 54 billion cartons, each one used once and thrown away. About 304 million electrical items were thrown out from American households in 2005, and two-thirds of them still worked; 3.16 million tons of obsolete electronics were discarded each year and only 13.6 percent were recycled.[34] The rest went into landfills. Of the 2.25 million tons of obsolete electronics in 2007, 82 percent were discarded … into landfills. By 2008, this figure had reached 3.16 million tons, but the recycling rate was still just 13.6 percent.

Growth like this is simply unsustainable, not only because resources are at some level finite, but because the negatives associated with this growth, including human misery, cultures ruined by stripped out rain forests, slums, and hunger, but also because of the depressive impacts on the biosphere's ability to supply the basics such as clean water, air, biotic cleansing functions, and soil, to say nothing of the international conflicts this type of resource depletion is already engendering. A cancer-like growth characterizes Hypercivilization and its radical disconnect with the past.

## *The Chemical Soup*

In the era of Hypercivilization humanity spewed toxic chemicals into the air, soil and water, and subsequently into our bodies. The U.S. Environmental Protection Agency reported that 90 percent of all Americans had measurable quantities of chemical such as toluene, styrene and ethyl phenol in their bodies.[35] Ninety-nine percent of pregnant American woman carry multiple man-made chemicals in their bodies, sharing that concoction through the umbilical cord. More than *80,000 chemicals* permitted for use in the U.S. have never been fully tested for toxicity to humans, let alone children or fetuses.[36] We are subject to a vast array of artifactual chemicals for which neither we nor the other organisms on the planet have had any evolutionary preparation. Many of these new chemical compounds were synthesized and released with no testing for toxicity or environmental impacts, especially as they combined and recombined with each other in the environment. It is a giant experiment and we humans are among the guinea pigs.

Beginning in the 1930s, the growth rates for consumption of such chemicals as ammonia, ethylene and chlorine began a steep rise,[37] as did those used in agriculture, including insecticides, herbicides, fungicides and synthetic fertilizers. In 2007, 887 million pounds of pesticides were sprayed on crops in the U.S., including 33 million pounds of organophosphates, one of the most toxic of the pesticides. The use of the herbicide glyphosate more than doubled from 85–90 million pounds in 2001 to 180–185 million pounds in 2007.[38] Total application in the U.S. since it was introduced as the active ingredient in Monsanto's weed killer, Roundup, is over 500 million pounds.[39] The World Health Organization has found that glyphosate is "a probable carcinogen and a team of international scientists recently reported that it can lead to the antibiotic resistance in common, disease causing bacteria."[40]

Methyl iodide was approved for use by the EPA and the states of Florida and California were considering its use even though it is a known carcinogen, a thyroid disrupter, can cause birth defects and late-term miscarriages, and may also be a neurotoxin. It is even used in labs to create cancerous cells.[41] The California regulatory agency approved using the pesticide at levels 120 times what its own advisory board of scientists said were acceptable levels of exposure, and against the warning of six Nobel Prize-winning chemists who had called for a ban on the substance.[42]

Many of the new chemical substances were micro-contaminants, man-made substances known as halogenated hydrocarbons. Ashworth, whose special concern is the Great Lakes in the U.S., writes: "It is this group which poses the greatest danger to living things. A pinch may kill; a microscopic speck may pose a substantial cancer risk. Thousands of pounds of them are released into the Great Lakes and their feeding waters each year."[43] These man-made substances are everywhere. PVC (polyvinylchloride) packaging is found in blister packs, toothpaste tubes, salad packs and bottle sleeving. According to the PVC industry, "Approximately 500,000 tons of PVC is used in packaging across Europe each year. Its major packaging applications are rigid film (about 60 percent), flexible film such as cling film (11 percent), and closures (3 percent)."[44] Much of this material is contaminated with heavy metals.

Many of these chemicals escaped accidentally into the environment, or, in a spasm of naiveté, were intentionally released, as in the case of PCBs.

These constituted a family of chemicals that come in over 200 types and were used in plastics, electrical insulators and hydraulic fluids and were originally thought to be inert in the environment. In fact, after flooding the environment with them, it was discovered that they were unusually toxic and persistent and damaged the vital organs.[45] Like many chemicals, they bioaccumulate. They are found in increasingly higher concentrations as one goes up the food chain. Samples taken in the North Sea found the concentration in the water to be 0.000002 ppm, but in marine mammals the concentrations had accumulated to 160 ppm, many orders of magnitude greater than that found in the water.

By the 1990s it was becoming evident that a particular class of chemicals routinely emitted to the environment was producing dire human health effects because they mimicked the female hormone estrogen, and functioned in the body as endocrine disrupters. These included lead, mercury and cadmium and the chlorinated compounds such as PCBs, dioxins, 2,4-D and 34 other pesticides. In January of 1994, some 300 scientists gathered at the National Institute of Environmental Health Sciences to exchange the results of research on endocrine disrupters. Theo Colburn reports: "Large numbers and large quantities of endocrine-disrupting chemicals have been released into the environment since World War II."[46] These chemicals confuse the glandular systems of both wildlife and humans. In 1996, a particular class of chemicals distributed ubiquitously in the environment was found to be hormone mimickers capable of producing physical deformities across a wide range of animal species.[47] These 51 synthetic chemicals trigger biological reactions that affect the brain, nervous system and reproductive systems. They regulate sperm production, cell division, and aid in "sculpting the development of the brain."[48] Examples were dioxins, PCBs, nonylphenol added to PVC plastic, and Vinclosolin, a common fungicide whose residues are often found on fruit, and bisphenol-A, which leaches from commonly used plastic drinking bottles and the plastic lining of food cans. While the latter has been eliminated from baby bottles and "sippy cups," it remains ubiquitous and is found even in cash register receipts and children's fillings. It is linked to cancer, birth defects, and behavioral problems. Trace amounts have been found in the urine of 90 percent of Americans.[49] These chemicals have been one cause of the new extinction crisis Hypercivilization has unleashed on the biosphere.

## The Sixth Extinction

On February 9, 1921, the last European bison was shot in the Bialowieza Forest in Poland.[50] The passenger pigeon in the U.S. had already been extinct for seven years, the last individual of a population once numbering 3 billion having died in 1914. Hypercivilization has set off an extinction crisis the likes of which have not been seen in 65 million years.

There have been five prior great episodes of extinction, all of them long before we humans had evolved and so are attributable to natural causes like climate change or giant asteroids hitting the earth. Now, in the era of Hypercivilization, we destroying species at hundreds of times the historical rate that has prevailed since the last one.[51] The famous paleontologist, Richard Leakey, warns in his book, *The Sixth Extinction*, that we will lose fifty percent of the world's species over the next hundred years, threatening the entire fabric of life on the planet.[52] By the end of the twentieth century, Earth was losing 140 species a day (thousands of invertebrates a year just from the clearing of the tropical rain forests),[53] prompting John C. Ryan to comment on the "massive bleeding of life from the planet."[54] This goes unseen by nearly all humans, caught up as they are in the urban-based consumer culture. Even if they knew, they might well say: "So what. Who needs these obscure creatures anyway?" The answer is, "We do."

Visualize the biosphere is as a hammock in which we humans are lying. The hammock is made up of thousands of strands each tied into one another to form the web that holds us up. We lie in this hammock getting fatter and fatter, our weight continually increasing through a burgeoning population and the weight of our energy use which skyrocketed in the twentieth century, all stressing the strings. What is more, the body economic exudes poisons that dissolve them—the thousands of toxic materials we throw into the biosphere every day. The stress on the little strings that make up the web has increased exponentially and they are breaking. But here is where the metaphor fails because in a simple backyard hammock the ground is a couple of feet beneath us and if it breaks we might get some bruises but we'd probably be unhurt. In the real world, our civilization, our lives depend on this web of life. Underneath the biosphere's hammock is only a dead planet floating in the black emptiness of space. We are carrying out a reckless experiment easily as dangerous, if not more so, than human-induced climate shift which, as we shall see, is one more cause of the extinctions. The others are habitat destruction and dumping toxins. Too many people with too much power are tearing at the web of life.

As of this writing, the die-off data are appalling. Here is a handful of examples.

- Lions have fallen from 250,000 in 1975 to a mere 30,000 as a result of habitat loss, prey loss, poaching, conflict with humans, disease and inbreeding. The international campaign to save lions known as World Lion Day warns us: "the extinction of the African lion is imminent."[55] In fact, most of the big cats are on the way out.

- Unlike the prior extinctions, ours also involves plants—the base of the food chain. One in five are now threatened with extinction.[56] Some scientists fear we will lose two-thirds of the world's 300,000 plant species by 2100. In the U.S., 29 percent of plant species are already threatened.

- Birds are also in trouble. Bird Life International reports: "One in eight of the world's bird species is deemed globally threatened and the fortunes of 217 Critically Endangered species are now so perilous that they are at risk of imminent extinction. Some of these species have not been sighted for many years and may already have succumbed."[57] A third of American bird species is in rapid decline and half of Europe's birds are threatened with extinction. The National Audubon Society reports that for some common widespread species in the U.S. that were generally thought to be stable, populations have in fact decreased by as much as 80 percent since 1967.

- "Malaysia faces the extinction of 45 bird species in the next five to ten years if it fails to introduce protected areas and breeding programmes for endangered species...."[58] As of 2015 at least a fifth of Malaysia's mammals faced extinction, including the Malay tiger.[59]

- Globally, 847 species of reptiles, or more than 20 percent of evaluated species, are endangered or vulnerable to extinction, according to the International Union for Conservation of Nature's Red List. The situation is even worse for amphibians. More than 1,900 species of frogs, toads and salamanders—fully 30 percent of the world's amphibians—are at risk of dying out.[60]

- The Institute of Biological Sciences at the University of Malaysia reports that half of her riverine fish species are at risk of extinction.[61] In Africa's once biodiverse and highly prolific Lake Victoria, the introduction of giant Nile Perch in 1962 has driven 300 other species to near or total extinction, while contributing to the deforestation of the region as fuel needs for drying fish shot up dramatically.[62]

- In the United Kingdom more than a quarter of the bird population is threatened.[63] The modernization of farming with its pesticides and destruction of hedge rows is even threatening the iconic nightingale.

- Air pollution in Poland's Ojcow National Park has killed off 43 plant species.[64]

- Snails are a bellwether species. Typical is the case of Hawaii, where snails have crashed; "only 15 of the 325 species recognized in Hawaii can still be found alive, and … the rate of extinction in the state has been as high as 14 percent per decade."[65] But it's not only snails that have crashed on the island; birds, moths and insects once present are also gone.
- Insect declines are particularly alarming. A recent German study found that "measured simply by weight, the overall abundance of flying insects in German nature reserves had decreased by 75 percent over just 27 years. If you looked at midsummer population peaks, the drop was 82 percent."[66] Other observers report we are experiencing a little-noticed insect holocaust world-wide.
- In the Southeastern U.S., it's not only snails that are disappearing. In an article titled *The Southeast Freshwater Extinction Crisis*, The Center for Biological Diversity reports: "Thanks to pollution, development, logging, poor agricultural practices, dams, mining, invasive species and other threats, extinction is looming for more than 28 percent of the region's fishes, more than 48 percent of its crayfishes and more than 70 percent of its mussels. As just one example of a Southeast waterway in peril, the Coosa River is the site of the greatest modern extinction event in North America, where 36 species went extinct following the construction of a series of dams. Overall, the Mobile Basin is home to half of all North American species that have gone extinct since European settlement."[67]
- The island of Madagascar, off the southeast coast of Africa, once supported 8,500 species of plants in a forest cover that was virtually total. Some 6,500 of these species were unique to the island. By the end of the century only 10 percent of the forest cover remained, 90 percent was gone.[68]
- In 1970 there were 4.5 million African elephants. By 1995 only 610,000 were left and by 2012 only 400,000, and poachers were killing a hundred a day.[69] Elephants are a keystone species, meaning their existence is crucial to the whole ecosystem which, without them, will change, dramatically affecting the survival of other species. Bears are also a keystone species. "Over the last 200 years nearly 99 percent of the Great Bear's territory has been lost. During this period, grizzly numbers have dropped from 50,000–100,000 to around 1,800 today. As a consequence, the grizzly now teeters on the brink of survival in the Lower 48 states."[70]
- Ninety percent of the world's remaining Black Rhinoceri were killed by poachers between 1970 and 1995, leaving only 3,000 of the

animals in the world, in spite of the fact that they were protected by law.[71] Of the several species of rhinos that numbered 500,000 in 1900, only 29,000 are left.[72]

- "Of the 207 species of turtle and tortoise alive today, 129 of them are listed by IUCN as vulnerable, endangered, or critically endangered."[73] In the Black, Caspian and Azov seas in the former Soviet Union, 90 percent of the major commercial fish species were gone.[74]

- In *Coming Soon: A World Without Penguins*, Carol Wellner writes that "The threat of extinction hangs over penguins."[75] Thirteen of the eighteen species are already endangered.

- Critical domestic plant species that will be needed as agriculture adjusts to global warming have also gone out of existence due to the ubiquitous spread of the handful of Green Revolution varieties. For example, in Indonesia, 1,500 local varieties of rice disappeared between 1977 and 1992.[76]

These are appalling statistics but by the time you are reading this they will be outdated. Things will have gotten worse.

Species extinction is due primarily to habitat loss. As cities and roads and airports expand, as mono-cropping industrialized agriculture expands, as the irrational growth economy roars on, the highly biodiverse wild areas are lost. Even the preserves are proving too small and to widely separated to protect the wild species within them.

Who gives us the right to cause the extinction of species? The extinction crisis is not only a matter of obvious prudence, as in "take care of nature so that it can take care of you," and "stop using chemicals and practices that are killing off the honeybees or you will lose half your food production because there are no pollinators." Who gives us the right to starve polar bears into extinction? To eliminate the intelligent and emotionally empathic elephant, or the magnificent tiger? Or any species? Who do we think we are?

But extinctions, exponential population growth, urbanization, consumption and the chemical soup do not exhaust Hypercivilization's assault on the biosphere. We turn now to the assault on forests, our waters, air, and to nuclear contamination, war and climate deterioration, completing our brief tour through the global environmental crisis of Hypercivilization.

# 2

## Assault on the World's Forests, Soils, Water and Air

In the era of Hypercivilization Earth's ecosystems came under withering attack by the combined force of too many humans wielding overwhelmingly powerful technologies. As Lester Brown pointed out, "The global economy is ... expanding at a robust rate. But the ecosystem on which it depends is not expanding at all."[1]

### Degrading Earth's Forests

Forests are our very breath. They convert our exhaled $CO_2$ into the oxygen we need for life, supply critical wood and fiber including paper pulp, and critical habitat that supports Earth's biodiversity. They act as sinks where rainfall can slowly settle into the groundwater table replenishing aquifers, combat erosion, and act as carbon sinks. They actually increase the moisture in the air around them. But in the era of Hypercivilization we have seriously degraded our forests.

"Before humans invented agriculture there were 6 billion hectares of forest on Earth. Now there are 4 billion, only 1.5 billion of which are undisturbed primary forest. Half of that forest loss occurred between 1950 and 1990."[2] It has proceeded apace since then. In the U.S. forests are clear cut to feed the demand for paper and for export to Japan. In 1949 the cut from the entire U.S. Forest Service lands had been 2.6 billion board feet. In 1988, 16 billion board feet were cut from Washington and Oregon alone. Ten square miles were being logged every month, often on extreme slopes.[3]

Properly managed, forests are renewable, but humans have been managing them improperly for millennia. In the era of Hypercivilization, degrading practices have greatly accelerated, especially in tropical forests which are the most biodiverse places on Earth. Even attempting to selectively log in those forests is highly damaging because the tree stands are so dense and tied together with vines that bringing down one tree often brings down or damages seventeen others. Nevertheless, smash-and-grab logging is being carried out in the world's tropical forests to supply the tremendous demand in the northern hemisphere, especially from the U.S., China and Japan. By century's end, more than 40 percent of the rain forest was down and a straight-line projection showed it would all be clear cut by 2135, a too-conservative projection since the rate of cut was growing. As of this writing, only 5 percent of the world's tropical rainforests are protected by parks and reserves and those are frequently poached. After decades of handwringing about rampant destruction of forests almost everywhere, investigators have recently demonstrated in extraordinary detail that much of this logging is blatantly illegal. In Romania, Brazil, Poland, Honduras, Peru and elsewhere, black markets and corrupt governments are conniving with timber companies to despoil protected areas, according to Alexander von Bismarck, executive director of the Environmental Investigation Agency. Richard Conniff sums up the practice in an article, "Amid the Plunder of Forests, a Ray of Hope." The illegal wood is everywhere. "Ask Donald Trump. According to Mr. von Bismarck, doors manufactured a few years ago with mahogany

A Clear Cut: Replacing natural forests with single-age, single-species monocultures such as paper pulp trees and palm oil plantations, or clearing for cattle ranches and cropland, has made major reductions in the quality and extent of the world's original forest cover, leading to erosion, siltation, increased use of pesticides, and major losses of biodiversity. Cattle ranches are the most impactful as they trade a carbon sink for a carbon spigot (Rich Carey, Shutterstock).

stolen from a UNESCO World Heritage Site in Honduras ended up in Mar-a-Lago."[4]

Far too many examples abound. The island of Madagascar has lost 93 percent of its tropical forest. Forests are often cut over to make way for cattle ranches to serve the export market of meat eaters in the First World, probably the worst possible transformation, as it is a double increase in greenhouse gases. Costa Rica was a case in point.

> Much of the forest was cleared in Costa Rica in order to expand cattle ranching for beef export. Many of the new pastures proved unsustainable. Within a few years they were grazed down, eroded, and abandoned. On steep hillsides and in heavy rains there were landslides which destroyed roads and villages. Silt from the eroded lands filled up reservoirs behind power dams or washed into the oceans where it buried and killed coral reefs and destroyed fisheries. The land will bear the scars for a long time from Costa Rica's few decades of intensive beef production.[5]

Often the sustainable practices of indigenous peoples were replaced by massive harvesting funded by the World Bank and other international financial institutions. Rainforest Action Network describes heartbreaking cases. "When the small farmers complain to the government that their land has been stolen, they are shown papers that prove that the land their families had farmed for generations was now owned by Cargill or ADM/Wilmar."[6] In *Plundering Paradise: The Struggle for Environment in the Philippines*, Broad and Cavanagh write: "We are entering a country of environmental ruin, a country where the lives of peasants, fishers, and others are being altered drastically by the sudden human devastation of millennia-old environments."[7]

By century's end, Cuba had less than 2 percent of its original forest cover.[8] On the Atlantic Coast of Brazil, once considered the most species-rich habitat in the world, only 5 percent of the original forest remained. Already by 1980, fuel wood demands were exceeding sustainable yield levels in 11 of 13 West African countries. By the 1990s, China's cut exceeded regrowth by 100 million cubic meters a year. In British Colombia, Canada, the 1989 cut was 30 percent above sustainable yield; in the western U.S. in the 1980s, 25 percent above on private lands and 61 percent on government land.[9] Often this cut is on extremely steep slopes. The best estimate of the rate of clearing world-wide was 135 acres per minute.[10]

Fragmenting forests by driving roads into them or leaving only islands of trees accelerates extinctions. Complex forest ecosystems are degraded by overharvesting, and, when replanted with even-age stands of a single species, the result is an ecological desert where few species other than the trees exist. This was proudly called "industrial forestry."[11] One of the worst examples is replacing complex native forests with palm oil plantations. The oil "is used in half of all the products commonly found in supermarkets including soaps, shampoos, candy, cosmetics and even biofuel. Imports to the U.S. tripled between 2005 and 2009."[12] Each item carries the cost of a complex ecosystem destroyed. Also, as we shall see when we consider climate change, fire is an increasing danger to forest ecosystems. Finally, forest degradation is not only a prime feature of Hypercivilization, it leads directly to another—soil erosion.

## Degrading Earth's Soils

In the era of Hypercivilization humanity failed to manage its natural inheritance of fertile topsoil. Productive soil is not just a medium to hold up plants; it's a very complex community of organisms—millions of bacteria, decomposers, aerators like earthworms and burrowing animals, etc. Nearly all our food comes from the soil. Healthy soil also acts as a carbon sink. Topsoil is made very slowly, in some cases only an inch every thousand years. The per capita amount of arable soil is shrinking in the age of Hypercivilization due to four factors: the explosion of population, erosion, desertification, and spreading urbanization. The quality of the soils is declining due to salinization of irrigated lands, compacting by heavy equipment, and pesticides which kill off the complex community of soil organisms. Drought is causing deserts to advance, and climate change is melting the glaciers which feed irrigated food-producing lands, preparing the way for a Malthusian nightmare in places like Pakistan, Bangladesh, Southeast Asia and elsewhere where population boomed, dependent on irrigating naturally dry lands. In some of these areas the soil has become waterlogged from too much irrigation, and crops drown. Existing irrigated land is being abandoned as fast as new irrigation projects come on line.[13] Finally, the failure to eliminate nuclear weapons leaves open the possibility of their use in a war which would spread lethal radiation over millions of square miles of soil, rendering it unfit for crops or anything else.

Most of the world's agricultural systems rely on six to eight inches of topsoil, but that is disappearing. Worldwide land degradation accelerated in the twentieth century. Lester Brown of World Watch observed: "Each year, some 6 million hectares of land are so severely degraded that they lose productive capacity, becoming wasteland."[14] In the last 50 years nearly two billion hectares of land were seriously degraded by human activities.[15] Already by 1993 the United Nations estimated that 70,000 square kilometers of farmland were being abandoned each year because of soil loss, an area five times the size of the state of Connecticut. The UN also reported that in the 1980s, the amount of arable land per person in the world declined by 19 percent.[16] Since then we have added almost two and a half billion more people. In 1996, biologist Charles Southwick stated that "to date, one third of the world's arable land has been lost."[17] By the last decade of the century desertification had claimed seventeen percent of the earth's land surface.[18]

In the U.S. the famous Dust Bowl of the 1930s severely damaged huge areas of soil. Drought hit there again in 1988 and in 2012 a great drought and fires struck the Midwest and the Southwest United States. More than half the counties of the U.S. were declared disaster areas.[19] Eighty-six percent of the nation's prime corn and soybean growing areas were affected. Already by 1982, the U.S. was losing 6.4 billion tons of soil a year from its agricultural land. In some places in Iowa the remaining topsoil was only a quarter of an inch in depth.[20] Where soil is left bare over the winter on farms in Missouri, each acre loses 41 tons a year. Even when cropped with corn, an acre loses almost 20 tons per year.[21] In hilly country the rate can rise to 200 tons per year. Droughts also hit North and South Korea, China and Russia in 2012.[22] Soil degradation and loss were occurring almost everywhere in the developing countries. In Ethiopia, over a million people had to abandon once-fertile land that had turned into "stony deserts" in the 1980s.[23] Here, as elsewhere, part of the cause is overgrazing—too many people reliant on too many animals. Soil erosion and desertification are like robbing the bank and condemn future generations to food insecurity.

In general, overuse, misuse and/or drought contribute to desertification, which then becomes self-fueling. The loss of plant cover reduces the supply of moisture that is evapotranspired to the air, which in turn reduces cloud cover, which both reduces the chance of rain and allows more desiccating sunlight to fall on the land, thereby increasing soil temperatures to the point at which soil microorganisms necessary for nutrient recycling die. This cycle continues until the area has become a desert.[24] Desertification now affects the southwestern parts of the United States, northern Mexico, North Africa, the Sahel, large parts of southern Africa … and parts of Australia. Between 1925 and 1975 the Sahara Desert grew by 250,000 square miles along its southern edge—in parts of the Sudan the desert boundary moved south by about 120 miles between 1958 and 1975—and it was moving north and encroaching on about 250,000 acres a year. In Chile the Atacama Desert is advancing by almost two miles a year.[25]

Along with other forms of pollution, degrading soils also contributes to the degradation of Earth's waters.

## The Global Water Crisis

Water is the medium of life. Plants of all kinds, both terrestrial and aquatic, form the base of the food chain and need clean, healthy reliable water. Our human bodies are primarily water and some carbon. Without water we die in a few days. In the brief era of Hypercivilization, we have squandered and degraded our billion-year inheritance of good water.

Aquatic ecosystems are the ultimate sink. Almost all pollutants end up in the water. The world's oceans, of which Lord Byron once said: "Man marks the earth with ruin—his control stops with the shore,"[26] are now a universal trash dump. Healthy oceans provide habitat for crucial protein, especially along the continental shelves and coral reefs. These are degraded by eroded soils, industrial pollutants, radioactive material, and the ocean is filling up with a phenomenal amount of plastic. Among of the worst offenders are plastic water bottles. "According to a 2001 report of the World Wide Fund for Nature (WWF), roughly 1.5 million tons of plastic are expended in the bottling of 89 billion liters of water each year."[27] Americans alone use 1,500 plastic drink bottles every second and 17 million barrels of oil are used in their production. The need to drink bottled water due to pollutants further burdens the Earth with the environmental costs of extraction, burning of fossil fuels, and litter—and much of it comes not from pure springs, but from city taps.

Much of our garbage ends up in the oceans where complex, shifting currents such as, the Pacific Gyre and other concentrations of floating and suspended plastic covers 40 percent of the ocean's surface.[28] The UN predicts that by 2050 the weight of plastic in the ocean will equal the weight of the fish.[29] Some of it is easily seen plastic bags and bottles, but a lot of it is small bits called microplastic, broken down from larger materials.[30]

> There is … growing evidence of microplastics being eaten by a range of important species at the bottom of the food chain (including seafood species such as mussels and langoustines). This is particularly worrying given that these particles are thought to attract and concentrate background pollutants such as DDT, and concerns are now being raised about their potential to provide a vehicle for these harmful chemicals to enter the food chain.[31]

Microplastics are also turning up in the Great Lakes in the United States at concentrations three times higher than in the oceans.[32] Some commercially available skin cleansers actually have "microscrubbers," which are minute polyethylene beads that go down the sink and into our waterways.[33] One has to wonder, where did the inventors and marketers of microscrubbers think they would end up? But of course, that is not a question commonly asked by such people in the era of Hypercivilization. Two other sources of microplastic in our waters are "nurdules" (raw, preproduction plastic bits), and tiny fibers from synthetic clothes that enter from home washing machines. Other things from tennis balls to televisions make up these enormous garbage patches. Plastic bags and pieces of fish net entrap and kill larger marine mammals such as sea turtles and even whales. The problem is not confined to great water bodies. Plastic microfibers are found in 94 percent of drinking water in the U.S.[34] None of it belongs in the waters, but at this point there is no feasible scheme for cleanup, let alone money for it.

One would think that fifty years of environmental regulation would have protected our waters, but it is surprising to read the following statistics.

- In 2010 alone, Industrial polluters dumped 226 million pounds of toxic chemicals in to U.S. waterways.
- According to American Rivers, "eighty percent of sampled streams in a national clean water survey contained drugs, hormones, pesticides, or other chemicals. Three-quarters of those streams contained more than one chemical."
- Those chemicals cycle back around to drinking water. In 2009, the Environmental Working Group published a study showing that 252 million Americans were served by drinking water systems that exceeded recommended safety guidelines for some contaminate at least once during the study period.
- The story is similar worldwide with 2 million tons of human waste dumped directly into waterways every single day. In developing countries, 70 percent of industrial waste is dumped into rivers and streams that communities depend upon for drinking, washing, watering farms and supplying livestock.
- Asian rivers may be the most polluted in the world. "They have three times as many bacteria from human waste as the global average, and 20 times more lead than rivers in industrialized countries."[35]

A 1995 estimate by NASA said that 542 million gallons of oil went into the seas, only 37 million of which were from big spills. Most of the rest came from drainage off the land, including used motor oil and industrial spills, especially in port cities.[36] In addition, rain falling on CAFOs pollutes ground and stream water with urine and fecal matter from hogs and cattle. The seas are polluted.

> Most of the major seas of the world, including the Baltic, Bering, Black Caspian, China, Java, Mediterranean, and Yellow, have serious ecological problems with declining fish populations. The Baltic Sea has high levels of toxic organochlorines, the Black Sea has excessive algal and bacterial blooms with anoxic conditions, the Bering Sea is depleted from over fishing, the South China and Yellow seas are plagued with oil and heavy metal contamination, and the Caspian Sea receives over 100,000 tons of waste water and petrochemical wastes each year. All of these, and several other seas, have sharply declining sea food harvests.[37]

The Bay of Naples, which at the beginning of the twentieth century produced "elegant fish of large size," by the end of the century yielded up only anemic little fish of less than 10 inches.[38] The Baltic Sea off Gdansk, Poland, was an ecological disaster area. The Vistula River pours in water so polluted it is not fit even for industrial uses, much less drinking, as it contains nitrates, phosphorous, mercury, lead, cadmium, and chlorinated hydrocarbons. The sea bottom there is an ecological desert where no life exists. The European Union reported that in 2014 the Baltic was also heavily polluted with manure running off from pig and cattle CAFOS, resulting in heavy loads of nutrients that cause toxic algae blooms and anoxic conditions, rendering it "one of the world's most polluted" oceans.[39]

It is surprising how much water is required just for our daily diet. According to the 2014 water footprint measurement methodology approved by the International Organization for Standards (ISO 14406: 2014), vegetables and grains take more than 1,500 liters (396.3 US gal) of water, and a meat rich diet takes 3,400 liters (898.2 US gal). It is astonishing how much water is required to produce various common foods. A cup of tea—8.6 gallons; pint of beer—42 gallons; glass of wine—31.7 gallons; glass of milk—52.8 gallons; 2.2 pounds of beef—3,962 gallons; 2.2 pounds of poultry—1,585 gallons; 8 ounces of M&Ms—304 gallons; and a jar of pasta sauce—53.4 gallons.[40]

The amount of fresh water on the planet is limited. No new water is being created. Ninety-seven percent of the Earth's water is salt and only three percent is fresh, two thirds of which is locked up in glaciers. Only 0.007 percent of the planet's water is available for its 7.6 billion people.[41] Without water, we humans can die of thirst in as few as three days. Much of it is used over and over again, mostly for mining and for industrial processing and cooling, for food prep, and to carry away human wastes. Skyrocketing per capita rates of use combined in the era of Hypercivilization with skyrocketing populations and the destruction of water absorbing wetlands that replenish aquifers to create a world-wide freshwater crisis.

## Mining

Mining is the most abusive use of the land. The ores constitute only a fraction of the earth that has to be moved resulting in huge dumps of "overburden" which carries material toxic to life including heavy metals, sulfuric acid and other nasty compounds. Frequently it is dumped into watersheds and even into streams. Hypercivilization is characterized by its unprecedented, indeed colossal rates of extraction and resultant land and water pollution from mines. The copper mines at Butte, Montana, were a typical case.

Anaconda Copper Company owned Butte and all its mines, including the hundreds of miles of shafts and tunnels, some going a mile below the surface, and its open pit mines which were the largest on earth. Dan Baum and Margaret L. Knox write: "Butte is saturated with a toxic legacy. Slag heaps and waste rock piles the size of small mesas tower everywhere. Swamps of bleached white mine tailings and layers of poisonous flue dust blanket old smelter sites, vacant lots and backyards."[42] The smelter soot is contaminated with metallic poisons, copper, manganese, arsenic and lead. These, "are blown around by the wind, picked up on shoes, ingested, inhaled and absorbed through the skin."[43] The trail of mine residues stretches a hundred miles along the Clark Fork River. The Berkeley Pit at Butte, which closed in 1982, is a mile across and the water in it, now as caustic as battery acid, has risen to a depth of 800 feet. Water (5.5 million gallons) contaminated with heavy metals and acids and other toxins were seeping into the pit each day, threatening Butte's water supply and the Clark Fork River. In just four western states almost 100,000 old mines have left toxic wastes needing to be cleaned up. The corporations responsible are mostly long since bankrupt and have disappeared. And the same processes have gone on all over the world in the era of Hypercivilization.

## Iron and Coal

The Mesabi Iron Range in northeast Minnesota produced millions of tons of ore, and the resulting overburden has turned the region into a moonscape of sterile rock piles that will last until the next Ice Age sweeps them away. From 1955 to 1980 Reserve Mining dumped 47 tons a minute of pulverized waste rock containing asbestos-like fibers into Lake Superior at Silver Bay, coating the lake floor with a grey, clay-like substance and clouding the once pristine water at the western end of the lake. The microscopic fibers found their way into the drinking water of Duluth and other shoreline towns. In 1980, the courts finally succeeded in getting the practice stopped. The iron itself ore went down the lakes to steel mills in Chicago, Cleveland and Pittsburgh. The coal for steelmaking came up the Ohio and Erie Canal.

Coal mining has and continues to destroy Appalachia, where poor farmers were conned into selling off their timber and then selling off the mineral rights to their land. Once a beautiful mountain wilderness, the area has ended in what Stewart Udall, former U.S. Secretary of the Interior, has called a "pathetic and disturbing story ... a tragic tale of the abuse and mismanagement of a resource heritage, and the human erosion that is always the concomitant of shortsighted exploitation ... the story of what happens when men betray their responsibility as land stewards."[44] As the slopes were clear cut, mountain families began farming them until the great deluge of 1927 came roaring down the mountains. Forty years later, Harry Caudhill would write: "The yellow hillsides now lay bleak and dead in a state of sterility from which they have not recovered to this day."[45]

The mining companies bought up the land for fifty cents an acre, using the "broad form deed" which left almost nothing to the land owner. It:

> ...authorized the grantees to excavate for the minerals, to build roads and structures on the land and to use the surface for any purpose "convenient or necessary" to the company ... to utilize as mining props the timber growing on the land, to divert and pollute the water and to cover the surface with toxic mining refuse. The landowner's estate was made "perpetually servient" to the superior or "dominant" rights of the owner of the minerals. And, for good measure, a final clause absolved the mining company from all liability to the landowner for such damages as might be caused "directly or indirectly" by mining operations on his land.[46]

Corporate capitalism at its best.

By 1914 the deep-shaft mines were in place and "the vast, backward, Cumberland Plateau was tied inseparably to the colossal industrial complex centering in Pittsburgh...."[47] The mountaineers found they had left their near-wilderness subsistence poverty for the grim and grimy coal towns where the company owned the houses, the stores, the taverns and the churches. The coal was mechanically sorted and graded by giant machines that conveyed and shook it, raising huge dust clouds that hung perpetually over the coal towns and penetrating every nook and cranny. Accompanying it was the oily black smoke coming from the fires that burned spontaneously and out of control in the ever-larger piles of discarded slate and low-grade coal—some were hundreds of feet high and hundreds of yards in length. Houses were heated with coal as well, augmenting the sooty atmosphere which everyone breathed, and which peeled the paint from the houses and caused the lungs and eyes to burn.

The worst was yet to come for Appalachia—strip and augur mining. In the deep shaft mines the companies did not take out the coal that was near the sloping surface of the mountain for fear of cave in. Strip mining allowed them access to this coal. Bulldozers would shove the trees over the mountain side. The men would then dynamite the rock overburden and push it over, leaving sheets of sterile material covering the original mountain soils. Then they dynamited the coal and hauled it away, leaving a flat plateau of rock where the coal had lain, and "high walls" at the back, 50 to 90 feet high and running in some places for a mile or more. Some mountains suffered several of these operations at varying heights and then, in a literal crowning blow, the top of the mountain would be bulldozed off to get at the seams lying up there. Finally, giant augurs would work into the remaining coal that lay too far under the surface for stripping, thus destabilizing the mountain above and causing it to collapse. The erosion was astronomical, silting up streams, covering bottom land farms, and releasing sulfuric acid into the watershed killing off the life in the streams.

Perhaps no other practice more perfectly exemplifies Hypercivilization's willingness to create massive, permanent desecration of the natural world for minimal short-term gain reaped by a few than does Mountain Top Removal, or MTR. It is a method of strip mining thin seams of coal in Appalachia by clear-cutting the hardwood forest cover (often just burning off the bulldozed trees), drilling deep holes for explosives and then blowing the tops off of the mountains. As much as 800–1,000 vertical feet of overburden is then stripped off and often dumped over the side, filling valleys and clogging the headwaters of streams. It is often described as "strip mining on steroids."[48] At least 500 mountaintops have been blown off. Earthjustice reports that "over 2,000 miles of streams and headwaters that provide drinking water for millions of Americans have been permanently buried and destroyed. An area the size of Delaware has been flattened."[49] The forests would have sequestered 3.2 million tons of $CO_2$ per year. But coal grew exponentially. Between 1920 and 1970 coal-fired power plants in the U.S. increased their consumption tenfold.[50] By 2015, coal still provided 33 percent of electrical energy generation in the U.S.,[51] although the industry appeared to be finally dying back as major companies faced bankruptcy. An ironic note is that MTR accounted for only 5 percent of total coal production in 2001; creating irreversible harm for very little gain in energy production. And coal smoke, along with other pollutants, fouled the air over the world's cities.

## Degrading the Air We Breathe

In the era of Hypercivilization, humans degraded the planet's atmosphere in three ways. First, they added harmful substances that were either previously not present or they added natural substances in quantities that natural systems could not tolerate, altering the chemical composition of the earth's atmosphere. There were several leading types of air pollution including particulates (smoke from combustion); dusts from agriculture and milling; and metal working which produced dusts such as powdered lead and beryllium, etc. There were the aerosols; sulfur dioxide and nitrous oxides (the source of acid rain); carbon monoxide from auto exhausts; ozone; hydrocarbons; hydrogen fluorides and sulfides; ammonia and phosgene; arsines; chlorines, aldehydes; benzene and benzo-a-pyrene. These substances ranged from being merely irritating to the respiratory tract and eyes to being toxic to plants to causing leukemia and cancer.[52] In addition there were huge releases of airborne radioactive substances from coal plants and nuclear meltdowns. Second, some of these chemical compounds raised the temperature of the atmosphere as a result of what came to be called "greenhouse gases," altering the climate with far reaching consequences we will examine in a later chapter. Third, they degraded the protective ozone layer. All of these developments had negative effects on human health and well-being. The situation was well summed up by astronaut Paul Weitz who commented, during a 1983 space flight, "Unfortunately, this world is becoming a grey planet...."[53]

Air pollution crises in large urban centers were typified by the infamous photo-chemical smog in Los Angeles, but deadly episodes occurred in other cities, especially before the change from coal to oil and natural gas for heating and industrial use. In Donora, Pennsylvania, a severe episode occurred in 1948, sickening 14,000 people and killing 20. In 1952, the London Smog caused 4,000 deaths.[54] While the cities of the rich First World were able to make the switch to cleaner burning fuels, most cities of the Global South were not and continued to produce extremely noxious local atmospheres.

The worst air pollution is in the rapidly developing areas of the world such as China. Burning high-sulfur coal and low-grade gasoline are resulting in foul and highly dangerous air in the cities. The Chinese were burning 6 million tons of low grade coal each day to power their economic miracle. The World Bank reports that 16 of the world's 20 most polluted cities are in China and only a third of the 340 Chinese cities they monitor meet China's own pollution standards.[55] The World Health Organization estimates that "the suspended particulate matter in northern China is 20 times the safe level."[56] The air quality of Beijing is 16 times worse than New York City. Sometimes you can't even see building a few blocks away and blue sky is a rare sight. "In Shanghai sometimes you can't see the street from the 5th floor window. Fresh air tours to the countryside are very popular."[57]

As the skyrocketing rates of combustion of fossil fuels accelerated in the twentieth century, the sulfur dioxide and nitrous oxides emitted from the smokestacks and tailpipes mixed with rainwater and fell as sulfuric and nitric acid. Between 1966 and 1993, "Norwegian soils became markedly more acidic, often changing by one full pH unit."[58] This tenfold increase in acidity slowed rates of decomposition and of nutrient cycling and reduced net primary production. In some cases the acidic rain killed the mycorrhizal fungi that interact symbiotically with the roots of trees. Acid rain can also precipitate out essential nutrients such as phosphorous. When taken up by the plant, acidic rainwater will degrade the chlorophyll, reducing the plants photosynthetic capacity. Acid rain also releases potential toxins in the soil including lead, aluminum, cadmium, copper, zinc and arsenic.[59] In the U.S. crop losses from air pollution were estimated to be from 5 to 10 percent of the total, ranging in value from $1.9 to $5.4 billion.[60] In some areas, where low clouds hang over high country forests, the trees were getting bathed in an acidic water vapor with a pH of 3.5, one hundred times normal. Even when the acidity and released toxins did not kill plant life outright, they increased its susceptibility to insect predation.

In California's San Bernardino and San Gabriel mountains, lying 129 miles east of Los Angeles, pine forests were suffering from that city's acid smog.[61] In Switzerland, 10 percent of the spruce and 25 percent of the fir had died from air pollution by 2000 and in Germany 43 percent of oak, 50 percent of beech, 51 percent of Norway spruce and 87 percent of white fir had been damaged.[62] Germans began to speak of *Waldsterben*—forest death. One third of Germany's forests were damaged at an economic loss of $800 million.[63]

Lake ecosystems were also highly sensitive to acid precipitation. By 1975 America's Adirondack lakes above 610 meters had a pH of less than 5.0 and 90 percent of them had lost their fish. These lakes looked exceptionally clean and pure but, in fact, were sterile. In Minnesota and Wisconsin 80 percent of the lakes were threatened by acid rain. By the 1980s the rain in parts of Pennsylvania in the U.S. was 1,000 times more acid than would occur naturally.[64] In Canada, 14,000 lakes lost their fish to acid rain in the second half of the twentieth century, partly as result of the 12 million tons of sulfur dioxide deposited annually from U.S. industries.

A second global level degradation in the earth's atmosphere caused by the activities of Hypercivilization was the partial destruction of the ozone layer by chlorofluorocarbons and other chemical emissions. Near the earth's surface ozone is a pollutant. High above the earth in the stratosphere it provides a shielding layer that protects living organisms from too much damaging ultraviolet light. The ozone layer had been intact and functional for eons until the advent of Hypercivilization. In 1928, chemists at General Motors invented a gas composed of chlorofluorocarbon or CFCs. They were eventually used as propellants in spray cans, as refrigerants, as a foaming agent in the manufacture of plastics, as a fumigant in granaries and the holds of ships, and later in the manufacture of Styrofoam and as a cleaning agent for circuit boards. Their use took off after World War II. Between 1958 and 1983, average production grew 13 percent a year. They are highly persistent in the upper atmosphere, lasting on average 100 years.[65] CFCs react with ozone in the upper atmosphere, breaking it into oxygen and a molecule of chlorine monoxide. In a second reaction, the chlorine atom is freed and can then destroy between 1,000 and 100,000 ozone molecules. To make one Styrofoam cup requires 10,000,000,000,000,000,000 molecules of CFCs.[66] Several other Hypercivilization chemicals also assist the CFCs in destroying ozone: the widely used dry cleaning agent, carbon tetrachloride—now banned in the U.S., methyl chloroform and the bromines and halons found in millions of home fire extinguishers.[67]

By 1980 U.S. satellites were observing a large hole in the ozone above the South Pole but their computers had been instructed to ignore such an impossible anomaly, treating it as a false reading. It was not until 1985 that the British Antarctic Survey at Halley Bay confirmed a huge hole in the ozone. Between 1991 and 1996, ozone over the Antarctic dropped by 50 percent and a hole had opened up over the Arctic as well.[68] Holes and thinning were occurring elsewhere as well, including over Norway and North Dakota. And the problem was developing more rapidly than scientists were at first able to get a hold of. By 1987 depletion was at levels originally predicted for 2020.[69] By 1989, the world had lost between 1 and 3 percent of its ozone and would continue to lose much more due not only to the fact that CFCs and other destructive chemicals persist for up to a century, but also to the fact that many nations had not yet phased them out and continued to produce and release these substances.

Declining ozone will result in depressing biota across a wide spectrum. The U.S. Environmental Protection Agency predicted declining yields of wheat, rice, corn and soybeans as a result of increased UV. It was also expected to have negative effects on other flora and marine food chains, beginning at the bottom at the level of phytoplankton. And, finally, rising levels of UV have already induced higher rates of skin cancer in Australia.[70] Excessive levels of UV also cause eye damage in humans and can kill smaller, more sensitive organisms.

Surprisingly, world leaders moved quickly in a bi-partisan and international effort under the auspices of the United Nations to develop the Montreal Protocol on Substances that Deplete the Ozone Layer. It entered into force in 1989 and has resulted in the elimination of 98 percent of all ozone depleting chemicals, one of the most successful international treaties ever signed and adhered to.[71]

In the next chapter we will look at the environmental impacts of splitting the atom, genetic engineering, war and the increase in environmentally induced diseases.

# 3

## War's War
## on the Environment
## and Splitting the Atom

When we look at the history of Hypercivilization in the twentieth century, we seem to be seeing a species that had gone berserk, a civilization with a lust for power and wealth and an unwillingness to acknowledge the long term consequences of reckless actions taken to fulfill its momentary desires.

### War's War on the Environment[1]

"The brass appeared, as scheduled, in full uniform, medals and all, and we asked one question. 'Did the U.S. military deposit any toxic material in Love Canal?' [President Carter's Chief of Staff, Jack] Watson asked. 'No sir, we did not.' 'Thank you gentlemen,' replied Watson, and off they went.... [Later, after that was proven a lie and a resident asked why.] 'Because, Ma'am,' replied Col. Norris, 'we're in the business of protecting your country, not protecting the environment.'"[2]

Mere conventional war, both the fighting and the preparation, devastate the environment. War is toxic. In the era of Hypercivilization, armies and destructive weapons experienced exponential growth. By 2018 armies numbered in the millions of men, as did the dead in warfare. Modern weapons included highly accurate long range artillery, intercontinental rockets, huge air fleets dropping powerful explosives, automatic weapons that fired hundreds of rounds per minute, poison gas, chemical, biological and nuclear weapons.

Even without nuclear weapons it became possible to destroy whole cities from the air. In World War II nearly every major city in Japan and

Fractured Forest: Modern war devastates the environment. Here we see the remains of a forest ecosystem brutalized by machine gun fire and artillery in World War I. During the Vietnam War the U.S. sprayed toxic defoliants on the forests of Vietnam and Laos (National Archives 531005).

Germany was destroyed, releasing all their toxins into the air, water and surrounding landscape. All of this was supplied by an enormous military–industrial complex that polluted the lands and waters for miles around.

## Korea and Vietnam

Deliberate environmental destruction rose to new levels in the Korean and Vietnam wars. "Massive firepower was directed against landscapes … and against croplands, water supplies, and transportation routes … the environment became the target of destruction."[3] The U.S. led UN forces bombed the dikes in order to flood the rice fields, destroying the crops and causing starvation among the population. The Vietnam War was:

> the first time military technology was employed to destroy the environment of an entire country. Carrying a 20-ton bomb load into the stratosphere, a B-52 could strike from 30,000 feet without warning…. By war's end, these behemoths of the Strategic Air Command had dropped triple the total tonnage dropped in World War II.

Their ferocious carpet bombing left at least 26 million craters in a country the size of Washington.[4]

The chemical warfare defoliant known as Agent Orange was a 50-50 combination of 2,4,5-T and 2,4-D, phenoxy herbicides designed to kill plant life by causing it to grow out of control until it dies. The former contains TCDD dioxin, one of the most toxic substances known. It is carcinogenic, teratogenic and mutagenic. By 1966, the U.S. had destroyed 850,000 acres of rainforests and crops and by 1967 they were destroying 1.5 million acres per year. All told, 72.4 million liters of Agent Orange were misted down over the tropical forests and wetlands of Vietnam.[5] Seven million gallons of other chemical defoliants (Agents Blue and White) were also sprayed.[6] These biocides were sprayed over 8 million acres of forest and 3.8 million acres of cropland. Land destruction in Vietnam involved 80 percent of all forest land and over 50 percent of coastal mangrove habitats.[7]

Twenty-ton bulldozers scraped villages and rice paddies off the surface of the earth, destroying over one million acres. The U.S. forces drained 39,000 hectares of sedge marshes and then torched the dried plant material with flame-throwers. They napalmed mangrove swamps, killing off half of these nurseries of the sea in the region. In total, some "5.4 million acres of tropical forests had been reduced to blackened rubble" leaving "almost lifeless Savannahs."[8] Wildlife was also badly impacted by the war. The damaging effects lasted for decades.

## The Persian Gulf War

In the Persian Gulf War of 1991 the Allied Forces literally attacked everything from the air, flying 2,555 sorties each day and detonating 240,000 tons of ammunition. They struck the 28 oil refineries and oil tankers in port, destroyed the grain fields, the irrigation systems, the chemical and biological warfare plants, the four nuclear power plants, the water treatment and pumping plants, the sewage treatment plants, the power stations, the roads, bridges, hospitals, and fertilizer factories. The Iraqis fired the oil wells to keep them from the enemy. The pollution and destruction of wildlife was astronomical.

The best estimates are six to eight million barrels of oil released each day directly into the Gulf, a warm, shallow sea that hosts a variety of ecosystems and species including the rare dugong (sea cow), the hawksbill and green turtles, and some rare fish species. It is a major flyway for migrating birds numbering between two to three billion each autumn. Six to eight million barrels a day went up in smoke.[9] Total amounts are estimated at 1.5 to 2 billion barrels lost into the sky and an equal amount onto the ground.[10] Massive oil lakes formed on the land, confusing migrating birds, which landed in them and died by the thousands. Oil spills in the Gulf killed invertebrate life in the broad intertidal flats, where rich life had existed before, with up to 100,000 tiny creatures per square meter. "After the oiling the count dropped in some areas to a dozen and even zero."[11]

It took nine months to put the fires out. The smoke from the burning oil refineries and wells traveled immense distances. By mid–March toxic smoke clouds were stretching from Romania and Bulgaria to Afghanistan and Pakistan. There were black rains in Iran and black snow fell on the Himalayas, 2,000 kilometers away. While the fires were burning they were emitting about one to two million tons of carbon dioxide, nine thousand tons of sulfur dioxide and about five thousand tons of soot each day.

The attacking forces also used depleted uranium weapons, which despite the name are radioactive. Tens of thousands of remnants from these weapons remain scattered in the desert. When humans are exposed to depleted uranium it can damage the kidneys and lungs, is deposited permanently in bones and can be released into the placenta. The stealth bombers used ozone-depleting chemicals to avoid detection by radar while, on the ground, the heavy tanks and tracked vehicles compacted fragile desert soils. Similar stories could be told about the Serbian War, Afghanistan, and now Syria and Yemen, but this is only part of the tale.

## Preparing for War Degrades the Environment

We need to look at the impact of war preparation, for war is now a heavily industrialized, highly profitable enterprise. William Thomas, author of *Scorched Earth: The Military Assault on the Environment,* writes: "In every country except Costa Rica, the military industry dwarfs secondary sectors—public health, education or environmental protection."[12]

> War making is the world's biggest business. The U.S. military is itself a major multinational with assets equal to half of all U.S. manufacturing corporations combined. Nearly 40 percent of industrial plants and equipment are devoted to military manufacturing; about 30 percent of all U.S. industry output was purchased by the Pentagon in 1989.[13]

Environmental damage comes from the pollutants associated with the manufacture of war material, aerial bombing practices, dumping of military waste, and contamination from military base operations. Practice bombing takes place in wilderness areas, and it craters the land and leaves both unexploded ordnance and toxic materials from them and the bomb casings and contents. It creates sonic booms, which causes behavioral changes in wildlife and stresses migratory birds. Between 700,000 and one million military aircraft sortie each year.

The military uses incredible amounts of fossil fuels. During the Vietnam War, B-52 bombers guzzled 3,612 gallons of fuel per hour while in flight. Each year, military flights contribute more than 10 million tons of carbon monoxide, nitrogen oxides, hydrocarbons, sulphur dioxide and soot and water vapor into the atmosphere, more than 60 percent of it at altitudes 9 kilometers above sea level, exacerbating the atmospheric effects.[14]

In the 1990s the aircraft carrier USS *Independence* required 100,000 gallons a day to keep it on station. Steaming from its U.S. port to the First Gulf War it burned 2 million gallons of fuel. The mechanized units of land armies required whole fleets of fuel tankers to follow them at close range. During that war a single armored division required 600,000 gallons of fuel per day to operate in the field.[15]

## Dumping

After World War II tons of military toxins were dumped in the ocean. Examples included 15 boxcars carrying 400 tons of mustard gas, intact bombs and other ammunition dumped in the Pacific Ocean off Vancouver Island. For 20 years the biologically rich breeding grounds off the Farallon

**Military Pollution: B-52 strategic bombers taking off from Barksdale Air Base in 1986 on a practice scramble to drop nuclear weapons. Until recently the military was exempt from environmental protection restriction and remains one of the largest polluting sectors (U.S. Defense Imagery/Wikimedia Commons).**

Islands were used as a repository for more than 1,000 barrels of nuclear and chemical waste, as well as explosives. "Today the sediments and surrounding waters contain plutonium, cesium and heavy metals which accumulate in higher concentrations as they move up the marine food chain into human tissues."[16] In the shallow waters between Scandinavia and the Baltic republics the victorious allies dumped 5,000 tons of old munitions and 350,000 tons of poison gases.

The manufacture of weapons and munitions contributed huge amounts of toxic materials to the environment, especially in the U.S. and the former Soviet bloc. The Department of Defense produces millions of tons of waste water each year including fuels, heavy metals, pesticides, solvents, acids, cyanide, explosives, TNT, PCBs, nerve gas and other chemical warfare agents. Many of these were leaking and contaminating groundwater off base. The U.S. Department of Defense itself identified 17,482 toxic sites on 1,855 bases in the country. Ninety-seven of these were so heavily polluted as to have been placed in the EPA's Superfund list.

But first, we turn to the opportunity costs of war making and preparation. How much money is involved? In 2017 the nations of the world spent on their militaries one trillion, six hundred eighty-six billion dollars.[17] In fiscal year 2015, U.S. military spending was projected to account for 54 percent of all federal discretionary spending, a total of $598.5 billion.[18] Add to that $65 billion on veterans affairs for past wars. The U.S. amount was more than the next ten nations combined. For comparison, U.S. spending on energy and environment was only $39.1 billion. The total annual UN budget was $5.4 billion with the U.S. contributing only $1.2 billion. Meanwhile, the U.S. is going ahead with a program started by President Obama to "modernize" our nuclear weapons, at a cost of $1.46 trillion.[19]

## Splitting the Atom—Nuclear War and Nuclear Power

On July 16, 1944, the United States detonated the world's first atomic bomb at a place in the New Mexico desert aptly named the *Jornada del Muerto*, the Journey of Death. When project director J. Robert Oppenheimer saw the fantastic explosion he quoted the *Bhagavad Gita*: "I am become death, destroyer of worlds." Physicist Kenneth Bainbridge was less literary; he said, "Now we are all sons of bitches."[20] Humans had learned to split the atom, first for unbelievably destructive bombs and then for boiling water.

Fissioning the atom was the ultimate Faustian bargain. Fissioning (and fusing) the atom releases what physicists call the "strong force," resulting in astronomical amounts of energy. It occurs naturally only in the interior of the sun. In order to do it on Earth, we need to enrich uranium and doing so creates substances that are so radioactive they must be kept from human contact for tens of thousands of years.

It is almost impossible to comprehend the environmental consequences of a nuclear war.

The bomb that destroyed 100,000 people in Hiroshima released the energy of just one gram of uranium. It was an unleashing of "the basic power of the universe." Jonathan Schell writes:

> The huge—the monstrous—disproportion between 'the basic power of the universe' and the merely terrestrial creatures by which and against which it was aimed in anger defined the dread predicament that the world has tried, and failed, to come to terms with ever since.[21]

The bombs that destroyed Hiroshima and Nagasaki were tiny compared to those that were developed later. Measured in tons of TNT equivalent, Hiroshima was about 12,500 tons, or 12.5 kilotons. The big bombs of World War II, called "blockbusters," carried about 8,000 pounds, or four tons of explosive. The Hiroshima bomb was 3,125 times as powerful and had the added lethality that came from radiation. The temperature on the ground underneath the fireball was 6,000 degrees Kelvin, hotter than the surface temperature of the sun. (Water boils at 315 degrees Kelvin.)

Very quickly the U.S. and the USSR developed thermonuclear weapons. Thermonuclear weapons are far more destructive. One-megaton and five-megaton bombs (a megaton equals a million tons), and even 10- and 20-megaton bombs were produced. All the conventional explosives expended in World War II are estimated to have been about three megatons over six years. Thus a single five-megaton nuclear bomb carries more destructive power than all of World War II. The first of the immediate effects of a nuclear blast is a blinding white light which can sear the eyeballs. This is followed by the "thermal pulse," a super-intense blast of heat (6,000–7,000 degrees Kelvin) which can last several seconds as the fireball climbs in the air. The third effect is the blast wave, which can flatten large concrete and steel buildings. The fourth effect is EMP, the release of an electromagnetic pulse, which knocks out all electrical and electronic devices. A 20-megaton bomb detonated 120 miles above Omaha, Nebraska, would disable all electrical devices in the United States, bringing the economy to a dead standstill.[22] The fifth effect is radiation, the release of materials which are sending out millions of invisible "bullets" that penetrate concrete, glass, wood, and the soft tissues of humans and animals. Radiation is a biocide that depresses life at all levels, depending on the dosage one receives. Large doses, such as would be experienced by anyone on the perimeter of the blast or immediately downwind, can kill rapidly, as the internal organs disintegrate and hemorrhage. Lesser doses, miles and even hundreds of miles from the blast, can make one violently ill and over time cause cancers. Effects on wild animals and on plants are seldom even discussed. The radiation would spread downwind and eventually circle the earth and would remain in the soil and water for tens of thousands of years. An attack on a metropolitan area would also result in the release of all the toxic materials in the city into the environment, an unprecedented amount of pollution.

The huge fires that would burn uncontrolled, all the fire departments having been vaporized or crushed, would release dense clouds of smoke that would cover the northern hemisphere and would constitute the next threat, nuclear winter. In the 1983 a team of scientists led by Carl Sagan and Richard Turco modeled the atmospheric effects of setting all the major cities of the northern hemisphere on fire, the famous TTAPS study which was subjected to peer review of approximately 100 other scientists. Sagan then teamed with Paul Ehrlich and 26 other scientists in a more comprehensive study, appropriately titled *The Cold and the Dark: The World After Nuclear War*.[23] The TTAPS study employed a range of megatonages and seasonal scenarios and attack strategies. In many of the scenarios, temperatures would drop sharply in much of the northern hemisphere. The model predicted readings well below zero in July in Kansas. Thus plant-killing temperatures and darkness that blocked photosynthesis would kill off much of the flora, already weakened by radiation and a general hyper-toxification of the biosphere. The nuclear war which the superpowers of the age of Hyper-civilization had prepared would, if unleashed, be the ultimate environmental disaster.

A more recent review of the effects of nuclear winter point to massive famine following even a limited nuclear war between India and Pakistan, in which only weapons amounting to some 2 megatons would be detonated. Dust and soot released into the atmosphere would result in significant declines in the production of wheat, rice and corn worldwide over a period of ten years, resulting in the potential starvation of over two billion people.

## Nuclear Power

After World War II scientists came up with the idea of using nuclear reactions to boil water to superheated steam in order to power electrical generating plants. They gifted the world with these monstrous devices, calling it "Atoms for Peace." Utilities claimed the power was going to be too cheap to meter.

Nuclear Power is unacceptable for six reasons. First, it is not clean power—it is the dirtiest of all power sources because of the radioactive materials left over will be dangerous contaminants for ten times longer than civilization has existed. All nuclear reactors produced plutonium 239 with a half-life of 24,000 years. Uranium 238 has a half-life of 4.5 million years. In spite of this, before starting out to develop nuclear power no attention was given to the waste problem, and it remains unsolved. Another dirty aspect is that fuel production is tied to fossil fuel contamination of air and water and adds to climate deterioration. The mined uranium itself is far too weak to generate a chain reaction and has to be highly concentrated before it is nuclear fuel, a process powered by coal-fired electric generating plants and then trucked to the commercial reactors, resulting in still more greenhouse gas emissions. Further, the places it is mined (mainly on Indian Reservations), are badly contaminated with radioactive dust.

Second, there is no safe place to store the radioactive wastes, so they are stored on site where they have to be kept cooled for years and later put in metal casks with a supposed life of 120 years. They then must be transferred to new casks. The Nuclear Regulatory Commission is hoping to develop a storage device that would last three hundred years, or one eightieth the half-life of plutonium, but some casks have already been discovered to have defective welds and cracks.[24] No underground site that does not have the potential to leak some of these wastes into aquifers has ever been found.

Third, a nuclear power plant is a prime target for terrorists. Blowing up the reactor, or even the stored wastes, would contaminate a vast surrounding area. Fourth, nuclear power plants have disastrous accidents—witness Windscale, Three Mile Island, Chernobyl, and Fukushima among the many other meltdowns. Fifth, nuclear power is incredibly expensive when we assess all the real charges, including the fuel cycle, the insurance (paid by the government since no private insurance company will insure a nuclear plant), the cost of perpetual storage, and the fact that the plants have a lifetime of less than a hundred years. The contaminated plant then has to be decommissioned, taken apart and buried at a cost no one has yet calculated. In fact, by 1996 some 86 reactors had already been closed after having an average service life of less than 17 years.[25] Lastly, operating plants frequently have spills or releases of radioactive material.

## Contamination from Weapons Production

Contamination occurs from testing weapons, from the enrichment and bomb making plants, from reactor meltdowns, and from general operating leaks. From 1945 to 1970 nuclear weapons were tested above ground, pouring radioactive material into the atmosphere, and which then settled out on the land and waters, contaminating people, soils, gardens, etc. From 1970 to 1997 they tested underground (but still experienced some leakage to the atmosphere), adding up to 1,909 nuclear tests. The U.S. tested in the desert southwest and in the South Pacific.

At Bikini and Eniwetok atolls 66 nuclear blasts were set off so that "few of the islands comprising Eniwetok Atoll are habitable, because of the subsurface radioactivity. The food chain is contaminated."[26] The March 1, 1954, "Bravo" test of a 15-megaton hydrogen bomb shot its fireball twenty miles up into the air and its radioactive cloud contaminated 7,000 square miles of ocean dotted with numerous islands, one of which was Rongelap. By afternoon radioactive ash began to fall on the island, piling up in shallow drifts. By nightfall the people were ill, vomiting and experiencing diarrhea. Three days later the U.S. Navy took them off the island. Later they were returned to the island, where the food supply was now contaminated. Twenty-five years later one-third of the original, exposed group was dead of cancer and the women were regularly miscarrying or delivering babies lacking bones and with exposed brains, the infamous "jellyfish" babies.[27] United Nations studies conclude that the 250 warheads detonated over the islands in the Pacific resulted in approximately 150,000 deaths of indigenous peoples from radiation-induced diseases, as well as obliterating six islands from the face of the Earth and rendering 12 other uninhabitable. The International Physicians for Social Responsibility (winners of the Nobel Peace Prize for 1985), estimate that the atmospheric testing will cause 2.4 million cancer deaths.[28]

## Contamination at the Bomb-Building Sites

Devastating nuclear contamination occurred at the manufacture and testing sites where the former Soviet Union and the U.S. made the atomic and hydrogen bomb warheads and where they dumped their wastes. The Fernald, Ohio, nuclear weapons plant stored hundreds of tons of radioactive wastes in leaking containers from which radon gas escaped into the atmosphere. Unlined pits were also used as dumping grounds for liquid wastes, which later leaked into the aquifer that supplies Cincinnati with its drinking water. The pits also overflowed into Paddy's Run creek and thence into the Great Miami River and into the Ohio-Mississippi system. Fernald engineers also constructed a secret pipe that illegally pumped wastes directly into the Miami. At the Mound, Ohio, plant so much tritium and plutonium leaked into the aquifer that as of 1990 it was being pumped out in large quantities and dumped into the river, in an effort to dilute the contamination in the aquifer. At the Savannah River nuclear site, some 35 million gallons of radioactive waste were being stored in decaying containers, and radioactive cesium-137, cobalt-60 and europium-152 were found in river water 150 miles downstream from the plant.[29] But it was the Hanford, Washington, site that was the flagship of nuclear pollution. Estimates to clean it up ran to $100 billion and doing so is expected to take 50 years.[30]

Peter Goin described the "Burial Gardens" at the plutonium finishing site at Hanford:

"At the Hanford Nuclear Reservation the nature and scale of ground contamination are beyond easy imagination. The yellow ropes and posts are everywhere. (Yellow is the color universally used to denote contamination.) Burial signs, "gardens," and radioactive notices are commonplace."[31] Much of the hot waste is temporarily contained in steel tanks. "Radiation levels in some of Hanford's readily accessible tanks are high enough to melt monitoring cameras."[32] The 570-square-mile "reservation" at Hanford contains an estimated 30 million cubic feet of radioactive waste and 3 billion cubic feet of radioactive and chemically contaminated soil. The Government Accounting Office believes that more than 750 sites on the base violate the Comprehensive Environmental Response, Compensation and Liability Act of 1980.[33] For its entire life, Hanford drew water from the Columbia River, ran it through the reactor cores to cool them, and then discharged in back into the river. Some 20,000 curies of radioactive waste went directly into the river and bio-concentrated in the salmon to 170,000 times river water levels, with a consequent high mortality of the fish.[34] In 1991 the Columbia was called the "most radioactive river in the world"[35] (although this was before knowledge about the Techa River in Russia was available). One glass of water per day per year from the river, 30 miles downstream at Pasco, yields 12 times the background radiation received by an average adult in a year. Something above 500,000 curies of radioactive iodine were released into the atmosphere between 1944 and 1957, including at least one deliberate release (in 1949), compared to 24 curies released at the Three Mile Island nuclear reactor meltdown. Goin writes:

> In a typical eight-hour shift at Hanford, more than 340 gallons of liquid high-level waste combined with hazardous chemicals were deposited in drums, more than 55,000 gallons of low-to-intermediate level wastes were disposed to "cribs" or "trenches," and more than 2.5 million gallons of potentially contaminated cooling waters were disposed to ponds.... Radioactive "swamps" were created when cooling water was collected in basins. Still other wastes, including plutonium, were injected deep into the ground.... The cumulative volume of liquid wastes discharged to the environment from Hanford surpassed 200 billion gallons, enough liquid to cover Manhattan to a depth of over 40 feet.[36]

Hanford has already released twice the radiation that escaped from the nuclear meltdown at the Soviet plant at Chernobyl.[37]

In the Soviet Union there was similar contamination. At Mayak, where they made nuclear bombs, they poured millions of tons of cesium, strontium and other highly radioactive wastes directly into the Techa River. Thousands of people living downstream received average doses of radiation four times greater than those received by the victims of Chernobyl. Radiation from this dumping was traced through the river system all the way to the Arctic Ocean. "Acute" radiation continues in nearby farms.

## Leaks and Meltdowns

While proponents of nuclear reactors say the risk of a meltdown is low, the consequences are catastrophic when it happens. "A large reactor contains as much radioactivity as the fallout from several thousand Hiroshima bombs. A large reprocessing plant can contain dozens of times more."[38] A nuclear chain reaction must be controlled or it will run away, resulting in astronomically high temperatures. The core of the reactor must be kept immersed in constantly circulating water and the radioactive fuel rods must be damped by inserting control rods that will absorb the radiation. A cooling-loss accident, either through a leak, pump failure, or being shut off by mistake, will rapidly raise the heat in the core. If it happens before it is noticed, or just too quickly for response, the fuel rods will melt and distort, preventing the damping rods from being lowered in, and the result is a meltdown. Unchecked, the core will melt through the steel and concrete floor into the ground and a steam explosion will blow off the containment dome, spreading radiation downwind. Stored spent fuel can also overheat and a steam explosion can blow it up. Both a meltdown and steam explosion happened at Fukushima, but first, we should look at a few examples of leakage.

The Muhleberg reactor near Bern, Switzerland, leaked cesium 137, contaminating Lake Biel, which is used for drinking water for the town of Biel. The Palisades reactor, built in 1971 along the shores of Lake Michigan, has been leaking into the lake for years. In Mexico, the Laguna Verde reactors on the Gulf of Mexico north of Veracruz have contaminated a 50-mile strip of ocean with cobalt, strontium and other radioactive isotopes.[39] In August 2011, fish taken from the Connecticut River downstream of the Vermont Yankee Reactor showed contamination by strontium-90. Tritium is found in the soils surrounding the reactor.[40] At the Indian Point Reactor, 30 miles north of Manhattan, the refueling-cavity liner—which has as its sole purpose preventing leaks after a seismic event—has been leaking for years at up to 10 gallons a minute. The El Dorado nuclear plant on the Canadian shore of Lake Ontario leaked waste and leachate for three weeks (March and April of 1980) at the rate of 40,000 gallons a day, elevating radium levels in that section of the lake to 40 times the safe drinking water levels.[41]

Reactor meltdowns are spectacular events.

## Chernobyl, April 26, 1986

During a planned test of the reactor, human errors caused a loss-of-coolant accident. Within moments the 2,000-megawatt reactor was on its way to meltdown, explosion and fire. The explosion blew off the reactor's 2000-ton steel cover plate and obliterated the containment structure. Radioactive material was blown high into the sky. The graphite in the reactor immediately ignited and the fire raged for ten days, lofting huge amounts of radioactive materials into the atmosphere. Some 7,000 kilograms, containing somewhere between 50 and 100 million curies, were released into the biosphere.[42]

A fortunate wind pattern blew the worst of the radioactivity away from the nearby city of Kiev, averting a catastrophe in the city of 2.4 million people. However, the cloud eventually dispersed and extended over virtually all of Europe as far north as the Arctic Circle, south to Greece, and west to the British Isles. A wide variety of radioactive materials, some 50 different isotopes, some with half-lives of up to 24,000 years were released, and were often borne to the ground by rainfall where, due to runoff patterns, they concentrated in hot spots from Lapland to the Mediterranean. They contaminated the food chain including reindeer in Lapland, fish in Italy's Lake Lugano, vegetables in Germany, and milk in many locations. Later some of these vegetables and the milk were sold to unwary persons in developing countries. The damaged site remains highly radioactive. It was entombed in a concrete sarcophagus, but the radiation has deteriorated that to the point where it has been covered with a second one. Forced evacuation removed 135,000 people, and the Soviet scientist in charge of the now-entombed and still dangerous radioactive blob at the bottom of the ruined reactor estimated that between 7,000 to 10,000 people died of radiation in the heroic effort to contain the catastrophe.

## Fukushima

On March 11, 2011, an earthquake and a 50-foot tsunami engulfed the Fukushima Daiichi nuclear installation on the coast of Japan, destroying the cooling apparatus and resulting in heavy damage to the four-reactor complex. Once the diesel generators were flooded and failing, the operators should have flooded the reactors with sea water, which could have cooled them quickly enough to prevent meltdown, but flooding was delayed because it would permanently ruin the costly reactors. Flooding finally commenced only after the government ordered it, and by then it was too late.[43] As the water boiled away in the reactors and the water levels in the fuel rod pools dropped, the reactor fuel rods began to severely overheat, and to melt down. In the hours and days that followed, Reactors 1, 2 and 3 experienced full meltdown.[44]

The damaged reactors and stored fuel rods continued to spew radiation; "scientists affiliated with the Nuclear and Industrial Safety Agency said the plant had released 15,000 terabecquerels of cancer-causing Cesium, equivalent to about 168 times the 1945 atomic bombing of Hiroshima...."[45] The International Atomic Energy Agency said a soil sample from Litate, a village of 7,000 people about 25 miles northwest of the plant, showed very high concentrations of cesium 137—an isotope that produces harmful gamma rays, accumulates in the food chain and persists in the environment for hundreds of years.[46] "By August, 2011, Fukushima was leaking radiation at one spot at 10,000 millisieverts per hour, a fatal dose for anyone coming into contact with the emissions. It could have been more since that reading is as high as Geiger counters go. (A dental X-ray is 0.005 of a millisievert)."[47]

"Radiation spewed for months from cracked reactors, waste storage pools, drainage ditches and broken ducts, poisoning soil, rainwater, tap water and the sea and foods. In the town of Okuma radiation was 25 times what the government deemed safe. Milk, eel, fish, tea, seaweed, flavorings, cabbage, spinach, beans, mushrooms, cauliflower, turnips, canola, broccoli, parsley and rice were all contaminated. Hay fed to cattle was found to have levels of radioactive cesium 250 times the official limit."[48] Tons of radioactive water were running into the Pacific Ocean, contaminating 600,000 square miles, where "concentrations peaked at 50 million times pre-existing ocean levels. A French study 'said that the amount of nuclear material called cesium 137 leaked by the Japanese plant, Fukushima, has proven to be the world's worst nuclear sea contamination event ever.'"[49]

## Conclusion

Not only is using radioactive material the most difficult way imaginable to boil water, it is a moral atrocity. What gives one people the right to get a few decades of electricity while condemning hundreds of future generations to the task of dealing with the deadly contamination? E.F. Schumacher summed up a common judgment on nuclear power reactors, writing:

No degree of prosperity could justify the accumulation of large amounts of highly toxic substances which nobody knows how to make "safe" and which remain an incalculable danger to the whole of creation for historical or even geologic ages. To do such a thing is a transgression against life itself, a transgression infinitely more serious than any crime ever perpetrated by man. The idea that a civilization could sustain itself on the basis of such a transgression is an ethical, spiritual and metaphysical monstrosity.[50]

When Pierre Curie (who along with his wife, Marie, discovered radium and set us on the path to nuclear war and power generation) gave his Nobel lecture, he asked a fateful question: "Is it right to probe so deeply into nature's secrets? The question must be raised here whether it will benefit mankind, or whether the knowledge will be harmful."[51] Today the answer is clear. Splitting the atom was one of Hypercivilization's major mistakes.

# 4

# Environmentally Induced Diseases and the Genetic Revolution

The revolution of Hypercivilization changed humanity's relationship to the biosphere on an even greater scale than did the invention of agriculture and cities. It magnified some of the older disease patterns, blunted others and created new ones. The most radical change was gene modification, and especially the technique known as CRISPR.

## Environmentally Induced Diseases

In the mid–twentieth century humans thought they were on the verge of eliminating infectious disease. In 1948, the U.S. Secretary of State declared that the conquest of all infectious diseases was imminent. They believed that with DDT, malaria could be completely eradicated. In 1967, the U.S. Surgeon General, William H. Stewart, announced that the time had come "to close the book on infectious diseases."[1] What actually happened was that some of humanity's oldest diseases spread more rapidly due to the concentration of billions of people in cities, the disturbance of wilderness areas, expanding agriculture, wars, and greatly increased global mobility. Easy air travel combined with dense urban populations made for pandemics. In Sao Paulo, one of Brazil's megacities with its 20 million people and dire slums, an epidemic of meningococcal meningitis involving 11,000 cases broke out in January–August 1974. In a pattern typical of the twentieth century, the bacterium had developed a resistance to the meningitis vaccine. It was especially severe among children, who would go from well to severely ill in a matter of minutes and within 12 to 20 hours would experience kidney hemorrhages, coma and death. Survivors often experienced brain damage, other long-term neurological disorders and, in some cases, loss of limbs. This epidemic seems to have originated in West Africa, making its way across the Atlantic, probably on airplanes. By 1975, 74 million people a year were traveling across international borders by air, and by 2017 the number rose to four billion, each a potential disease vector. The world had become a single unified and contiguous field for disease organisms.[2]

Some disease organisms became more virulent and others developed resistance to antibiotics due in part to factory farming. New diseases emerged, AIDS, Zika, and Covid-19. Increased air pollution from coal-fired power plants and diesel engines contributed to respiratory diseases, especially among children and the elderly. The World Health Organization reported that air pollution claimed seven million lives around the world in 2012, a third of them in the developing world, where rates of heart and lung disease have been soaring. The agency identified air pollution as the "world's single biggest environmental health risk."[3] It was also implicated in diabetes.[4]

## The Special Case of Warfare and Disease

Wars create a perfect environment for disease. As the scale of wars has grown, so too, have the casualties attributed to disease. In the American

**Incoming Aircraft: Disease vectors need no longer rely on birds and insects to spread them to receptive hosts, including humans. Thousands of aircraft take off and land somewhere else every day, making possible the rapid spread of pandemic disease (Nieuwland Photography, Shutterstock).**

Civil War, 357,000 soldiers died of disease. In the Spanish American War, the disease rate was ten times the number of combat deaths for the U.S. In the Philippine phase of that war it is estimated that several hundred thousand died from disease.[5]

The most common diseases associated with war are cholera, dysentery, plague, smallpox, typhoid fever, louse-borne typhus, scurvy, influenza, measles, and yellow fever.[6] Often civilians are more affected even than combatants. The most famous modern pandemic was the global outbreak of influenza after the First World War. J.S. Oxford "recently hypothesized that the pandemic commenced in the large British base camp at Étaples in Northern France during the winter of 1917, linking over-crowding in the many large camp hospitals with the putative sources of the virus, the camp piggeries and live geese, duck and chicken markets in nearby townships."[7] Whether it started there or in an army camp in the U.S., it moved around the world in five months, traveling with its human carriers on the crowded troop ships and railroad trains that brought the soldiers home from the battlefields. It swept along in three waves, getting more virulent each time. Twenty thousand people died in New York City, 500,000 in the U.S. and 21,000,000 world-wide. In some cases it could kill within 45 minutes of the first appearance of symptoms. The current war in Syria has seen the destruction of nearly all the country's hospitals and clinics, and healthcare is practically nonexistent while refugee camps swell to overcapacity, creating breeding grounds for disease.

## Disease and Development

Environmental diseases also arise from what we are pleased to call "development." The many-faceted environmental changes associated with development have brought on more than one new threat to humanity. The Green Revolution movement, with its massive use of pesticides and elimination of natural competition and predation, backfired on the malaria control program. The Indira Gandhi Canal, dug in 1958 to irrigate the desert in Rajasthan, resulted in farmers switching to high-water-demand crops, and many workers migrated into the area seeking work. "The canal—445 kilometers long—turned out to be an ideal breeding site for mosquitoes…. The heavy rains brought the farmers rapidly spreading cerebral malaria … causing the brain to hemorrhage."[8] In Africa, the Daima Dam, built on the Senegal River in 1985, irrigated 40,000 hectares of desert. The irrigation canals, which were the only source of water for drinking, bathing and laundry, became infected with diarrheal bacteria. The dam also halted the influx upriver of salt water from the ocean, which had previously kept snails in check, creating perfect conditions for an epidemic of snail-born schistosomiasis. Sixty percent of the population contracted the disease, with its bloody urine and stools, vomiting, diarrhea, coughing, and swelling of internal organs.[9] Worldwide, some 200,000 people die from schistosomiasis each year. Clearing the Trans-Amazon highway in 1976 led to an increase in malaria, from two- to tenfold at various locations along the road, as people came into contact with vectors previously confined to feeding on animal hosts. And in many areas of Asia, "a new epidemic of Japanese encephalitis was brewing consequent upon the expansion of paddy rice cultivation."[10] Meanwhile, the use of DDT led to drug-resistant strains of mosquitoes. By 1975, the worldwide incidence of malaria was 2.5 times what it had been in 1961.

Some of the new diseases are horrific, such as Marburg Disease. In 1967, 196 pharmaceutical workers in Marburg, Germany, came down with a ferocious and excruciatingly painful hemorrhagic disease. It attacked the nerves and resulted in horrendous symptoms, including a painful shedding of skin, massive hemorrhaging and, in many cases, death. Survivors suffered chronic hepatitis, impotence and for some, psychosis from the horrifying experience of the pain. It was traced to three shipments of monkeys from Uganda, where an epidemic had broken out among the primates in 1961. Nine years later cases were reported in Australia, and these were less understood than even the German incidents. Marburg Disease remains at large and its aetiology almost a total mystery.[11] It is only one of several such diseases, among them Bolivian Hemorrhagic Fever, which escaped the South American rainforest in 1952, also as a result of environmental changes. Clearing for corn plantations resulted in an exploding population of the Calomys field mouse, the carrier of the virus. The mice overwhelmed the town of San Joaquin and infected the people whose food supplies, drinking water and persons were exposed to the urine and feces of the mice. The Calomys mouse was also the vector in the outbreak of another hemorrhagic disease, Argentine Junin virus. At the same time, massive DDT spraying in the region killed the cat population, eliminating the most important predator of the mice. Some 20,000 people were affected, with high fever, headache, low blood pressure, vomiting, dehydration and internal hemorrhage and shock. One-third of them died, drowning in their own blood.[12] Mice were also implicated in the outbreak of the deadly Hanta virus in the American southwest.[13] Rodents carrying various Hanta strains and other viruses and bacteria are on the increase in urban areas of the world as governments cut funds for their control.

Another example of modern urban conditions contributing to outbreaks of disease is that of dengue fever. Incidences of dengue hemorrhagic fever and dengue shock syndrome were skyrocketing at century's end. The

Harvard Working Group reported that "Between 1986 and 1990, an annual average of 267,000 cases were reported, as compared with an average of 29,803 cases in previous [comparable] years."[14] The female *Aedes aegypti* mosquito breeds in urban conditions—small containers of water such as birdbaths, gutters, barrels, and old tires. So too does the Zika virus, which causes microcephaly in the womb. Areas at risk now include nearly all of South and Central America and Mexico, most of central Africa, Florida and Texas.[15]

The general result has been that exposure to pathogens has increased from a variety of sources including the anthropogenic eutrophication of coastal waters that has allowed increased toxic plankton blooms and increased bacteria and virus populations. At the same time various immunosuppressants have lowered resistance to disease around the world so that "Old diseases have come back, new diseases have appeared, and the public health system has been caught unprepared."[16] The Harvard Working Group on New and Resurgent Diseases concludes: "The emergence of new diseases has been greatly assisted by environmental degradation."[17]

The massive use of antibiotics for livestock and aquaculture contributed to the rapid development of resistant strains of disease organisms and vectors. It was not uncommon in various monocultures, such as the poultry industry—where 100,000 chickens can be raised in close confinement—for diseases to rampage through the stocks. Platt writes: "Monoculture is a major source of the drug resistance."[18] Even as diseases were breaking out of transformed wilderness areas and finding ready hosts in overcrowded urban areas without adequate sanitation or health care, scientists were proclaiming a grand new era of disease control as the result of breaking the genetic code. In the last decade of the twentieth century and the first decades of the twenty-first, great hope for medical miracles again promised that with genetic engineering, not only could disease be "conquered" but that a whole new paradisiacal biosphere would be created via human ingenuity. Others saw grave danger in such notions.

## Altering Life at Its Core: The Genetic Revolution of Recombinant DNA, CRISPR and the Age of the Anthropocene

Late in the era of Hypercivilization humans broke the genetic code and began to alter life in ways never before seen in the history of the planet. Jeremy Rifkin wrote in *The Biotech Century* that this revolutionary development is "likely to be the most radical experiment humankind has ever carried out in the natural world."[19] The new technology of recombinant

DNA allowed humans to combine genetic material across natural, species boundaries, "reducing all of life to manipulate-able chemical materials."[20] Including humans.

A radical new way of thinking about nature was being formulated as a rationalization for this power, a belief that living organisms are without boundaries and that nature is pure process. Thomas Eisner, biologist at Cornell University, argued that a biological species "must be viewed ... as a depository of genes that are potentially transferable. A species is not merely a hard-bound volume of the library of nature. It is also a loose-leaf book, whose individual pages, the genes, might be available for selective transfer and modification of other species."[21] Available to one species—us. It was, as Rifkin said, a "second Genesis."[22] Is it also the ultimate case of Pandora's box?

The terms genetic engineering, genetic manipulation, recombinant DNA, synthetic biology, gene drive, and the latest term, CRISPR-Cas9, describe similar and sometimes overlapping activities. Recombinant DNA, developed in the 1970s, is a process that involves transferring genes from one organism to another, for example, a wheat gene to a rice plant. CRISP-Cas 9, developed in the second decade of the twenty-first century, is a much more powerful technique for editing genes within any organism, but what they all have in common is the manipulation of the fundamental building blocks of life, i.e., genes.

By the 1970s and accelerating in subsequent decades, experiments and ideas that would have seemed fantastic to our grandparents began to make the news, including creating trees that glow in the dark.[23] A graduate student at the University of Wisconsin created recombinant strain of a bacterium that he called "ice-minus," which could be sprayed on fields of strawberries to retard the formulation of ice crystals.[24] Yet again we see an example of a project designed for a single end without first asking about its other ramifications in the web of life and if these might be harmful or even disastrous, another failure to exercise the precautionary principle. Subsequently there was a realization that, should it propagate freely in the wild, ice minus could play havoc with natural formations of ice. In May of 1997, Japanese researchers successfully planted an entire human chromosome into a strain of mice. Some scientists are even working on developing artificial wombs, propelled apparently by the views of the late Joseph Fletcher at the University of Virginia School of Medicine, who said: "The womb is a dark and dangerous place, a hazardous environment. We should want our potential children to be where they can be watched and protected as much as possible."[25] (Misogynistic patriarchy at its finest.) Others have projected the ability to grow headless human clones in artificial wombs, to be used as spare parts for the donors whose cells have been cloned to make the creature.

The ostensible purpose of genetic manipulation is to create organisms that are good for humans and to make a profit for the corporations that quickly came to control much of the research and the results. It is at this point that we leave objective science and that subjective value judgments came to enter into and to guide the science. This is not necessarily always a bad thing. The creation of virtually unlimited supplies of inexpensive insulin by means of gene manipulation has been an incredible boon to people with diabetes. But there is a dark side. Moreover, the welcome that genetic engineering has received is a part of a much larger and controversial cultural thrust.

In the revolutionary development of recombinant DNA, scientists isolate genes or parts of genes and combine them with other genes. An even further departure from traditional and naturalistic techniques for manipulating organisms has been the development of "synthetic biology," a term that seems to be an oxymoron. "Synthetic biology is the design and construction of biological devices and systems for useful purposes."[26] Of course, who defines "useful" is a key issue and useful for a specific purpose may well be highly detrimental for general ecological stability and human health, as we have witnessed far too many times with PCBs, DDT, et cetera, ad nauseam. While "traditional" genetic engineering takes genetic material from living organisms and inserts it into other living organisms, material that has evolved over eons, synthetic biology inserts man-made DNA into a living cell in order to rewrite or rebuild natural systems "to provide engineered surrogates."[27] It has been compared to building with Legos, that is, with mainly interchangeable parts: "synthetic biology puts these 'blocks' together from scratch to build an entirely new strand of DNA which is then placed into an empty living cell. These new cells can be 'built' to perform a number of functions...."[28] Traditional biology has approached the natural world as a system, as an integrated whole, as a way of thinking the world back together, seeing all the connections, whereas in synthetic biology the focus is disassembling and then reassembling, that is, "on ways of taking parts of natural biological systems, characterizing and simplifying them, and using them as a component of a highly unnatural, engineered, biological system."[29] One of the outcomes has been the creation of standardized DNA sequences called "Biobricks" (significantly, a trademarked name), "designed to be composed and incorporated into living cells such as E. coli to construct new biological systems."[30] One of the long-term goals is "to produce a synthetic living organism from parts that are completely understood."[31] Their purpose is "to construct new biological systems."[32]

The rapid and relentless development of genetic manipulation was and is driven by scientific curiosity, the laudable desire to improve human health, and corporate visions of great profit. In 1973 the first transgenic organism was made by inserting antibiotic-resistance genes into the *E. coli* bacterium, a project that seems stupifyingly reckless. Why would one want to make a disease organism resistant to antibiotics? A year later the first transgenic animal, a mouse, was created. Early on some saw the great commercial potential. Genentech Corporation was founded in 1976 and in 1978 produced genetically engineered human insulin. In 1980 a landmark ruling by the United States Supreme Court favored the commercialization of genetic engineering. In *Diamond v. Chakrabarty* the judges ruled that genetically altered life could be patented. In 1983 a biotech company, Advanced Genetic Sciences, applied for a license to field test it. The year of 1986 saw another major development, the first field trials of crops genetically engineered to be resistant to herbicides. China introduced GMO tobacco to the market in 1992. In 1994 a company called Calgene was approved to release for sale a GMO tomato engineered to last longer on supermarket shelves and the European Union approved a GMO tobacco resistant to the herbicide bromoxynil. In 1995 the U.S. Environmental Protection Agency approved the Bt potato, the first pesticide-producing crop approved in the U.S. Bt stands for *Bacillus thuringiensis*, a soil bacterium used as a pesticide. When ingested by insects it punctures their gut. By 2009 eleven different transgenic crops were being grown commercially in twenty-five countries. In 2010 the first synthetically created life form was created by scientists at the Ventner Institute, a bacterium appropriately named "Synthia."[33] Knowing what we now know about the huge microbial environment within and on us, how sensible was it to create a new bacterium without knowing the ramifications?

When we begin to remake human cells we are dealing with large numbers and with very complex interactions. The human genome as defined by the U.S. National Institute of Health, "is an organism's complete set of DNA, including all of its genes. Each genome contains all of the information needed to build and maintain that organism. In humans, a copy of the entire genome some 21,000 genes—more than 3 billion DNA base pairs—is contained in all cells that have a nucleus."[34] The base pairs are the genetic materials that are the building blocks of the DNA double helix. These make up what is sometimes called the blueprint for our bodies and design and regulate the way our bodies function.[35] Our bodies are even more complex because environment also has an influence on how our genes are actually expressed to shape our structure and our body's behavior. It is much the same for all other organisms. Tinkering with this is the business, the big business, of genetic engineering. The whole future of living beings is tied up in their current genetic structure and in the possibility of random mutations in the genetic material.

As a human embryo develops from a single cell, dividing and dividing, the new cells become differentiated, specialized as skin cells or heart muscle cells, and so on. The original stem cells had the capacity to become many different things and are called "pluripotent," i.e., endowed with many potentialities. Once specialized, say as heart cells or liver cells, they can become no other type, or so it was thought up until 2006 when Shinya Yammanaka learned to genetically reprogram adult skin cells to revert to a pluripotent state. Using these techniques, other scientists have created heart cells that actually beat in a lab dish, and they "brought neurons, bone, fat and blood vessels to life" outside a human body.[36]

Sensing danger, the United States Supreme Court outlawed the patenting of human genes in 2013. Myriad Genetics had isolated a human gene associated with hereditary breast cancer and had received a patent for it. Sued by researchers and doctors, the case reached the high court. The Justices said that something found in nature, as in the case of this gene, could not be patented because it was not an invention. Conservative Justice Clarence Thomas wrote: "A naturally occurring DNA segment is a product of nature and not patent eligible merely because it has been isolated."[37] This decision has restricted the commercialization of these techniques to some degree, but a gene that is isolated and then modified in some way can be patented.[38] Modifying humans in some ways using genetics is still illegal in the U.S.; however, in the summer of 2013 the government of the United Kingdom announced that they would permit scientists using in vitro fertilization to create babies with genetic material from three different parents, opening the way for the "first genetically modified humans who could pass down those genetic tweaks to their children."[39]

Because the potential for upsetting delicate ecological balances and compromising human health is enormous, many people are suspicious of these new genetic engineering techniques and their applications. Early on scientists themselves became concerned as a result of an experiment planned in 1971 to introduce SV40 DNA into an *E. coli* cell. This was of concern because SV40 is a monkey virus that can transform monkey cell lines—as well as human cell lines—into a cancerous state. These experiments were postponed and subsequently the National Institute of Health, which was sponsoring much of the rDNA research, set up an advisory committee, and a now-famous gathering of scientists and ethicists was held in 1975 at the Asilomar Conference Center in California to discuss the dangers of a possible escape. Federal regulation was rejected but the NIH did institute guidelines for the research, and later the Environmental Protection Agency and the Food and Drug Administration indicated that no commercial products would be approved unless they had followed the guidelines.[40] Nevertheless the whole field of genetic engineering is highly controversial, with some scientists and the gentech companies arguing they are harmless and necessary, while other scientists, environmentalists and public health advocates refute such claims as either self-interested or short-sighted.

In general, bioengineering is a radical departure from the human past and could bode well for disaster. Stanford University ecologist Steve Palumbi pointed out: "It's not like engineering a machine. If you make a really bad car, it won't remake itself. If you make a mistake with a living thing, you may be stuck with it forever."[41] And as one delegate to a conference on synthetic biology held at Cambridge University said: "There could be thousands of people making millions of invasive species."[42] The main conceptual problem is the same one found throughout innovations of Hypercivilization: it ignores context. In a review of Adam Rutherford's *Creation: How Science Is Reinventing Life Itself,* the review author quotes Rutherford as he describes synthetic biology as a discipline that "takes the principles of biology and reinvents them with the goal of engineering solutions to specific human problems."[43] It's the same old linear thinking that has characterized Hypercivilization for several centuries. Bioengineers see a single issue or problem and then want to design a single GMO to address it, but the ecological and evolutionary context is not amenable to such single shot approaches. As we are continually reminded by our mistakes, you can't do just one thing in nature. Anything introduced is going to interact with many other organisms and ecosystems, and often in ways no one could have foreseen. So while genetic engineering holds out some promise it could also radically upset the biosphere.

The chief concerns about genetically modified organisms fall under five headings: biohazards, accidental biological contamination of non-target organisms, ecological destruction, social justice, and terrorism. The case of the monkey virus was an example of a biohazard. A dangerous organism could escape from a lab and pose a significant threat to human health. It is highly likely that governments, which have long dealt in biological warfare agents (especially the U.S., Russia and China), are using these new techniques to produce very nasty biological weapons that could escape or be deliberately loosed in a desperate wartime situation. Furthermore, the science is not so sophisticated that terrorists could not make use of it. There is almost no control.

In 2014, a small group of hobbyist scientists began a project to use synthetic biology to create trees that glow in the dark.[44] Phosphorescence is a well-known phenomenon in nature, as in the insects commonly known as lightning bugs. Presumably these researchers will create an artificial gene for phosphorescence and transfer it to a compatible species of tree. This small group advertised in the web-based crowd funding site, Kickstarter, for start-up capital and received over $500,000.00, promising to deliver seeds to their 4,000 investors. They failed. Others are trying, and now tout it as a way to save energy by eliminating electric street lighting,[45] but the obvious foolishness of such an outcome—who would want glowing trees all around the landscape?—demonstrates the twofold danger associated with such a project. First the uncontrolled, unmonitored release of genetically modified seeds into the world by a group of amateurs could wreak ecological havoc. Second is the apparently growing number of so-called communal laboratories springing up around the nation as the techniques of genetic manipulation become cheap enough that people can start tinkering in their garages. One of the stated goals of the glowing trees group is "to publicize do-it-yourself synthetic biology 'to inspire others to create new living things.'"[46] Their research will be done at "Biocurious," one of the new communal labs that describes itself as a "hackerspace for biotech research."[47] Their ecologically innocent mission statement reads: "We believe that innovations in biology should be accessible, affordable and open to everyone. We're building a lab for amateurs, inventors, entrepreneurs and anyone who wants to experiment with friends."[48]

If that doesn't strike terror into our hearts, I don't know what will. But it's not just bumbling amateurs we need to fear. It's the genetic scientists at big corporations and government funded universities. Genetically modified crops are now ubiquitous and there are experimental programs to make genetically modified biofuels. Because of the astronomical increase of mono-cropping in industrial agriculture so typical of Hypercivilization there has been a corresponding increase in the "pest" species that attack them. Herbicides such as the glyphosate that Monsanto markets under the trade name "Roundup," are sprayed all over the world to kill weeds, but they only work with crops that are genetically engineered to carry the Bt toxin. The crops are immune to it but it kills all other plants. The obvious longstanding concern is that through aerial drift, many plants beneficial to insects such as the milkweed that monarch butterflies and other pollinators feed on are killed in the process. The GMO crops themselves are highly controversial. In fact, no long-term studies have ever been done on the safety of GMO food, and what studies have been done have been primarily conducted by the corporations that profit from this technology. Even the World Health Organization requires only ninety days of study to conclude that a GMO food is safe, a standard which, had it been applied to cigarettes, never would have detected any human health risks from smoking. Meanwhile evidence of harm continues to mount: "Lab mice fed just a 33 percent GMO diet begin developing aggressive cancers (particularly breast cancer), liver failure, and kidney failure … 50 percent of the males and 70 percent of the female animals on the GMO diet succumbed to early death at an age equivalent to 40 to 50 human years."[49]

Then there is the problem of GMO crops escaping. In three recent cases, GMO crops have escaped the fields where they were planted. An Oregon farmer found GMO wheat growing in his field, a crop that Monsanto had experimented with but that had never been approved for sale, and GMO rice has escaped and contaminated many fields in the U.S.[50] In eastern Washington, a shipment of alfalfa that was supposed to be free of GMO strains turned up contaminated. And in Oregon, Monsanto's field trials of over a decade ago "allowed GE grass to escape and turn up in the Oregon wilderness despite continuous efforts to eradicate it."[51] The Roundup-ready bentgrass had never been approved for commercial release and was being field tested in Eastern Idaho.[52]

The applications of the new technology can apparently be endlessly dreamed up scientists, from those who want to create a bacterium that could turn moon dust into bricks so that we can build a civilization on the moon, to the more prosaic exploring algae for biofuel. In searching to create a super-algae that can be used to make biodiesel, over 4,000 strains have been engineered in a hundred or more projects. The algae projects, funded in part by big oil companies, raise a particular alarm. While not refusing to carry these forward, one scientist said that "efforts to genetically engineer algae, which usually means to splice in genes from other organisms, worry some experts because algae play a vital role in the environment. The single-celled photosynthetic organisms produce much of the oxygen on earth and are the base of the marine food chain."[53] And at a meeting of the President's Bioethics Commission, ecologist Allison Snow pointed out that a: "'worst-case hypothetical scenario' would be that algae engineered to be extremely hardy might escape into the environment, displace other species and cause algal overgrowths that deprive waters of oxygen, killing fish."[54]

## The CRISPR Revolution

> "CRISPR gives us the power to radically and irreversibly alter the biosphere that we inhabit by providing a way to rewrite the very molecules of life in any way we wish."
> —Jennifer Doudna, *A Crack in Creation*[55]

The latest development in the field of genetic engineering is known as CRISPR-Cas9, a way of editing genetic sequences without needing to cross species. Far faster, cheaper, and effective than recombinant DNA, CRISPR is a huge advance over prior techniques of genetic manipulation. Developed primarily in the research lab run by Jennifer Doudna at the University of California at Berkeley, "CRISPR" (pronounced *crisper*) stands for

"Clustered Regularly Interspaced Short Palindromic Repeats. These are segments of <u>prokaryotic</u> <u>DNA</u> containing short repetitions of base sequences. Each repetition is followed by short segments of '<u>spacer</u> <u>DNA</u>' from previous exposures to a bacteriophage virus or <u>plasmid</u>."[56] Using CRISPR, we can far more efficiently alter an organism's germ cells so that the new trait is inheritable through all subsequent generations and can, when coupled with an older technique called "Gene Drive," spread quickly throughout a population. In an article titled *Tweaking Life*, Nathan J. Comp writes: "CRISPR/Cas9 is the biochemical equivalent of a magic wand."[57] Doudna writes that CRISPR gives us "The ability to control the future of life."[58] The world has changed.

A recent article in *National Geographic* pointed out: "CRISPR places an entirely new kind of power into human hands. For the first time, scientists can quickly and precisely alter, delete, and rearrange the DNA of nearly any living organism, including us."[59] Comp's analogy is helpful

CRISPR: Gene editing is the revolutionary technique of altering an organism at the genetic level, making it possible to introduce genes from one species to another. Beginning with the development of recombinant DNA in the 1970s, a difficult and costly process whereby genes from one organism could be transferred to another, creating a new creature, the cheaper and easier technique CRISPR-mediated genome editing developed by Jennifer Doudna and colleagues in 2012 can simply alter the genes of an existing organism without recourse to a different species. It opens up a huge array of ethical issues and questions of simple prudence; it may be as big a scientific revolution as splitting the atom (Vchal, Shutterstock).

here: "If the accuracy of transgenics was that of a free fall bomb, CRISPR/ Cas9 is a laser guided missile capable of editing."[60] Note the military language.

With this new technique, the DNA of an organism can be cut and new genetic material precisely inserted to alter not only the organism itself, but all its descendants, driving it into the entire population. To take a favorite example, a gene for sterility could be engineered into mosquitoes that would result in the radical diminution of their numbers or even in making them extinct. Genetic modifications could be introduced into the sperm or egg of human parents—for example, genes for making their children more muscular or taller or smarter, or just about anything. The Army is wondering about the possibility of "making" bigger, stronger, more aggressive soldiers. In November of 2018, Chinese scientist He Jiankui claims to have altered genes in seven embryos, one of which has come to term. He was denounced by the Chinese government, which said he had broken the law, and in general by scientists and ethicists world-wide.[61]

Nevertheless, some observers speculate and even rhapsodize about the possibilities. *Smithsonian* magazine recently carried an article with the frightening title, "Kill All the Mosquitoes?!," where we read, "you could, in theory, wipe out an entire species of mosquito. You could wipe out every species of mosquito." And even more alarming: "We can remake the biosphere to be what we want, from wooly mammoths to non-biting mosquitoes."[62] Other alluring hints are made in these two articles about eliminating AIDs and curing cancer, Alzheimer's, Tay-Sachs disease and many others. Jennifer Doudna is excitable to the point of giddiness about the potential for eliminating genetic diseases, and who would not be? Some even think it could be used to save endangered species by inserting lost DNA into gene pools that had become so small as to not be diverse enough to survive. Others are already working on "de-extincting" the woolly mammoth. In short, there is a ring of Paradise just around the corner in all this talk. Who could be against it? Both of the magazine articles raised some doubts, but the overall tone was excitedly optimistic, unlike that of critic, Gris Anak, who wrote in his blog: "We didn't start tinkering with the nuts and bolts of genetics until the 70's and historically it's been a pretty exclusive thing; you needed a specialized lab with hundreds of thousands of dollars of equipment and a staff of highly trained scientists to make the most basic changes.... But [with] CRISPR that exclusivity proviso goes out the window ... any halfway decent lab can use it to alter the DNA of anything."[63]

"Genome editing started with just a few big labs putting in lots of effort, trying something 1,000 times for one or two successes," says Hank Greely, a bioethicist at Stanford. "Now it's something that someone with a

BS and a couple thousand dollars' worth of equipment can do. What was impractical is now almost every day."[64] The developers of CRISPR immediately saw the good it could do, especially with regard to controlling genetic diseases, but they were naïve in releasing it to the entire world. Now, in fact, even high school students can perform this trick for only $130 dollars for a CRISPR kit.[65] In fact, even a computer can do it.[66] We have to wonder, how many garage labs will be or already are experimenting with the future of the biosphere?

In our race to make a perfect world for ourselves, we forget how many mistakes we have already made and that we are not the only beings on the planet—we are totally integrated with countless other species, a fact which ought to make us cautious. Tinkering with them, we tinker with our life support system. CRISPR raises profound ecological and ethical questions and issues of who we are as humans and what is our rightful place in the Creation. While recent human activity has led to unprecedented rates of extinction, we haven't set out to do that deliberately. It's been a tragic and unintended consequence of population explosion, habitat alteration as we destroy wild and semi-wild places, and of side effects of technologies that aim at some other end. With CRISPR we can eliminate and alter them intentionally. Should we?

What about unintended consequences? One scientist points out that if we eliminated the malaria borne mosquito another would quite likely take its place, perhaps one that carries Dengue fever or some other disease. And what about the bird and bat species that rely on that particular hatch of mosquitoes, and so on? One thing we do know for sure is that there are countless and sometimes hidden connections in the biosphere. The strength of engineering in general is that is focuses on solving one specific problem in order to achieve one specific end, like figuring out how to bridge a river. That's also its profound weakness since you can't do just one thing in nature. Furthermore, an engineering approach does not ask whether that river should be bridged, or what will be all the environmental and social changes that will follow on bridging it.

What about accidents? Kevin Esvelt, an ecologist at MIT, points out: "But as soon as you're thinking of a gene drive technology you have to assume whatever you're making will spread once it gets out of the lab. Human error will win out, if not deliberate human action."[67] Some observers worry that this technique could escape the control of carefully regulated scientific labs and find its way into the hands of amateurs working in unregulated and unsupervised garage labs or places like Biocurious making ecologically destructive organisms. This is not a fantasy according to the *National Geographic* article: "Do-it-yourself biology is already a reality; soon it will almost certainly be possible to experiment with a CRISPR kit in the same way that previous generations of garage-based tinkerers played with ham radios or rudimentary computers. It makes sense to be apprehensive about the prospect of amateurs using tools that can alter the fundamental genetics of plants and animals."[68] Comp quotes Dustin Robinson, a professor at University of Wisconsin, referring to the desire to establish rules and regulations governing the irresponsible use of the technique, as was done with recombinant DNA experiments: "Rubenstein says with so many more labs across the globe it wouldn't be hard for poorly funded labs or rogue geneticists to cross ethical lines."[69] And Robert Streiffer, a bioethics professor at the University of Wisconsin, concludes: "We're not going to be able to put a lid on it as much as people might like."[70] All of this needs to give us pause. Comp believes that "It's the end of the biological world as we know it."[71] And, "this may well be a watershed in human history—the world before CRISPR/Cas9 and the world after."[72]

## The Age of the Anthropocene?

Some people are talking about a fundamental shift in our way of conceiving humanity's relationship to nature, arguing that we have left all natural history behind entered a new geological age where, for the first time, we humans will take control of the natural world and reshape it at its most fundamental levels of biology and geology. Advocates of employing the new powers of science to create a new Earth and a new human nature were calling this the "Anthropocene," the age of humans.[73] Using genetic engineering we will create new life forms and, combined with geoengineering or "terraforming,"[74] we will create a benign environment for the expansion of the human enterprise on Earth. We hear in these voices secular echoes both of the old Christian idea of original sin and the Enlightenment idea of the perfection of humanity. For example, scientist Robert Sinsheimer betrays a dark view of human beings and his carefree desire to "improve" ourselves, writing: "For the first time in all time, a living creature understands its origin and can undertake to design its future.... Man is too clearly an imperfect, a flawed creature.... We now glimpse another route—the chance to ease the internal strains and heal the internal flaws directly—to carry on and consciously perfect, far beyond our present vision, this remarkable product of two billion years of evolution."[75]

Recombinant DNA, synthetic biology and CRISPR are all underlain by a peculiar mind-set, the engineering mentality. While many good things come from engineering, it is a mechanistic science of analysis, of deconstruction into parts and reconstruction in order to achieve specifically human ends. It is highly useful in building bridges, safe airplanes and skyscrapers. But applied to the natural world it ignores the lessons of evolutionary biology and ecology, is always innovative and pragmatic, and operates on a short-term horizon, seeing and treating the world as a machine rather than as an organism. In a machine the whole is made up of the sum of its inert parts and can be taken apart and reconstructed. But no machine is as complex as a simple tree frog. The entire natural world is too complex, too incompletely known and too mysterious to approach with such a blunt instrument.

One of the disturbing aspects of these advocates is the nature of their discourse—the very words they use in discussing genetic manipulation. It is the language of power over, of dominator society, and includes militarized language. Jennifer Doudna, the primary developer of CRISPR, is an example. In *A Crack in Creation*, she speaks of "command and control," and, of "targets," "biological mastery," "gene knockout, or KO," the "power" to "rewrite the code of life" and that we can "bend nature to our will," "primary authority over life's genetic makeup."[76] She writes of insects as "winged pests" that could be "wiped out" with this new "best weapon" against this "pervasive threat."[77]

It's a blatantly anthropocentric view of the world. Their whole approach privileges a species that is already out of control. In *A Crack in Creation*, Doudna hardly mentions ecosystems, the biospheric context of human civilization. This approach fails not merely on the basis of morality, i.e., its presumption that whatever humans do is OK, but on the basis of prudence, i.e., that we humans at least need to protect ourselves from the folly of wrecking the natural base of life which supports us. It is hard to believe how ecologically naïve these folks are.

Typical of these advocates for the Anthropocene, Doudna argues that we have been reshaping nature all along, citing the breeding of cattle and plants. However there are natural limits to this sort of traditional selective breeding and planting. In the main, one could not cross species boundaries. But, even more importantly, this argument that humans have been altering the genetic make-up of plants and animals since the dawn of agriculture, and therefore genetic manipulation is no different, fails because it does not take into account that vast increases in scale (both in speed and in the ease with which we can now alter germ lines by CRISPR), creates a difference in kind. To illustrate, one fisherman relieving himself in a trout stream will not affect its ability to produce trout, but 10,000 will destroy it. Genetic manipulation is not the same thing as historic breeding practices. It is an unprecedented revolution.

To believe that humans should attempt to "create a new heaven and a new earth," as the Bible has it (Rev. 21:1) seems to many a dangerous illusion. Believing that we have at last achieved the status of the overlords of nature, which we will now recreate to suit our own convenience, is colossal hubris. Recreate, or wreak havoc, as we have already done with far lesser technologies? All over the world soils are being destroyed, deserts are spreading, beneficial insects attacked, waters polluted, aquifers drawn down, species are disappearing forever, the oceans are inundated with plastic and becoming acidified, and the climate is destabilized. The web of life is being torn apart and most people are not even noticing. And now we want to define ourselves as the managers of evolution, as its ascending arrow? To the contrary, are we not the ascending arrow of destruction? Without reflection, it seems, we may be slipping into a collective form of insanity. Whether we think it so or not, we are not gods. If anyone doubts that they need only look at our track record. The arrogance of Age of the Anthropocene is just the final illusion of Hypercivilization, preceding the stage of breakdown.

We are a clever species. Look at all the marvels we have made. But we are not a wise species, not yet anyway. We have made some profound mistakes, like splitting the atom. It produced a bomb that may have ended a terrible war a few weeks early in 1945, but it has left us with the potential for nuclear holocaust and has spread around the world deadly radioactive toxins that will last for ten times as long as civilization has yet existed. We all recall the nuclear power industry telling us how safe its technology was but we also remember Chernobyl and Fukushima. They are still pouring radioactivity into the environment. But we don't have to look to something as exotic as nuclear weapons and nuclear power for evidence of our incapacity to handle complex technologies safely. Simply look at our mistake with fossil fuels. We didn't know we'd alter the climate, sending it spiraling into profound changes that threaten global well-being. What about DDT and the other hard pesticides? What about our simple inability to control our species reproduction until now we are facing critical water shortages, food shortages, a massive loss of biodiversity, soil exhaustion, polluted oceans, and on and on. Our history as planet managers leaves a lot to be desired. What make us think we can handle CRISPR? We have been wrong so many times before and going forward with genetic engineering beyond the lab could lead to mistakes on a colossal scale.

Henry Greely, a Stanford law professor and bioethicist asks: "Do we want to live in nature or in Disneyland?" He goes on to give us some wisdom: "I would want there to be some consideration and reflection, and a social consensus before we take that step."[78] We could easily be on a slippery slope. If mosquitoes, then why not rats, then raccoons (they're pesky), and other species that someone doesn't like? Who gets to decide? We certainly don't want pharmaceutical corporations deciding on the basis of their idea of the good—namely bottom-line profit. We don't want some government bureaucrat deciding. Eric Lander, the prominent scientist who was Director of the Human Genome Project, says: "And you'd better be able to say that society made a choice to do this—that unless there's broad agreement, it is not going to happen. Scientists do not have standing to answer these questions."[79]

# 5

## Climate Change
### *The Crowning Blow*

"Climate is an angry beast, and we are poking at it with a stick."[1]—Colombia University scientist Wallace Broecker

"We've already done deep and systemic damage."[2]
—Bill McKibben, 350.Org

*"Have you looked into the eyes of a climate scientist recently? They look really scared."*[3]—Michael Pollan, "Why Bother?" in *Drawdown*

### *Human-induced Climate Deterioration*

A stable civilization requires a stable climate. Climate change is just one more element in the "perfect storm of environmental problems now confronting humanity."[4] In the last 250 years we have been changing the climate for the worse by burning of fossil fuels, putting ever-increasing concentrations of carbon dioxide in the atmosphere where it acts like the windshield on an automobile, trapping the sun's heat, warming the atmosphere, soil, and water bodies. Dr. Jerry Malhman, a scientist at the National Center for Atmospheric Research, testified before Congress: "Our burning of fossil fuels ... is the indisputably direct cause of the ever-increasing concentrations of carbon dioxide in the atmosphere. This added carbon dioxide acts directly to warm the planet. There is no scientific controversy about these facts."[5]

The warming produces intense and more frequent storms, deadly heat waves, droughts, melting of the sea ice, ice caps, and glaciers, rising sea levels, and acidification of the oceans. These all then have cascading adverse effects on plant and animal species and extremely negative social and economic impacts. Deforestation, industrial agriculture and soil deterioration also contribute to the build-up of $CO_2$ in the atmosphere and cattle contribute methane, a greenhouse gas 25 times as powerful as $CO_2$. The rapidly increasing global population, and even more rapid increase in meat consumption, has resulted in increasing numbers of these animals and an increase in the amount of added methane. Humanity's herds were giving off 2.2 billion tons per year.[6] A rice paddy can give off 115 million tons of methane per year. Fracking is another source of methane release.

The mean temperature of the atmosphere is the result of a balance between the amount of solar energy that is reflected into space and the amount trapped by the upper atmosphere. Earth is a natural greenhouse. Without the naturally occurring layer of carbon dioxide and other gases the earth would cool to about minus 18 degrees centigrade. Our position vis-à-vis the sun is crucial. Our nearest neighbors, Venus (700 degrees) and Mars (minus 100 degrees) are, respectively, far too hot for human life and far too cold. The livable temperatures on earth are the result of a delicate balance of incoming solar energy and the amount that is re-radiated into space. What makes Hypercivilization such a radical discontinuity is the huge increase in the extraction and burning of fossil fuels. Also, the planet is running out of "sinks," natural features such as rainforests that trap carbon and keep it out of the atmosphere.

One of the disturbing revelations is the phenomenon of positive feedback. As the atmosphere warms, it triggers other changes in the biosphere which then further accelerate the warming. "Feedbacks are the guts of the climate problem."[7] An example of positive feedback is the shrinking of polar ice sheets which reflect sunlight back into space. As they melt they are replaced by dark water which absorbs solar energy. The larger the area of absorbent warmer water becomes, the more it undermines the remaining sea ice from below, exacerbating the melting so that the water absorbs yet more heat and melts yet more ice and the Earth warms even more, and on and on. The result—the hotter it gets, the faster it gets hotter. Thus the climate does not change at a gradual, even pace but can "lurch" in a step-wise process.[8] Another example is the melting of the permafrost around the Arctic Circle releasing methane that has been trapped for thousands of years. By 2013 methane concentrations in the atmosphere were at their highest in 800,000 years and had spiked from 1950 levels of approximately 1,100 ppb to 1,800 ppb.[9] By 2014, craters were appearing in Siberia, caused by methane bubbles pushing up through the permafrost, which had been warming and softening for the previous twenty years. A crater in the Yamal peninsula, thirty meters across and over seventy deep, registered methane at 53,000 times the normal concentration in air. The summers of 2012 and 2013 in Russia were 5 degrees centigrade warmer on average and the permafrost thawed and collapsed, releasing the potent heat-trapping gas.[10]

Coal is the worst source of carbon dioxide, releasing twice the amount of carbon per unit of energy as natural gas, but that should not make us tolerant of any fossil fuel. All forms contribute $CO_2$ to the atmosphere. In the automobile-mad North, cars are a major source of emissions of $CO_2$ in the atmosphere.

> A gallon of gas weighs about eight pounds. When it's burned in a car, about five and a half pounds of carbon, in the form of carbon dioxide, comes spewing out the back.... And no filter can reduce that flow—it's an inevitable by-product of fossil-fuel combustion, which is why $CO_2$ has been piling up in the atmosphere ever since the Industrial Revolution.[11]

Since 1880, when records began to be kept world-wide, the amount of $CO_2$ in the atmosphere has increased by 43 percent and the earth has warmed 2 degrees Fahrenheit as an average over the surface of the planet.[12] This may not seem like much, but it is high enough to cause ice fields to melt and

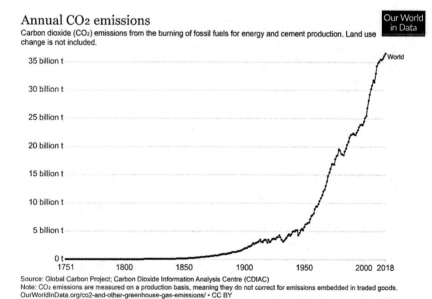

**Annual CO2 emissions**

Carbon dioxide (CO₂) emissions from the burning of fossil fuels for energy and cement production. Land use change is not included.

Our World in Data

Source: Global Carbon Project; Carbon Dioxide Information Analysis Centre (CDIAC)
Note: CO₂ emissions are measured on a production basis, meaning they do not correct for emissions embedded in traded goods.
OurWorldInData.org/co2-and-other-greenhouse-gas-emissions/ • CC BY

**Global Carbon Emissions: The sudden, human-induced increase in $CO_2$ and other greenhouse gases is a shift of geological proportions in the history of the Earth and brings many other changes in its train, most of them threatening to the stability of civilization (G. Marland, T.A. Boden, and R. J. Andres. 2003. "Global, Regional, and National $CO_2$ Emissions." In *Trends: A Compendium of Data on Global Change. Carbon Dioxide Information Analysis Center*, Oak Ridge National Laboratory, U.S. Department of Energy, Oak Ridge, TN).**

the oceans to begin rising at an accelerating rate, as far a foot since 1900. If the current warming trends continue the overall rise could be as high as 8 degrees Fahrenheit, making the planet unsuitable for much of humans and other life forms. While climate shift has occurred in the past, changes of this magnitude required thousands of years. We are now adding about 33.5 billion tons of $CO_2$ to the atmosphere each year, an enormous amount which has increased 120 percent just since 1971; we are increasing the carbon dioxide "at a rate 1000 times faster than in previous global warming periods."[13]

The earliest warning came over a hundred years ago when Svante Arrhenius performed the first calculations of a possible effect on the atmosphere of increased carbon dioxide production. His work was ignored. Then a 1957 article in the journal *Tellus* warned: "Human beings are now carrying out a large-scale geophysical experiment of a kind that could not have happened in the past nor be repeated in the future."[14] We are living in a new world. Dr. James Hanson, former NASA climatologist and considered the Dean of climate scientists, states the situation quite starkly: "The greatest danger hanging over our children and grandchildren is initiation of changes that will be irreversible on any time scale that humans can imagine."[15]

In late 2014, the Intergovernmental Panel on Climate Change, an organization of thousands of scientists paying attention to tens of thousands of climate studies, issued a "dire warning":

> The gathering risks of climate change are so profound they could stall or even reverse generations of progress against poverty and hunger if greenhouse emissions continue at a runaway pace, according to a major new United Nations report … the overall global situation is growing more acute as developing countries join the West in burning huge amounts of fossil fuels…. Failure to reduce emissions … could threaten society with food shortages, refugee crises, the flooding of major cities and entire island nations, mass extinction of plants and animals, and a climate so drastically altered it might become dangerous for people to work or play outside during the hottest times of the year. Continued emission of greenhouse gases will cause further warming and long-lasting changes in all components of the climate system, increasing the likelihood of severe, pervasive and irreversible impacts for people and ecosystems….[16]

Furthermore, climate change is already here. "The group cited mass die-offs of forests, including those in the American West; the melting of land ice virtually everywhere in the world; an accelerating rise of the seas that is leading to increased coastal flooding; and heat waves that have devastated crops and killed tens of thousands of people."[17] To this list we can add more and more intense storms and a longer and more intense forest fire season.

$CO_2$ is not emitted evenly over all the world. In 1998 the average person in the Third World was releasing 0.1 ton of carbon each year, and the average person in the rich Global North was releasing 3.5 tons, or 35 times as much carbon.[18] In a given year, an American will use 500 times as much energy as a person from Mali over her lifetime. That means that even though India and China will add almost eighteen times as many people over the next decade as the northern hemisphere, the northerners will nevertheless add more greenhouse gases to the atmosphere.[19] Still, population increase anywhere in the world adds to the net loading of the atmosphere with greenhouse gases. However, the tremendous economic growth in Asia, and especially in China, which has now become the leading net emitter of $CO_2$, is in a substantial part due to the demand pull from the developed nations. America, Europe and Japan are importing record numbers of goods made in China. Economists call the emissions associated with this trade embedded emissions.

The carbon content of the atmosphere had increased from 280 ppm about 1860 to 360 ppm by 1989, with the annual increment being about 5.5 billion tons. By July 2012 the number was 399 ppm[20] and had reached 400 over the Arctic.[21] By March 2019 it was 414.84.[22] This is the highest level in 800,000 years, and guarantees elevated concentrations for a long time. If we add in the other heat trapping gases we are emitting the figure is the equivalent of 480 ppm of $CO_2$.[23] These gases do not just go up and disappear. "Once emitted, a single molecule of carbon dioxide can remain aloft for hundreds of years ... the effects of today's industrial activities will be felt for the next several centuries, if not thousands of years."[24] Thus we can't just put off halting emissions at some time in the future and then quickly fall back to safe levels. That is why most scientists believe that 360 ppm is the maximum limit beyond which dangerous and even catastrophic changes in the biosphere and human civilization will result, such as the drying up of the Amazon rain forest, a major carbon sink and provider of oxygen and biodiversity.[25] However, if we continue the current rate of growth, the IPCC estimates we could reach a level of concentration of 650 to 1,200 ppm.[26] That is a recipe for a climate disaster threatening the very existence of modern civilization.

Temperatures are steadily rising. "Averaged over all land and ocean surfaces, temperatures have warmed roughly 1.33°F (0.74°C) over the last century, according to the Intergovernmental Panel on Climate Change...."[27] By 2014 it was 1.4 degrees Fahrenheit. The twentieth century was the warmest in the last six hundred years and the decade of the 1980s the warmest in the century with the 1990s even warmer than the 1980s.[28] "The first decade of the twenty-first century was the hottest on record in the lower 48 states of the U.S."[29] The last 5 years have been the hottest on record. Bush reports that "Temperatures are now higher than at any time in the past 1,000 years."[30]

## *The Great Melting*

Snow cover, glaciers, sea ice and ice sheets are melting all over the world. Already by 1992 Meadows and Jorgen Randers reported in *Beyond the Limits*, that "Long term studies of Canadian lakes are showing an increased ice-free season of three weeks, which is changing the relative populations of aquatic species."[31] The *Huffington Post* reports scientists who say: "We've documented in the mountains of the U.S. West that the spring runoff pulse now comes between one and three weeks earlier than it used to 60 years ago and that's because of warmer temperatures tending to melt that snowpack earlier and earlier."[32] In the northeastern U.S., frost-free season is now beginning eleven days earlier.[33] Migrating birds are arriving in earlier in the spring. And the freezing level in the atmosphere, that is, the height at which air temperature reaches 32 degrees Fahrenheit, has been gaining altitude since 1970 at the rate of 15 feet a year.[34]

Ohio State University researchers report: "in 2005 a survey of mountain glaciers around the world found that most of those … had been shrinking since 1900. Some that had survived for many thousands of years were vanishing, a striking sign of unprecedented climate change."[35] In Southwest Asia, the receding glaciers pose a serious threat to high density populations. Already melting faster than anywhere else, the glaciers in the Himalayas are disappearing, which will affect two billion people in India, China, Nepal and Pakistan who depend on the meltwater of the Brahmaputra, Indus and Ganges Rivers.[36] Half the Alpine glaciers in Europe have been lost in the last century, more than two thirds in Glacier National Park, and there has been extensive melting in Muir Glacier in Alaska.[37] By 2014, Greenland's glaciers were accelerating their descent into the ocean at two to three times the 1990 speeds. In 2012 the Peterman Glacier calved off a 43 square mile iceberg, and two years earlier, an area four times the size of Manhattan floated away to melt in the ocean.[38]

The West Antarctic ice sheet is disintegrating as warmer ocean waters melt it from underneath and topside meltwaters pour down crevasses and further destabilize it. Scientists had thought it would take over a thousand years to melt, but new research suggests it could go in a matter of decades and by itself raise sea level by three feet by 2100. Combined with the other melting going on, the total rise could be six feet. "That is roughly twice the increase reported as a plausible worst-case scenario by a United Nations panel in 2013…. The long term effect would likely be to drown the world's coastlines, including many of its cities."[39]

The seas are already rising. On Cape Cod the superintendent of the National Seashore reports: "We're retreating."[40] Beach access parking lots have had to be moved back 125 feet. Flooding is happening in Miami, Florida, and Norfolk, Virginia. The Director of the Florida Center for Environmental Studies reports, "At the spring and fall high tides, we get flooding of coastal areas…. You've got saltwater coming up through the drains, into the garages and sidewalks and so on."[41] The rising temperatures also swell the volume of the ocean and it expands into coastal areas, as the people of Norfolk discovered. They are spending $1.2 million to raise street levels and redirect storm sewers so that the high tides no longer put 2 to 3 feet of water on some of their streets. Norfolk has seen the seas rise 14.5 inches since 1930. Rising seas also threaten coastal wetlands, which act as buffers against storm surges and as carbon sinks and nurseries of marine life. The U.S. is already losing 80,000 acres of these wetlands each year.[42] What is more, the oceans will continue to rise for the next 500 years due to $CO_2$ already aloft.[43]

Island nations are in trouble. The Pacific Island nation of Kiribati is already experiencing threatening seas. Bloomberg News reports:

> A recent study found that the oceans are absorbing heat 15 times faster than they have at any point during the past 10,000 years. Before the rising Pacific drowns these atolls, though, it will infiltrate, and irreversibly poison, their already inadequate supply of fresh water. The apocalypse could come even sooner for Kiribati if violent storms, of the sort that recently destroyed parts of the Philippines, strike its islands.[44]

Bangladesh is in trouble, as noted in a *New York Times* article in 2014:

> Dr. Pethick found that high tides in Bangladesh were rising 10 times faster than the global average. He predicted that seas in Bangladesh could rise as much as 13 feet by 2100, four times the global average. In an area where land is often a thin brown line between sky and river—nearly a quarter of Bangladesh is less than seven feet above sea level….[45]

Sea level rise there is expected to flood seventeen percent of Bangladesh and displace eighteen million people in the next forty years. Storm surges are eroding coastal lands and salt water intrusion is ruining fields people depend on for food, resulting in miserable conditions and, in many cases, migration into urban slums.[46]

Global warming may also affect ocean currents and dramatically change weather patterns in Europe and elsewhere. In the North Atlantic, cold salty water sinks and flows southward toward the equator where it warms, picking up equatorial heat. Then the warmer water rises and flows back north, bringing tropical heat with it. This is known as the Atlantic Meridional Overturning Circulation, or the AMOC, and includes the Gulf Stream. Southern heat is also carried northward up the Atlantic coast of the U.S. and then on around to England, making it possible for palm trees to grow on the Cornish coast and keeping Europe much warmer than it would be otherwise. A similar conveyer belt of north moving heat exists in the Pacific. Globally this circulation is known as the "thermohaline pump" and geological records indicate that has shut down abruptly in the past, profoundly altering weather and climate. The addition of lighter, fresh waters coming from the melting of the ice sheets and enhanced river flows into the ocean due to more rainfall, itself a product of warming, could conceivably shut it down.

## Acidifying the Oceans

The Center for Biological Diversity reports that "Ocean acidification could disrupt the entire marine ecosystem."[47] Carbon emissions are rapidly changing the pH balance of the oceans in which marine organisms have evolved over eons. The oceans are now absorbing about half the carbon emissions. As $CO_2$ dissolves in water it changes to carbonic acid. The oceans are now 26 percent more acidic than around 1750, but the future looks far worse. UN statistics indicate that by 2050 it could increase 150 percent, which is 100 times faster than any acidification trend change in the last 20 million years.[48] As the ocean becomes more acidic, the concentration of carbonate is reduced. Shellfish depend on carbonate to create their shells. Mussels, oysters, clams, lobsters, shrimp and crabs that need carbonate, especially at the juvenile stages, are all at risk. More ominously, smaller planktonic organisms like foraminifera and pteropods are also at risk. These form the base of the oceanic food chain and could cause severe damage to cascade up the food chain making for large reductions in fish populations. Berger writes that climate change thus threatens us with "an ocean food chain catastrophe."[49]

## Dying Coral Reefs

Another casualty of global warming are the coral reefs. Reefs are nurseries of the sea and harbor much of its biodiversity including many species of fish which provide much needed protein for millions of people. Already stressed by pollution and overfishing, now rising temperatures and too much sunlight getting through to them are causing the algae to over metabolize and create toxins. In turn the polyps react by getting rid of them and, losing their food source, the reefs die. The years 1993, 1998, 2010 and 2013–2017 were high stress years for the world's corals. Dead reefs collapse and the fish lose their habitat. The prognosis is bad. "The World Resources Institute report, 'Reefs at Risk Revisited,' suggests that by 2030, over 90 percent of coral reefs will be threatened. If action isn't taken soon, nearly all reefs will be threatened by 2050."[50]

## Accelerating the Great Extinction

We are in the midst of a Great Extinction unparalleled in millions of years, resulting from humanity's destruction of habitat, its radical chemistry, and polluting industries. On top of these assaults, human induced global warming brings its own particular species depressants. Many animal species and even more plant species will be unable to evolve rapidly enough to adapt. Adult phases of many animal species such as clams, oysters, barnacles, snails, move very little and will be outrun by climate change. Plants obviously can't uproot and seek cooler elevations. Climate change is also favoring pest species such as bark beetles and other invasive plants and animals which move northward and overwhelm existing ecosystems. These changes are coming too fast for successful adaptation of many native species. John Berger writes that "Earth's species are the end result of millions of years of evolution and thus it will take earth millions of years to replace the biodiversity that humanity appears likely to destroy through climate change and other means in a matter of decades."[51] And further: "Unless the accelerated extinction of species due to climate disruption is brought under control, slowed, and then halted very soon, so many other species will follow the first wave of casualties that hundreds of millions, if not billions, of people will be endangered or die as the ecosystems on which they depend for existence … are degraded."[52]

## Social Impacts

Climate change will impact almost every aspect of our economic and social lives, including the critical areas of food production, health, infrastructure damage from extreme storms and bigger fires, and will exacerbate conflict at all levels. It will bring difficult challenges to food production worldwide, including changes in the length of seasons to which agriculture will need to adapt, if possible, as well as drought, flood, extreme heat, pounding rains and hail, expanding ranges of pests and invasives, and more robust weeds. A descending curve of depressants on the food system will intersect with the rising curve of the number of food consumers by century's end. More forests will be cleared for farms, reducing carbon absorption and droughts will be longer and more severe.

# Drought

Paradoxically, climate shift that creates heavier precipitation events also results in more droughts, because more water is held in the atmosphere and less is available for the soil. John Berger points out that the ten year drought that hit the Sahel region of Africa in the 1980s impacted nine countries. It came back in 2011 and was so severe in Kenya that it killed 150,000 livestock and affected 23 million people.[53] The trend is clear. "Drought has also gripped parts of Alaska, Canada, the Mediterranean, eastern South America, eastern Australia, and large parts of Eurasia since the mid twentieth century. China, in February 2009, had the worst drought in 50 years."[54] Two thirds of Iran is turning to desert and the government is concerned that "the Iranian plateau is becoming inhospitable to human habitation."[55]

From Texas to California to Syria severe drought has already impacted agriculture. Texas experienced the worst ever October to May drought in 2011. Water levels plummeted in Lake Travis, which is fed by the Little Colorado River, and pitted the city of Austin against rice farmers near the Gulf of Mexico, as they rely on huge amounts of water from the river for irrigation. Over the last thirty years, water use in Austin has tripled due to burgeoning population, and, almost unbelievably, a utility wanted to build a coal plant on the river and use its water for cooling. Meanwhile, the oyster beds off the mouth of the river are in trouble due to the high salinity of the water during times of drought.[56] Negative changes radiate through the web of the biosphere. Drought also hit the ranchers. "The worst one-year drought in Texas history produced a statewide hay shortage that more than doubled the price of ... bales, forcing many ranchers to sell or even abandon all of their cattle and horses because they cannot afford to feed them."[57] An extreme heat event in Russia, where temperatures hit 100 degrees in the summer of 2011, caused wheat production to plummet, driving up prices worldwide.

In California, 2014 was the driest year on record; eighty-six percent of California was experiencing extreme drought.[58] The San Luis Reservoir had its lowest water level in 27 years, dropping to 10 percent of August's historical average.[59] Toxic blue-green algae develop in such conditions. In 2017, because of the drought, record wildfires destroyed vegetation and compacted soils so that when the deluging rains came, catastrophic mud slides followed. In Syria where wheat fields and sheep pastures supported thousands of farmers for centuries, four years of drought wreaked havoc. "Ancient irrigation systems have collapsed, underground water sources have run dry and hundreds of villages have been abandoned as farmlands turn to cracked desert and grazing animals die off. Sandstorms have become far more common, and vast tent cities of dispossessed farmers and their families have risen up around the larger towns…."[60] Some 1.3 million people there were impacted by the heat and erratic rainfall. Herders have lost 85 percent of their livestock. "'We saw whole villages buried in sand,' said Zaid al-Ali, an Iraqi-born lecturer at the Institut d'Études Politiques in Paris."[61]

Climate change will impact already stressed world fisheries. The IPCC predicts that:

- as sea temperatures change, fish numbers will change and fish will move to different areas
- some species will go extinct in particular areas
- predators and prey will move to different areas, disrupting food chains
- wetlands and other low lying habitats where fish reproduce will be covered by rising sea levels
- water in lakes will get warmer
- bad weather may stop fishers going to sea

These changes may affect fisheries worldwide, but the impacts will be particularly damaging for fishers in developing countries.[62] "Fishing communities in Bangladesh are subject not only to sea-level rise, but also flooding and increased typhoons. Fishing communities along the Mekong River produce over 1 million tons of basa fish annually and livelihoods and fish production will suffer from saltwater intrusion resulting from rising sea level."[63] Along the coasts of West Africa, according to a recent United Nations Environment Program report titled "In Dead Water," climate change has the potential to alter the ocean currents which flush and clean the continental shelves, processes critical to maintaining the fisheries. Meanwhile, the sea temperature around England has risen 1.6 degrees centigrade since 1980, four times as fast as the rest of the ocean and the result is a changing array of fish as more desirable species like cod move north and less desirable ones take their place. At the very least, there will be major disruptions of the global food system as crops and agricultural technology struggle to adapt to a changing climate.

## Human Health

Droughts leading to shortages of water will impact health. More dramatic is stress related to extreme heat waves, which began in the late 1990s when deadly urban heat waves struck Chicago. Phoenix, the fifth largest city in the U.S., will probably become uninhabitable: by 2050 the average number of hundred degree days will have risen to 132 per year and, due to droughts that desiccate the Colorado River, they will experience a short-fall of water amounting to 23.2 million acre-feet, four times the amount Los Angeles uses in as year.[64] The August 2003 heat wave in Europe killed 35,000 people.[65] Extreme high temperatures are especially dangerous for the elderly, children and people living in cities. Over the past ten years, 75 countries set record high temperatures while only 15 set record lows. In 2010, no countries set record lows. In the summer of 2015, a deadly heat wave, the worst on record, hit Pakistan. Government hospitals saw 1,150 people die. In 2016, 2,300 people died of a heat wave in India, followed by hundreds more in 2017 when temperatures hit 124 degrees Fahrenheit.[66] Dr. Peter Wilk, Executive Director of Physicians for Social Responsibility, writes:

> Global warming is one of the gravest health emergencies facing humanity. It's life-threatening and it's affecting us now…. In the United States, heat waves already kill more people during a typical year than floods, tornadoes and earthquakes combined.[67]

Robert Repetto, a United Nations economist says: "climate change will intensify smog, leading to 'increased outbreaks of asthma and allergies' and 'exacerbate vector-borne diseases such as Hanta virus, West Nile virus, Lyme disease and dengue fever.'"[68]

## Super Storms

Warming atmosphere can hold more moisture and creates higher winds resulting in superstorms with "heavier rains, more extreme floods, and more intense storms driven by latent heat, including thunderstorms, tornadoes, and tropical storms."[69] Bill McKibben stated the point clearly in a radio interview, saying, "what's happening is we're making the earth a more dynamic and violent place."[70]

In March 1993 a cyclonic blizzard including a line of severe thunderstorms moved from the Gulf of Mexico, passed over Cuba and Florida and up the East Coast of the U.S. "The storm stretched from Central America to Nova Scotia … winds reached hurricane force in the gulf region and well over 100 miles an hour in Cuba."[71] Blizzard conditions stretched from Texas to Pennsylvania; seventeen inches of snow fell on Birmingham, Alabama. Hurricane Katrina, which devastated New Orleans in August 2005, affected 90,000 square miles of the United States, scattered refugees far and wide and killed 1,833 people. The storm surge was 27 feet.[72] Costs were estimated at $148 billion. These severe storms bring torrential rains and can produce baseball-sized hail driven on 100-mile-per-hour winds. Katrina destroyed hundreds of millions of trees over an area of 5 million acres in Alabama.

These dead trees will release an estimated 105 million tons of $CO_2$ to the atmosphere in a tragic feedback process, further warming the globe.[73] In 2012, Superstorm Sandy became the largest Atlantic hurricane in history with winds swirling in a 1,000 mile diameter circle, deluging 24 states as far inland as Michigan and Wisconsin, destroying or damaging hundreds of thousands of homes and businesses and caused $68 billion in damage.[74] The storm left 8.5 million homes without power. The National Oceanic and Atmospheric Administration believes that the number of severe class 4 and 5 hurricanes will continue to increase.[75] Tropical hurricanes, known as typhoons, are equally destructive and deadly. Tropical cyclones have killed nearly 175,000 people since the year 2000.[76] Typhoon Haiyan, the strongest tropical storm in history, hit the Philippines in 2013 with wind gusts of 235 miles per hour. The storm killed 6,800 people and devastated the islands.[77] In the spring of 2014, a massive storm hit the Gulf Coast. The National Weather Service estimated that an astonishing 26 inches of rain fell around the Alabama-Florida state line causing heavy flooding. The rainstorm came after an outbreak of deadly tornadoes that had killed dozens of people a few days prior.[78] Megafloods have occurred in Queensland, Australia, Sri Lanka, Vietnam, the Philippines, Brazil and Colombia.[79]

In August of 2016 devastating floods hit Louisiana, which received more than 31 inches of rain in a week. According to a NOAA spokesperson: "The flooding in Louisiana is the eighth event since May of last year in which the amount of rainfall in an area in a specified window of time matches or exceeds the NOAA predictions for an amount of precipitation that will occur once every five hundred years, or has a 0.2 percent chance of occurring in any given year."[80] In 2017, Houston and Haiti were devastated by deadly hurricanes. Projected negative impact on the economy for Houston is $196 billion.[81] In Haiti there simply isn't much left. Hurricane Florence in August of 2018 lingered over the Carolinas, dumping record rainfall that led to devastating floods. Hurricane Michael, in November of 2018, actually accelerated as it came ashore at Panama Beach in the Florida Panhandle and then failed to break up as it traveled hundreds of miles inland, laying waste to towns and villages.

A bi-partisan report led by former Secretaries of the U.S. Treasury Henry Paulson, a Republican, and Robert Rubin, a Democrat, pointed out two consequences of global warming and rising seas: "More than a million homes and businesses along the nation's coasts could flood repeatedly before ultimately being destroyed. Entire states in the Southeast and the Corn Belt may lose much of their agriculture as farming shifts northward in a warming world."[82] Paulson said: "I feel as if I'm watching as we fly in slow motion on a collision course toward a giant mountain. We can see the crash coming, and yet we're sitting on our hands rather than altering course."[83]

## Forests and Fires

Fires have been increasing in frequency since the 1980s and are occurring four times as often and burning six times the land area and lasting five times as long.[84] The firefighting budget of the U.S. Forest Service has exceeded a billion dollars a year. Large fires of more than 10,000 acres are seven times more common than forty years ago and the fire season is two months longer. In 2012, 9.3 million acres burned in the U.S. Fifty-one of those fires were larger than 40,000 acres and 14 burned over 100,000 acres.[85] In August 2016, the Blue Cut fire that raged near San Bernardino, California, drove 82,000 people from their homes.[86] In December of 2017 five major fires raged around Santa Barbara, California, including the Thomas Fire, which grew by 50,000 acres in a single day and burned 250,000 acres. Nine thousand firefighters struggled with the blaze and 100,000 people were evacuated. Several thousand homes were destroyed.[87] The big fires also impact wildlife, killing them outright, damaging or destroying habitat and resulting in major stress for survivors trying to recover. In November of 2018 deadly fires struck California, including the Camp Fire, which burned out the entire town of Paradise, destroying 12,000 homes and as of this writing resulting in 87 deaths. In short, a hotter world will more easily ignite.

## Climate Economics

The up-front costs of climate mitigation are high. The costs of doing nothing, or too little, are even higher and the longer we wait for significant action the higher they get. There is a plus side to the ledger. Mitigation will create new jobs, improve health and rebuild decaying infrastructure. The only people opposed are the dirty energy companies and members of Congress who depend on them for campaign contributions which can only be characterized as legal bribery.

Had we moved earlier the costs would have been negligible. A U.S. Department of Energy study "projected in 2000 that carbon emissions could have been reduced to 1990 levels by 2010 at roughly *zero net cost.*"[88] A 2006 study led by former World Bank chief economist Sir Nicholas Stern concluded that the cost of unmitigated global warming would amount to "losing at least 5 percent of global GDP per year, now and forever," and if a wider range of impacts that had not been included in the models on which the report was based, it could rise to 20 percent per year.[89] Looking back on the report in 2013, Stern said he had "underestimated the risks."[90] The International Energy Agency estimates that every year of delay in taking serious action to cut emissions adds half a trillion dollars to the inevitable cost.[91]

By 1998 major insurance companies in the U.S. saw payouts for storm related damage go from $16 billion in the decade of the 1980s to $48 billion in the period 1990–94.[92] Analysts at the International Institute for Applied Systems Analysis have said that the range of permissible emissions from 2015 onwards is not 2,390 billion tons, but rather about half of that at 1,240 billion tons. "In effect, that halves the levels of diesel and petrol available, coal for power stations, and natural gas for central heating and cooking available to humankind before the global average temperature … reaches the notional 2C mark, long agreed internationally as being the point of no return for the planet."[93] However, Robert Pollin, author of *Greening the Global Economy*, also believes it could still be done at reasonable cost; that we can achieve the IPCC's twenty year emission goals by investing no more than 1.5–2 percent of global GDP in energy efficiency and low emission renewables, an amount between $1.3 and $2 trillion, and he points out that already in 2011 we were investing $227 billion in renewables and over $150 billion in efficiency.[94] And, he points out, investments in efficiency pay for themselves in about three years so what is needed is just up front capital.

## Migration and International Conflict

Rising sea levels will inundate coastal areas and low-lying islands where one tenth of the human population lives and irrigation rivers will dry up leading to significant and unwelcome migration by millions of people and a very probable rise in violence. The U.S. military establishment takes this threat seriously, asserting that climate change poses an "immediate threat" to U.S. national security.[95] "There may be hordes of climate refugees, fleeing homes on islands and coasts made uninhabitable by climate change—anywhere from 25 million to 1 billion people by 2050, according to the International Organization for Migration."[96] The drying up of the Himalayan glaciers that feed the irrigated agriculture of the region is critical and is "'threatening the food security of an estimated 60 million people' in the Indus and Brahmaputra basins."[97] The water there is already 100 percent used. The Pentagon, itself a major contributor to global emissions, calls climate change a "threat multiplier,"[98] in part because the U.S., Russia, Norway and Canada are scrambling to claim oil and gas under the now opening up Arctic sea and are moving military assets into the region. Already there are thousands of climate migrants at the southern border of the U.S.

## Climate Deniers, Inaction, Obfuscation

Any progress on shifting toward a carbon neutral economy was thwarted for decades by intentional misinformation promulgated by conservative think tanks and others largely funded by a handful of industry sectors including fossil fuels, transportation, electric utilities and cement. Science denying conservatives had come to dominate and control much of government in the United States, like Senator Ron Johnson of Wisconsin, who said: "I absolutely do not believe in the science of man-caused climate change. It's not proven by any stretch of the imagination." Johnson told the *Milwaukee Journal* that the climate change theory was "lunacy" and blamed changes in the Earth's temperature to "sunspot activity or just something in the geologic eons of time."[99] Industry front groups with misleading names, such as the "Greening Earth Society" (a creation of the Western Fuels Association, a $400 million coal producing co-op), carried on public relations campaigns.[100] The Center for the Study of Carbon Dioxide and Global Change issued a report claiming that "CO$_2$ *enrichment* [italics added] brings growth and prosperity." It was authored by Craig D. Idso, former Director of "Environmental Science" at Peabody Energy Company, the world's largest coal producer.[101] In 1989 the fossil fuel industry hired the public relations firm of Burson-Marsteller to set up a front group called the Global Climate Coalition, which was given approximately a million dollars to fund its disinformation campaign. The group met in the offices of the National Association of Manufacturers, a notoriously conservative and anti-government lobby opposed to any regulation of corporations.[102] In 1997 the public relations firm Sandwick Public Affairs set up the Global Climate Information Project and spent three million dollars on ads designed to instill public fears that the Kyoto Protocol would add a half-dollar tax to the price of gas and raise prices on everything else.[103] The *London Times* reported that "ExxonMobil gave £1 million to fund 'organizations that campaign against controls on greenhouse gas emissions,'"[104] having known about its dangers and covered them up almost 40 years ago.[105]

In the face of overwhelming scientific consensus these organizations continued to reiterate ten myths which have confused much of the public. These were: that the evidence for climate change is contradictory and uncertain; carbon dioxide is not a powerful greenhouse gas; burning fossil fuels is not the source of global warming; earth is in a natural warming cycle; the world is really cooling and not warming at all; the warming is slight and the dangers are exaggerated; climate change is slow and we can wait to see if there are any bad effects; global warming is a good thing; combatting climate change is too expensive; and doing anything places an unfair burden on developed nations.[106] None of these are true and all have been refuted at length. The media, in a misguided attempt to tell both sides of a story, has unwittingly portrayed the deniers as if they were equal in number and validity to the climate scientists. In fact only a handful of scientists have been climate deniers and these have close relationships with industries that have funded their research and publication.[107] Many well-funded conservative think tanks, including the American Enterprise Institute, the Cato Institute and the Heartland Institute, also attacked the scientific evidence on global warming.

For a long time many of the world's governments were loath to acknowledge the greenhouse effect because it would have meant abandoning their very popular, centuries-old policies of supporting economic growth via cheap energy. Through the 1980s, they temporized in the face of growing evidence, suggesting always more study. Finally, in 1992, the United Nations held an "earth summit" at Rio de Janeiro and succeeded in getting an international treaty drafted called the United Nations Framework Convention on Climate Change. Its objective is to "stabilize greenhouse gas concentrations in the atmosphere at a level that would prevent dangerous anthropogenic interference with the climate system."[108] One hundred and ninety-six nations signed on. But this agreement set no limits on greenhouse gas emissions nor did it have any enforcement mechanisms. It was no more than an agreement to talk. By 1997 they had developed the Kyoto Protocol, which did set limits for the developed countries, specified to commitment periods. The first of these was from 2005 to 2012 and the second 2012 to 2020. Not all of the parties agreed to the second commitment and the United States never ratified the Kyoto Protocol. In 2001 Republican President George W. Bush withdrew the U.S. signature, claiming the science of global warming was not conclusive and to limit greenhouse emissions would hurt the U.S. economy. He also argued that the treaty was unjustified because it did not include developing nations, even though they were responsible for only a small fraction of emissions. His vice president, Dick Cheney, forced the Environmental Protection Agency to delete sections on climate change from its reports in 2002 and 2003 and tried to control or censor congressional testimony by federal employees in order to dis-inform or confuse the debate by minimizing threats to the environment. In 2012 conservative Prime Minister Stephen Harper withdrew Canada from the Protocol.

In his first term, President Obama demonstrated very little leadership on climate or any other environmental issue.[109] In his second term he tried hard but was thwarted at every turn by a conservative Congress. He outraged the Republicans by saying in his State of the Union address, "If Congress doesn't act, I will."[110] He began issuing executive orders including a "Clean Power Plan" to shut down the nation's dirtiest coal plants. The

Administration was promptly sued by 27 Republican State Attorneys General and the plan stalled in the courts. However, his move was enough to restore U.S. credibility on climate matters and, sending his Secretary of State to China, they negotiated an agreement that broke the 20-year deadlock and led to the Paris Accords of 2015.

## *The Paris Accord: Too Little, Too Late*

The Paris Accord, established under the auspices of the United Nations Framework Convention on Climate Change, was signed on December 12, 2015, and opened for signature on Earth Day, April 22, 2016. It is based on the nearly universal consensus among scientists that the world is warming as a result of human generated $CO_2$ and other gases and that an uncontrolled warming beyond 2 degrees centigrade above preindustrial levels will bring catastrophic consequences for the biosphere and human civilization. A commitment was made by 195 nations to reduce their greenhouse gas emissions to reach the goal of no more than a 2 degree centigrade (3.6 degrees Fahrenheit) increase in global atmospheric temperature above preindustrial levels and hopefully, no more than 1.5 degrees. It also relies on increasing carbon sinks such reforestation, or stopping clear cutting to slow further rises in temperature. The Accord went into effect on November 5, 2016, as the ninety-second signatory ratified it. The Accord gets the world off the starting line, but it's a long way to the finish. The plans so far submitted do not get the world to the 2 degree goal. "The new deal will not, on its own, solve global warming."[111] In an article titled "Fighting Climate Change? We're Not Even Landing a Punch," the author reported: "when experts tallied the offers made in Paris by all the countries in the collective effort, they concluded that greenhouse gas emissions in 2030 would exceed the level needed to remain under 2 degrees by 12 billion to 14 billion tons of $CO_2$."[112] Furthermore, the accord legally binds nations only in that each must submit an emissions reduction target and meet every five years to show how well they are doing. What is not legally binding is the target: "there is no legal requirement dictating how, or how much, countries should cut emissions."[113] Much greater reductions in emissions will need to be made in the near future, but as of 2019, the nations are not even doing enough to meet the goal of 2 degrees.

In October of 2018 the world's chief climate monitoring agency, the Intergovernmental Panel on Climate Change, concluded alarmingly that the negative effects once assumed to be associated with 2 degree rise could begin at a 1.5 degree rise. They issued a dire warning, namely that the world has only twelve years to make far-reaching and unprecedented changes in the global economy in order to sufficiently curb greenhouse gas emissions or be confronted with inevitable, serious and adverse effects of global warming.[114]

The Paris pledges remained voluntary because of the Republican-controlled U.S. Congress. Had the targets been legally binding they would have constituted a treaty which, according to the U.S. Constitution, would have to be approved by the Senate, an impossible event. But even this minimal amount of progress was threatened by U.S. President Donald Trump.

## Hope Dashed—The Tragedy of Trump

The election of Donald Trump derailed the progress of the United States on climate and on many other environmental issues. Trump is a climate change denier and generally disdainful of science-based evidence. He filled his cabinet and lower positions in government with like-minded men and women, often drawn from the ranks of corporate polluters. He purged the Whitehouse webpage of any mention of climate change. He announced the U.S. withdrawal from the Paris Agreement. He cancelled President Obama's move to close the dirtiest coal plants, promised a revival of the coal industry, issued executive orders opening the Atlantic and Pacific seaboards and public lands, including the Arctic National Wildlife Refuge to oil and gas drilling, ordered the go ahead of the North Dakota Access and Keystone pipelines, and met with auto manufacturers to roll back auto efficiency standards. His energy policy was "energy dominance," that is, for the U.S. to be the largest producer of fossil fuel–based energy in the world. He pledged to increase the size of the military, which is one of the chief emitters of carbon, and to vastly ramp up the nuclear arsenal which, if only a tiny fraction is used, will be the most environmentally devastating act in human history, and these items hardly exhaust the list. In general he attacked environmental safeguards across the board.

His initial appointments included the head of Exxon Mobile, Rex Tillerson as Secretary of State; Scott Pruitt, who had sued the EPA 14 times as the leader of State Attorneys General on behalf of the fossil fuel industry, to head the Environmental Protection Agency; Ryan Zinke, a western anti-environmentalist and former oil company board member, to head the Department of the Interior; former Texas governor Rick Perry, a climate change denier, to head up the Department of Energy which, as a former presidential candidate he had vowed to abolish. Trump's first choice to head the President's Council on Environmental Quality was Kathleen Hartnett, formerly a senior fellow at the Texas Public Policy Foundation (funded by the Koch Brothers and the Exxon). She believed carbon dioxide is harmless "plant food." She equated belief in climate change to "paganism." And she called solar and wind power "unreliable and parasitic." Hundreds of lower offices in the federal government were staffed with climate change deniers. President Trump governed as if the richest one percent are the appropriate governing class. He had no concept of the necessity of a healthy environment as the base on which civilization rests, and his sole criterion of good was maximizing unfettered economic growth in such a way as to benefit that class. It was a return to the mentality and practices of the nineteenth century.

The ideology of endless economic growth makes it very difficult to radically reduce carbon emissions. McKibben reports that international climate negotiators said that "the momentum of growth was so great that making the changes required to slow global warming significantly would be like 'trying to turn a supertanker in a sea of syrup.'"[115]

One would think that the dire warnings in the latest IPCC report would have had some impact, but even this report appears to have watered down the case for immediate, broad scale action. In a starkly critical review of the report, the Arctic News Blogspot slams the IPCC's latest report.

> the IPCC bends over backward to make it look as if temperatures were lower than they really are, in an effort to make it look as if there were *carbon budgets* to be divided, and polluters should be allowed to keep polluting until those budgets had run out. This is like saying that drug junkies who cause damage and are deeply in debt, should be handed over more *OPM* (other people's money, in this case the future of all people and other species).[116]

The critique provides a very compelling argument, backed by sophisticated data, that the IPCC report has ignored evidence that the temperature has already risen higher than they indicate and that their focus on average temperatures ignores both the fact that it's the peaks that kill and that it's in the Arctic where the highest temperature changes are found, which means that Arctic melting will accelerate global heating more rapidly than they have accounted for. The conclusion is that there is no carbon budget left and that we must move immediately to a post-carbon economy. No more coal or oil should be burned. As Wallace-Wells says in *The Uninhabitable Earth*, "It's worse, much worse, than you think."[117] Richard Heinberg, an analyst at the Post Carbon Institute, lays out our challenge: Now we're at the point where we must finally either succeed in overcoming growthism or face the failure not just of the environmental movement, but of civilization itself.

How did it all come to this? Do we face global collapse?

# 6

## The Prospect of Collapse

"A new study sponsored by NASA's Goddard Space Flight Center has highlighted the prospect that global industrial civilization could collapse in coming decades...."[1]
—Reported in *The Guardian,* March 14, 2014

So far we have looked at explosive rates of population growth, urbanization, consumption, the contamination of the earth with 90,000 chemical compounds, an extinction rate unseen in 60 million years, massive deforestation and accelerating rates of forest fires, deteriorating soils, droughts, depleted aquifers, disappearing wetlands, polluted waters, the devastation war wreaks on the environment, and the folly of nuclear weapons and nuclear power and, finally, at human-induced climate deterioration. These are the trends that make Hypercivilization a radical discontinuity from all prior history. We have never been here before.

For the last 200 years we have more and more absented ourselves from nature and built instead a mental and physical habitat I have called Hypercivilization. The results of that experiment show it to be a failure. Our thought systems and our technical and social systems have resulted in serious and massive degradation of the biosphere. Furthermore, the rate of negative change is speeding up due to the phenomena of mechanisms that trigger tipping points. If we wait much longer we may well undermine the natural base of civilization and usher in a new Dark Age. We are not doomed to planetary collapse, but it is at the end of the path we are currently racing along.

All systems are failing under the weight of human impact. The past weighs heavily on the present and threatens the future. There are many who are simply not aware of the facts, who education and the media have failed, or they were just not paying attention. There are the foolish who think that more of the same, that is economic growth, will somehow suddenly begin to produce results opposite of what it has always produced, namely depletion and toxification of the biosphere. There are some who think that God will not let us ruin the planet but have no plan for change. Then there are those self-styled "masters of the universe," the ones *Ecotopia's* author Ernest Callenbach wrote about in his very last piece, found on his computer after he died. "We live," he said, "in the declining years of what is still the biggest economy in the world, where a looter elite has fastened itself upon the decaying carcass of the empire."[2] These are the ones who know but just don't care because their fortunes need to be protected for as long as possible.

## Breakdown: Collapse Scenarios

It is very easy to think that things will go on pretty much as before. We know of course that change happens. My grandmother was born before the automobile, radio and heavier than air flight. She saw the moon landing on television. We have developed a cornucopia of new labor saving products made of new materials. We saw it all as "progress." DuPont's 1935 slogan, "better living through chemistry," caught the mood of a century.[3] This has produced a great irony. The failures of the future are embedded in the solutions of the past. Furthermore, we don't want to think about this. If you are comfortable, living in a developed country with an air conditioned home and a nice car and a good education, you don't want to contemplate that it could all change for the worse. But advanced civilizations have collapsed before by putting unsustainable pressure on the ecosystems that supported them. Jared Diamond, who studied many societies that collapsed in the past, writes: The path to "unintended ecological suicide" includes deforestation, habitat destruction, soil problems (erosion, salinization, fertility lost), over fishing and hunting, invasive species, human population growth, increasing per capita impact.[4] Another factor was growing inequality.

The ancient civilizations of the Fertile Crescent turned it into a desert and went away. The Minoan Civilization of ancient Crete collapsed, as did the Mycenaean Civilization of ancient Greece.

Roman civilization declined to the point where the baths and central heating were abandoned and depopulated cities left empty while public order disintegrated in invasions and wars. The Mayans went down as did the peoples of Easter Island and some thirty other societies. Collapse happens. None of these civilizations weighed on the global biosphere to anywhere near the extent that we do. Their leaders and peoples did not see the catastrophe coming in time, and when they finally did, they thought the solution was to do more of what brought them there in the first place. The 2016 election in the United States placed in office a perfect example of

**Civilizations Collapse: The great capital of the Southern Kingdom of ancient Egypt, it flourished for less than 200 years and is today a ruin. Civilizations come and go (isparklinglife, Shutterstock).**

this tendency. In 2016 a new administration came to office in the U.S. on the promise to extract and process even more coal, oil, and minerals and to grow the economy even faster than before, in other words an attempt to double down on business as usual, the race to run out, the futile attempt to preserve the old lifestyle of Hypercivilization. Jared Diamond, author of *Collapse: How Societies Choose to Fail or Succeed*, writes, "Failure comes from faulty group decision making" and he cites four aspects: (1) failure to anticipate a problem [some are hard to see coming; reasoning by false analogy, e.g., the Maginot Line]; or it comes too slowly, or "creeping normalcy" or "landscape amnesia"; (2) failure to perceive it once it arrives; (3) failure to even try to solve because even while the elites know it's real, they don't care as long as they get their short-term profits, or the "tragedy of the commons"; and (4) they try to solve it but don't succeed.[5]

Arthur Demarest, another archeologist who has studied many collapsed civilizations, observed:

> …that need and the capacity for continual growth is also a diagnostic of eventual collapse. No society can sustain unlimited growth—none ever has. History demonstrates that expectations of infinite growth lead to collapse. Unfortunately, millennia of evidence also indicates that needed attempts to stabilize such societies run counter to the expectations of the populace and of interest groups. For that reason, such attempts at stabilization frequently fail.[6]

There is also the tendency of elites to preserve themselves regardless of the common good. Writing on the response of leaders to impending catastrophe, Demarest points out:

When there is pressure for leaders to respond to problems or crises, they often simply intensify their efforts in their particular defined sphere of activity—even if that's not relevant to the real problem. To do otherwise requires taking on entrenched practices and asserting power in areas where it often will not be well received. And leaders tend to see major crises more as threats to their own position rather than as systemic challenges for the societies that they govern or the institutions that they manage.[7]

In general, leaders in such times are also crippled by the short-term thinking characteristic especially of capitalism and of democracy with its frequent election cycles.

It is short-term thinking and decision-making that is the most universal factor leading to collapse. Yet, such short-cycle evaluation is a fundamental characteristic, and a basic strength, of both democracy and capitalism. In democracy, the competition which "short-cycle evaluation" generates in government, and the limited tenure of leaders, assures responsiveness to the needs of the people and a protection against oppression. However, such competition, and the expensive campaigning that it entails, has led to an unrealistic evaluative process that consistently sets aside long-term problems and consequences in order to try to achieve some short-term successes, so as to survive reelection in two, four or six-year cycles.[8]

Our leaders are not so different when they think the solution to society's problems is more long-distance trade, more manufacturing, more economic growth, clinging to old and outmoded values and beliefs that once served but, in the changed world we now inhabit, inhibit rational decision making including free enterprise, individualism, militaristic nationalism. Diamond observes: "a common theme throughout history and also in the modern world ... the values to which people cling most stubbornly under inappropriate conditions are those values that were previously the source of their greatest triumphs."[9]

One might want to argue that these archeologists are examining long ago societies that do not have our sophisticated grasp of science and powerful technologies, but in fact what actually makes our situation different from those in the past is that we are far more powerful than they were. Our impact on the biospheric base is exponentially greater because there are so many of us, our per capita input and output are far greater, and our technologies are so incredibly transformative. We can alter the atmosphere, poison the soils, cut the forests, and degrade the oceans on a scale never seen before. We can and are destroying not only specific locales like the rain forests of the Amazon and northwest Canada, but the global commons as a whole. One might also want to argue that technology will save us, but there is no guarantee there. The NASA study reminded us: "Technological change can raise the efficiency of resource use, but it also tends to raise both per capita resource consumption and the scale of resource extraction, so that, absent policy effects, the increases in consumption often compensate for the increased efficiency of resource use."[10] Of course, we have innumerable examples of technological innovation making the situation worse.

The biosphere and the world we have made within it are composed of a web of relationships so complex that even now we don't know all of them. A single example of one simple linkage like electricity and water, which is in short supply all over the world, shows how vulnerable we are. According to the Institute of Electrical and Electronic Engineers:

> In almost every type of power plant, water is a major hidden cost. Water cools the blistering steam of thermal plants and allows hydroelectric turbines to churn. It brings biofuel crops from the ground and geothermal energy from the depths of the Earth. Our power sources would be impotent without water.... In the United States alone, on just one average day, more than 500 billion liters of freshwater travel through the country's power plants—more than twice what flows through the Nile.[11]

And they go on to point out that even simply charging an iPhone requires half a liter of water to flow through miles of pipes and pumps and heat exchangers in a power plant.[12] New technologies also bring new problems. In Iceland scientists have learned to sequester carbon dioxide as a rock. While it sounds like a magnificent breakthrough, to sequester a single ton requires 25 tons of water.

Civilization rests on the biosphere and the multiple services it provides. Ours can be likened to a fat man lying in string hammock and putting on weight all the time, using more and more resources as the knots begin to strain. But to make the situation worse he is also snipping at the strands, dramatically reducing biodiversity, depleting resources and clogging the biological and geological sinks. And so we head for collapse.

What is it that breaks down, either gradually or rapidly? A civilization needs to assure enough throughput from and to the biosphere to provide enough food, water, shelter, clothing and energy to sustain the population. If either the population or the per capita take is growing, or both, then it must provide increasing amounts of these things in like measure. But to do so it must protect the commons from degradation at both the local and global levels including the atmosphere and climate, water, soil, biodiversity and energy supplies. It must also provide security from disruptive social violence so that the biological and social mechanisms which provide sustenance can function and be taken advantage of. That means protecting the peace which must also be considered a global commons. Breakdown occurs when the social and biological mechanisms to provide these goods and services fail. It may be gradual decline or it may be a cascading collapse. In a highly complex, fragmented and highly interdependent global system such as ours, breakdown can begin at any one of a number of places and spread quickly. In this society almost no one of us produces their own water, food, clothing, shelter, health care, transport or the various types of infrastructure on which these depend. As individuals we have become so specialized as to be radically deskilled and totally interdependent, what Arthur Demarest calls "hypercoherence." He argues that it "…is one of the most dangerous threats to the long-term survival of our civilization. Hypercoherence is the close efficient linkage of all parts of the world economic, communication and transport systems," and goes on to point out:

> However, this strength is, again, one of the most common symptoms of impending collapse. Perturbations, even small ones, immediately radiate throughout the entire system. Today there are few, if any, refuges against international crises of any kind. Thus, our brilliant communication, information, and transport systems, which will be remembered as the hallmark of our age, are also a point of great fragility.[13]

The division of labor and its world-wide disbursement leave us highly vulnerable. Our food and clothes are produced by somebody else and travel thousands of miles to get to us; our energy sources can be half a world away and very few of us know how to fix the pumps that deliver the water or the dynamos that turn on our lights or the pipelines that bring us the natural gas that heats our homes. We are incredibly vulnerable to breakdown that can be caused by absolute shortages of supply (especially minerals), or by market prices that cannot be met for lack of money or capital to maintain and develop, or by the clogging of earth's sinks for waste, or its renewal mechanisms, or by technical breakdowns such as the New York blackout of 1977 which resulted in immediate rioting and looting. The meltdown of the nuclear reactors at Fukushima in 2011 caused a long delay in auto manufacturing in the U.S. for lack of parts. And in political terms, people need to believe that the power elites are presiding over a more or less fair distribution of resources, or revolution can be the result, which further destabilizes supply. In other words, a special feature of the contemporary system is that human society is grossly inflated or top heavy as never before in history. A huge and still rising population and a huge level of throughput per capita weigh on the biosphere and make the whole edifice precarious.

One of the bitterest ironies of modern times is that at last we are seeing a rising standard of living in the developing countries. Steven Radelet writes in *The Great Surge: The Ascent of the Developing World*: "We live at a time of the greatest development progress among the global poor in the history of the world…. The next two decades can be even better and can become the greatest era of progress for the world's poor in human history."[14] This is a bitter irony because it is also propelling growing impacts on the biosphere. Of course we want everyone in India to at least have a refrigerator, but for everyone in India to have a refrigerator will require huge amounts of copper, steel and other minerals for the machine and the transmission lines from which it will get its power. It will require more mining, more carbon spewing trucks and power plants. Quoting *Big World,*

*Small Planet,"* by Johan Rockstrom, *New York Times* columnist Thomas Friedman writes "The good news is that in this period many more of the world's have-nots have escaped from poverty. They've joined the party. The bad news, says Rockstrom, is that 'the old party' cannot go on as it did."[15] It is tragic that the world's poor are joining the party just when it's about to wreck the house. This is an unspeakable social injustice and any plan to avoid collapse must include redress which can only be accomplished by large transfers of wealth from the overdeveloped world to these new arrivals, as well as a plan to reduce global population to sustainable limits, and huge investments in less damaging technologies. These may well be too much for the governing elites to grasp, much less to sell to the masses especially in the Global North. Certainly there is no sign that the current government of the U.S. has a clue about these matters and is, instead, intensifying the old strategies, the very program that Demarest says led other civilizations to breakdown and collapse.

What does breakdown look like? It depends on how far and how fast it happens and it depends on how many linkages there are in a society's interrelationships with other societies and with nature. The more there are the faster and more extensive it will be. However, we have no experience with breakdown of a modern, highly complex, world-spanning industrial society, so we find several different scenarios, none of which are comforting. In an article titled *The Imminent Collapse of Industrial Society*, Peter Goodchild paints a devastating picture of cascading disconnections.

The collapse of modern industrial society has 14 parts, each with a somewhat causal relationship to the next. (1) Fossil fuels, (2) metals, and (3) electricity are a tightly-knit group, and no industrial civilization can have one without the others. The decline in fossil-fuel production is the most critical aspect of the collapse.... As those three disappear, (4) food and (5) fresh water become scarce; grain and wild fish supplies per capita have been declining for years, water tables are falling everywhere, rivers are not reaching the sea. Matters of infrastructure then follow: (6) transportation and (7) communication, no paved roads, no telephones, no computers. After that, the social structure begins to fail: (8) government, (9) education, and (10) the large-scale division of labor that makes complex technology possible.

After these 10 parts, however, there are four others that form a separate layer, in some respects more psychological or sociological. We might call these "the four Cs." The first three are (11) crime, (12) cults, and (13) craziness, the breakdown of traditional law; the ascendance of dogmas based on superstition, ignorance, cruelty, and intolerance; the overall tendency toward anti-intellectualism; and the inability to distinguish mental health from mental illness. There is also a final and more general part that is (14) chaos, resulting in the pervasive sense that "nothing works anymore." ... If we look at matters from a more purely chronological viewpoint, however, we can say that there is a clear division into two time periods, two phases. The first phase will be merely economic hardship, and the second will be entropy. In the first phase the major issues will be inflation, unemployment, and the stock market. The second phase will be characterized by the disappearance of money, law, and government. In more pragmatic terms, we can say that the second phase will begin when money is no longer accepted as a means of exchange.[16]

Arthur Demarest points out: "Paradoxically, the key strengths of civilizations are also their central weaknesses. You can see that from the fact that the golden ages of civilizations are very often right before the collapse.... The ample record of failed societies chronicles systems at their peak of success, then rapidly disintegrating."[17]

A different scenario for is provided by Dimitri Orlov in *The Five Stages of Collapse*. He begins by pointing out three facts: that most resources necessary to run a global industrial economy have passed their peak supply; that the pollution consequent on operating such an economy requires ever more capital be devoted to mitigating it; and that as an economy grows so does all of its infrastructure: roads, bridges, airports, transmission lines, seaports, pipelines, etc. When shrinkage begins due to resources shortages all of this infrastructure still has to be maintained but there is not enough capital for the job. "At some point maintenance costs become unbearable and maintenance is foregone. Shortly thereafter they become nonfunctional and with them the rest of the industrial economy."[18] A case in point is the fact that in the U.S., which is a fantastically wealthy economy, one of every four bridges is either unsafe, deficient or obsolete.[19] Local roads are deteriorating rapidly, as are thousands of dams. A further problem Orlov points out is that in order to function, an industrial capitalist society must bet on the future. Financial institutions provide advance capital for trade on the assumption that the future economy will have grown enough to pay off the loans *with interest*. "This borrowing is used not just to finance expansion but to finance all of the shipments that make up global trade."[20] If the economy stops growing long enough, these turn into bad loans, some banks become insolvent and others will not honor their letters of credit. Global trade stops, supply chains dry up and the manufacturing process stops. So a capitalist industrial system must continue to grow or collapse, and the fact that resources are finite and the global sinks are filling up, most notably the atmosphere, and creating perturbations which will require more capital to be diverted to repairs of bridges and roads after record floods, indicates that this arrangement cannot survive indefinitely. It is what Orlov calls the "inevitability of a discontinuous future."[21]

He sees five stages of collapse and casts them in terms of a loss of faith. They are:

1. Financial Collapse: Faith in "business as usual" is lost. The future is no longer assumed to resemble the past in any way that allows risk to be assessed and financial assets to be guaranteed. Financial

institutions become insolvent; savings are wiped out and access to capital is lost.

2. Commercial Collapse: Faith that "the market shall provide" is lost. Money is devalued and/or becomes scarce, commodities are hoarded, import and retail chains break own and widespread shortages of survival necessities become the norm.

3. Political Collapse: Faith that "the government will take care of you" is lost. As official attempts to mitigate loss of access to commercial sources of survival necessities fail to make a difference, the political establishment loses legitimacy and relevance.

4. Social Collapse: Faith that "your people will take care of you" is lost, as local institutions, be they charities or other groups that rush in to fill the power vacuum, run out of resources or fail through internal conflict.

5. Cultural Collapse: Faith in the goodness of humanity is lost. People lose their capacity for "kindness, affection, honesty, hospitality, compassion, charity." Families disband and compete as individuals for scarce resources. The new motto becomes "May you die today so that I can die tomorrow."[22]

Orlov does suggest that this order might be inverted, or that all five "stages" might be more or less simultaneous. This is what Demarest calls the "grand systemic collapse."[23]

This is a dystopia so frightening that we don't even want to contemplate it, but so was the fall of Rome for those who experienced it. One question for us in contemplating such a scenario is, what role would ecological degradation play in triggering these stages in whatever order they might cascade? That the economic collapse might be built into global capitalism certainly weakens the system, but it did survive the Great Recession of 2008 when subprime loans proved to be bad. But all the while the infrastructure decays and more and more people understandably want to use more of the world's dwindling supply of resources to lift themselves out of poverty. Two related considerations come to mind: climate change and loss of biodiversity. As we need to adapt to climate change, more and more resources will need to be put into climate defense infrastructure such as sea walls and moving coastal populations, to say nothing of repairing existing structures damaged by weather extremes and paying for fire-fighting. Furthermore if resource wars expand not only over the remaining oil but water and various minerals, more and more capital will be put into fighting them which itself will destroy use value as cities and dikes and dams are bombed while mechanized warfare spews ever more carbon into the atmosphere. And climate change will set millions of people in motion that will make the migrations of the last few years seen paltry. Climate change will also exacerbate the already grave extinction crisis caused by habitat loss as cities and industrialized agriculture expand. No one knows at what point biotic simplification will undermine the biosphere's ecological services to the economy. Certainly the health of a complex community of soil organisms is crucial to food production, as is the ability of the ocean to produce protein, both of which are under threat. What is certain is that many of these phenomena are underway and that humanity has not taken corrective steps. It seems the runaway train is headed for the chasm where the bridge is out. But maybe not. In Part III of this book we will look at new values, ideas, technologies and institutions that, if adopted in time can in my view avert or at least mitigate collapse, but only if we adopt a wholly new mindset. Demarest reminds that the answer to our problems "must begin with ideology—with a change in general expectations."[24]

Collapse is the downside—the depressing news, but we do have hope, and here is what Jack Nelson-Pallmeyer has to say about it in *Authentic Hope.*

> Optimists expect everything to turn out nicely without any effort being expended toward that goal. Pessimists assume we're doomed and there's nothing to do about it but infect everyone else with despair. Hope is based on uncertainty. The central premise of authentic hope is that we have an opportunity and a responsibility to shape the quality of our *as yet to be determined* future. Authentic hope pays attention to problems as they actually are. The good news is that when we act constructively to engage any one of these problems we create favorable conditions for resolving the others.[25]

The knowledge we need is available to us, though not widespread enough and not understood by the governing elites. But before we can look at that, we need to go back in history to see just how this epic crisis of Hypercivilization came about. Without knowing the origins of the problem, we can't create the right solutions. We will look next at how humanity lived with nature before the rise of Hypercivilization, or The Old Way, and then the rise of dominator civilizations including our own.

# PART II

## The Rise of Hypercivilization

Analyzing the long history that preceded and then prepared the way for Hypercivilization deserves a library of its own. If we are to invent a sustainable alternative, we need to understand how Hypercivilization came about, but here we must be content with the merest sketch. First, we look at what came before it, not because we are going back to the remote past, but to see what we can learn from it for constructing Ecocivilization which we will take up in Chapter 8.

# 7

# "The Old Way"[1]

Before civilization, hunter-gatherer peoples had an enchanted view of nature and their place in it. They had crafted a symbiotic relationship with nature that was sustainable for over 100,000 years until it was rejected as "primitive" and "savage" and they were shoved off into remote places by dominator cultures. However, some of these peoples and their world view did survive into the present. The views of contemporary indigenous peoples, whose reverence for Earth was once disdained as savagery and superstition, are making a comeback, and looking more and more like wisdom.

By 35,000 BCE the advanced hunter-gatherers had crafted a sophisticated culture that included a tool kit well-adapted to interact with the rest of nature, amazing art on cave walls, and a complex spiritual life that anthropologist Elizabeth Marshall Thomas called "The Old Way."[2] Here we are in the presence of fully developed humans. A great deal of nonsense has been written about these peoples, including Thomas Hobbes's famous statement that those living in the what he called the "state of nature" had "no Knowledge of the face of the Earth; no account of Time; no Arts; no Letters; no Society; and which is worst of all, continuall feare, and danger of violent death; And the life of man, solitary, poore, nasty, brutish, and short."[3] This is what is known as the "hunter stereotype"[4] and no part of it is true.

We can construct a model of their mode of apprehending the world that both explains their behavior and, by contrast, illumines ours. According to historian Calvin Martin, "Hunters maintain that all of nature is empowered—is conscious, intelligent, sensate, and articulate."[5] They sensed themselves to be in it and of it, to be non-distinct from the great living or the "surround" as he calls it.[6] They communicated with the rest of the sentient world through visions and sacred song and story-telling. They lived in the here and now and did not know the terrors of a linear concept of time. Of course they took from what we call "nature" to eat, but hunting was not murder. They were not in competition with "nature." As Martin put it, "Nature conserves me—not I it. This [was] the underlying ethic."[7] Therefore

**109**

they consciously limited their impact on the earth. What we mistakenly call their "economy" was in fact their "spirituality,"[8] and the difference in these two concepts delineates the great gulf between the modern consciousness and that of the Old Way and many contemporary indigenous peoples.

Everything was imbued with spirit. Nothing is harder for the modern mind to comprehend. It is so "unscientific." Hunters learn the language of the other-than-human beings. And the way in which Hunters speak and sing about these others, the tribes of the winged peoples, the four-footed peoples, and so forth, is always in terms of care, respect, and humility. Not defining themselves as distinct from the surround but as integral to it, other beings were kin and this kinship bred "both respect and confidence."[9] They lived without fear, knowing that nature would support them as long as they behaved with respect and humility toward the other beings.

"The other-than-human persons, vegetable and animal, will give themselves to me as long as I refrain from overexploitation, as long as I treat their flesh and substance, including their remains, with respect and avoid all other forms of offense—this is the prevailing sentiment."[10] Perhaps it was this ethic among other things that made hunting and gathering the most sustainable way of human interaction with nature. We do not know

The Old Way: Although we don't know much about the religion practiced at Stonehenge, it has come to stand for the Old Way in which humanity and nature were seen as indissolubly bound together and where humans were not thought to be the masters of nature but members of a larger, sacred community that included the plants and the animals (PTZ Pictures, Shutterstock).

that any other way of life is sustainable over such a period of time. Martin's hypothesis sounds reminiscent of the orientalist Mircea Eliade, who points out that in the mythic perception of the cosmos, man "feels himself indissolubly connected with the Cosmos and cosmic rhythms...."[11] We live rather in the impersonal world of nature as perceived by modern science. "The fundamental difference between the attitudes of ancient man and those of modern man as regards the surrounding world is this: for modern, scientific man the phenomenal world is primarily an 'It'; for ancient—and also for primitive—man it is a 'Thou.'"[12] Nature was not a wilderness, but the benign surround that gave life.

Very gradually, almost imperceptibly, hunting and gathering gave way to farming in upland villages. It was a seamless shift and was probably never total until the industrial revolution drove people off the land. It is unlikely that the fundamental values and perceptions of nature and humanity's place in nature changed until the advent of the great river valley civilizations. But before that, between the hunters and gatherers and the rise of dominator societies there appear to have been "partnership societies." (The term is Riane Eisler's in *The Chalice and the Blade*.[13]) Agriculture developed first in the uplands in Neolithic villages before the militarized empires took over and revamped whole river valleys and denuded the forested hillsides. Imagine a scene from such a village.

*A party of women make their way quietly through the cool, fresh air of the pre-dawn dark to a slight rise of land. The place toward which they are headed is a small field, its surface of dew-moist, broken earth mixed with wood ash was laboriously prepared with bone hoes and digging sticks the day before. They arrive just before sunrise, kneel in the field and sing a hymn to the Earth, the Great Mother Goddess who showed them how to live like this by cultivating grain as farmers, and from whose womb their lives are nourished. As the sky brightens and the sun's rim begins to show and then climb above the horizon, they break into praise for the life-giving Father Sun who guards the plants. They make offerings of pure water and wheat cake. Then they bend to their labor, poking holes in the light soil with digging sticks, singing a rhythmic work song as they entrust the mysterious seed to the ground. It is 10,000 BCE in ancient Mesopotamia. With their hard work completed the women return to the village for a midday meal. They meet the men and boys coming in the other direction, sweating and tired, carrying their stone axes and small bundles of firewood. They have been felling trees which, when dried out, will be burned as the first step in preparing a new field. As they come into the village they are greeted by the old women, the very young children and the old men, and of course, dogs, goats, cattle and sheep. Each family goes to its own hearth to eat flatbreads, onions and peas and drink a pot of beer.*

This scene occurred daily over thousands of years in the ancient Near

East somewhere in the hills above the Tigris-Euphrates flood plain, as well as in Anatolia and Old Europe. These people (who were real and not fictional we—just don't happen to have a record of their names) were creating a new environment, a mixed landscape of wilderness and farms, a humanized environment. They were settled. They lived in that place as their parents and grandparents had before them. Their village was surrounded by small, irregular fields carved out of light woodland. A little gathering and hunting still took place as a supplement to the grains, vegetables, and small herds, but farming was now the dominant mode of interacting with nature. They could look out on their fields and see where their life came from. Unlike us, their asparagus was not grown in Brazil and then shipped to the E.U. for eating.[14] It is very likely that these lands which produced their life were managed as a commons, as the property of the community with a special sense of stewardship based on intimate knowledge of soils and seasons and governed by rules and norms that provided for cooperation and equal access and prevented free riders. It is highly unlikely that they were treated as a space in which a competitive free-for-all took place; in fact, it's unthinkable.

In many ways, the Old Way persisted in the new age. Hunting continued as a means of supplementing the food supply. The reverence formerly felt for wild species was maintained and extended to the domestic ones that had become the sustenance of life. In the large agricultural village of Catal Huyuk in Anatolia, for example, murals of bulls in black and red decorated the walls and the skulls and horns of bulls, covered by clay and painted with geometric designs, projected from walls and floors. Domestic sheep and goats were honored in art along with wild leopard and vultures. The Mother Goddess, enthroned between animals or shown in the posture of giving birth with her arms raised in benediction, may well have personified the life-giving grain.[15] The civilization of Crete was probably the last surviving example and it well documented in Riane Eisler's *The Chalice and the Blade*.[16] They were not matriarchies, but rather "partnership societies." The impact on the Earth made by these village farmers was for the most part benign. Their small fields actually increased edge habitat and probably the populations of some species of wildlife. In their beliefs, what we would call "nature" was the sacred source of life.

Five features made these civilizations sustainable. They numbers were few, their technology was organic, that is, biodegradable, and limited in power. There was a reverence for or at least a respect for nature. The world was a benign place of kindred spirits. The nature-based peoples perceived themselves as living within the larger community of inspired nature. Of necessity these peoples were frugal, and they lived within the limits of what nature could provide on a recurring basis, the annual cycles of harvest and growth. But they were conquered by a very different kind of civilization that had abandoned The Old Way.

## Dominator Societies

Hypercivilization is the latest phase of Western Civilization, which has its origins in the Semitic history of the Middle East. What arose there were the dominator societies including Hammurabi's Babylon, the ancient Hebrews, the Achaeans and Dorians in Greece, and the Romans. They overtook the old partnership societies. The new male elites imposed a harsh rule over humanity and nature. Their creation stories were not tales of a mother goddess giving birth to order and beauty, but of violent clashes. Their religious symbols were thunderbolts and weapons. The Scythians "made sacrifices to their sacred dagger, Akenakes."[17] The Mesopotamians believed that their male god, Marduk, created the universe by slaying the goddess Tiamat, hacking her to pieces and then constructing the world out of her body parts.[18] Wave after wave of these peoples came out of the desert fringes and off the Eurasian steppes, penetrating the Fertile Crescent and later into the Mediterranean. They developed highly regimented, centralized river valley civilizations based on a radical alteration of the environment. They were characterized by a male-dominated hierarchy including divine kings, harsh laws, slavery, patriarchy, and warfare. Deities were no longer benign mother goddesses but male warrior gods associated with the sky or volcanoes. Nature was a hostile realm to be conquered for the glory of the king as in the *Epic of Gilgamesh*, about a real king who ruled around 2,700 BCE.

As the story opens, the young Gilgamesh is raping the daughters of his subjects, who are praying to the Goddess Inanna for relief. She sends another strong man, Enkidu, to contend with Gilgamesh. Enkidu can be understood either as the symbol of nature or of hunting and gathering societies. He lives in the wilderness and the animals are not afraid of him and he is able to talk with them. To get him to come into the city, a priestess is sent to him and she lies with him, after which he is unable to communicate with the animals. He has fallen from grace and leaves the wilderness. In the ensuing fight, Gilgamesh triumphs but does not kill Enkidu. Instead they become fast friends:

> Together they went on a quest for cedar wood in the far mountains. The forest was a sacred grove protected by the wild giant, Humbaba, and his defeat and death at the hands of the two heroes was a symbol for the subjugation of the wilderness by the city. Gilgamesh promptly cut down the cedar trees and carried them off to Uruk to use in building a palace for himself. The proper effort of mankind toward wild things, in the view of the Mesopotamians, was to domesticate them.... Animals that could not be

domesticated were hunted mercilessly; Gilgamesh is said to have killed lions simply because he saw them "glorying in life."[19]

Another king, Gudea, the ensi of Lagash around 2200 BCE, is also recorded as going into the mountains to collect timber. Deforestation and the extincting of species through over-hunting were characteristic of the dominator societies. These were also the first urban societies.

> The attitude of city-dwelling people toward the natural environment shows a striking change from that of the hunter-gatherers, early farmers, and herders. It is as if the barrier of city walls and the rectilinear pattern of canals had divided urban human beings from wild nature and substituted an attitude of confrontation for the earlier feeling of cooperation…. It was only through the conquests made by the gods, and the constant labor of their human followers, that the natural chaotic state of the universe could be overcome and order established … the order of the city, with its straight streets and strong walls….[20]

The cities made huge demands on distant uplands and mountains, exploiting mines, creating surpluses, and trading them over long distances. They knitted together whole resource areas by means of empire. Society was reorganized for what historian Lewis Mumford considered the first great attack on nature.[21]

Their power came from extensive and complex irrigation systems that increased food production and hence led to a larger population, which could be managed for warfare. But in the end they failed because they deforested the hills and the erosion silted up the irrigation systems. At the same time, salinization turned the soil into unproductive hardpan where the water sat, creating malarial marshes. The blowing sand buried their cities. It was a systemic failure but they set the pattern for the later taker civilizations that succeeded in the Mediterranean world, first the Greeks and then the Romans.

## The Ancient Greeks

Theirs was a complicated society exhibiting both dominator characteristics and a mystical reverence for nature in the myths of Pan, from which we get our term pantheism, and in the mystery religions, as well as being the culture that birthed the very first materialist philosophy in the works of Thales and the atomist, Democritus. Humanism, an ideology that privileges humans over other creatures, was first formulated by Protagoras (480 to 410 BCE), who stated that "man is the measure of all things." The Greeks also gave us the idea of hubris, that overweening pride that leads to a tragic fall. The myth of Prometheus, who stole fire from the Gods and gave it to humans, is not so much about the origins of fire-using among humanity as it is about upsetting the natural order and the trouble that inevitably follows. Prometheus was severely punished by the Gods, bound to a rock where a vulture repeatedly tore out his entrails. It is instructive that the image of Prometheus Unbound would become a symbol of modern, industrial progress in the Age of Hypercivilization.

But in the end, the Greeks overtaxed their soils. The once beautiful mainland and the islands, including Crete, were forested and wild. But overpopulation and foolish grazing practices with goats and sheep have left most of Greece and the islands rocky wastelands. The soils of Attica were intact during Homeric times, when population densities were low and people practiced a mixed economy of hunting, herding and farming. We know that in the Odyssey we read about green pastures for grazing and forested hills and running streams. By the fourth century BCE Plato had noted the environmental decline without ambiguity.

> What now remains of the once rich land is like a skeleton of a rich man, all the fat and soft earth having wasted away, only the bare framework is left. Formerly, many of the present mountains were arable hills, the present marshes were plains full of rich soil; hills were once covered with forests, and produced boundless pasturage that now produce only food for bees. Moreover, the land was enriched by yearly rains, which were not lost, as now, by flowing from the bare land into the sea; the soil was deep, it received the water, storing it up in retentive, loamy soil; the water that soaked into the hills provided abundant springs and flowing streams in all districts.[22]

No more. "Today the home ... of the wily goat king–turned seafarer is a rock-filled desert with bare hills and desiccated fields furrowed by brooks gone dry."[23]

## The Roman Debacle

Rome was the dominator civilization par excellence of the ancient world, a society built on slavery, warfare, exploitation of distant provinces and brutal "entertainment" in which not only did humans kill each other for sport, but tens of thousands of animals like lions and elephants were rounded up and slaughtered as well. Rome contributed to the further exhaustion of Mediterranean soils, and all the while inequality continued to rise and the state was either engulfed in chaos of civil wars or stifled with dictatorship, until it was finally overrun by "barbarians" in the West.

The decline of Rome and the end of the ancient world in the West was an incredibly complex series of events which can be best understood as a systemic biosocial failure. Much of the land Rome came to control in its thousand-year history was already old and worn out, including the Greek peninsula and the man-made deserts of ancient Near East. Some of the land they came to control, especially in Italy and northern Africa, they themselves wore out. In other areas they created malarial swamps. "The Pontine marshes supported sixteen Volscian towns in the seventh century BCE. Five hundred years later they supported only mosquitos."[24] Hannibal's troops had brought malaria with them as they invaded Italy in the Punic wars.[25] Rome's port at Ostia silted up so badly that the seashore was removed several miles to the west.

By around 400 CE, we see a class of world managers, the Roman administrators, trying to organize and regulate widely scattered, powerful private land owners and great areas of public domain so that the whole system would continue to produce food and defend itself. They failed. They failed in the sense that depopulation, desiccation and desertification took place in the dry regions, and regional extinctions and wild reforestation occurred elsewhere. They were unable to manage this vast system. It was overheated and inflated and for a time could thrive and give a sense of prosperity but it was probably doomed to inevitable failure. The Roman people lived at a level of consumption that was far above their soil's carrying capacity and eventually beyond the capacity of the soils of Europe, the Mediterranean and North Africa and the ancient Near East. Their economy was based largely on robbery and exploitation. One must agree with historian Frederick Cartwright when he asks us to "Imagine Rome as a bloated spider sitting in the center of its web."[26]

## The Ambivalence of the Judeo-Christian Tradition

The Judeo-Christian tradition provided another building block of Western Civilization.

The ancient Hebrews were an archetypical dominator people whose angry warrior god of the mountains ordered the wholesale slaughter of the nature-worshipping peoples in the land of Canaan.

> In passage after passage of the Old Testament we read how Jehovah gives orders to destroy, plunder, and kill—and how these orders are faithfully carried out.... In Numbers 31, for example, we read of what happened after the fall of Midian. Having slain all the adult males, the ancient Hebrew invaders "took all the women of Midian captives, and their little ones." And now they were told by Moses that this was the command of the Lord: "Kill every male among the little ones and every woman who hath known a man by lying with him, but all the women children that have not ... keep alive for yourselves."[27]

Cutting down sacred groves was a particular passion of the invaders. It is attested to time and again in the Old Testament.

> You shall tear down their altars, and break their pillars, and cut down their Asherim ... lest you make a covenant with the inhabitants of the land, and when they play the harlot after their gods and sacrifice to their gods ... and their daughters play the harlot after their gods and make your sons play the harlot after their gods [Exodus 34:13–16].

In first Kings a contest is arranged between Elijah and the prophets of Baal. After winning the contest the Hebrews seized the prophets of Baal and drowned them in the river (I Kings, 18).

In the Book of Genesis, humans are given dominion over nature:

> Then God said, Let us make man in our image; and let them have dominion over the fish of the sea and over the birds of the air, and over the cattle and over all the earth, and over every creeping thing that creeps upon the earth [1:26].
>
> Be fruitful and multiply, fill the earth and subdue it and have dominion over the fish of the sea and over the birds of the air and over every living thing that moves upon the earth [1:28].
>
> And the fear of you and the dread of you shall be upon every beast, and upon every fowl of the air, upon all that moveth upon the earth and upon all the fishes of the sea: into your hand they are delivered [9:1–4].

God is not in nature and not of nature but over and above it, and nature is an It, not a Thou. The Earth was thoroughly demythologized. Nevertheless, these nature-hating trends by no means exhaust the Hebrew tradition. Ultimately God is the owner of nature and humans are just "sojourners in the land," temporary guests (Leviticus 25:23). And God has made creatures who have no utility for humans, sometimes because they make him laugh, as in Psalm 104.

> Leviathan whom you made to amuse you.
> All creatures depend on you....
> with generous hand you satisfy their hunger.
> You give breath, fresh life begins,
> You keep renewing the world.

## The Christian Middle Ages

Christianity grew out of this ambivalent matrix of Jewish scripture and Greek philosophy and carried the two seemingly contradictory attitudes toward nature through the next two thousand years. It probably started out as a movement to reform Judaism but quickly became a separate religion, since most Jews did not recognize Jesus as Messiah. It developed slowly, gaining adherents primarily in the cities of the Empire. Our word "pagan" comes from the Latin, *paganus,* which means rural peasant, i.e., one not yet converted to Christianity and therefore a "worshiper of false gods."[28] By the end of the fourth century the Church had developed a complex theology and institutions of governance. It seems that the religion of Jesus became a religion about Jesus and, for the faithful, he was their guarantor of an eternal after life in a realm beyond Earth which itself was regarded as a mere testing ground. Christianity had become the official religion of the Roman Empire and its successor States in the Middle Ages.

Christians made short work of temples devoted to nature worship. "To a Christian," writes Lynn White, Jr., "a tree can be no more than a physical fact. The whole concept of the sacred grove is alien to Christianity and to the ethos of the West. For nearly two millennia Christian missionaries have been chopping down sacred groves, which are idolatrous because they assume spirit in nature."[29] He writes further that "By destroying pagan animism, Christianity made it possible to exploit nature in a mood of indifference to the feelings of natural objects," and "Christianity is the most anthropocentric religion the world has seen."[30] Historian Victor Ferkiss concurs: "Unlike the pagan religions, Christianity desacralized nature…. But the cult of saints, as one historian argues, 'ousted spirits from the material objects of nature and liberated mankind to exploit nature freely.'"[31]

The world view of the church was entirely non-scientific. Christian thinkers rejected ancient materialism and damned its chief author, Lucretius. Writing in the *Norton History of the Environmental Sciences*, Peter Bowler says, "we are now dealing with a time in which the study of Nature was almost by definition subordinated to the worship of God. The early fathers of the Church were suspicious of ancient learning because it tended to deflect people's attention away from salvation."[32] The goal of life was virtue leading to heaven, not the "laying up of treasures where moth and rust corrupt" (Matt. 6:19–20). The rejection of ancient materialistic philosophy led to "the decline and virtual disappearance of science," according to historian Url Lanham.[33] But the beauties of nature were still to be enjoyed and nature was seen as the Codex Dei, the Book of God wherein his Creation revealed his majesty.

After Rome, much of Europe became reforested and provided the retreats from civilization for the burgeoning, world-denying monastic movement. The monks did not go to the wilderness to enjoy it *per se,* they went to escape the sin of the cities, and their main work was clearing the forest because the central sacrament required wheat and vines for the bread and wine. Ironically, many of their establishments eventually attracted urban development around them.

Unable to control nature by means of technology or the meager social organizations that medieval people were able to construct, and being but sparsely provided with companions in a seemingly hostile world, they turned to an other-worldly religion. Humanity was weak. Nature was strong, the reverse of our situation. Over the centuries our forebears managed to extricate humanity from that situation. In the process, many attitudes, techniques and forms of social organization were hammered out. These were functional in that they worked to increase humanity's control over the environment, but their very success makes them dysfunctional today. We have been like a person pushing a large boulder up a hill, learning new ways of pushing, making new tools to help and all the while gathering strength and speed as we neared the top, proud of what we have learned and developed. Suddenly we were over the crest in a rush and in danger of a rapid uncontrolled descent down the other side.

Scientific materialism, capitalism, and the Industrial Revolution were just around history's corner and together would constitute Hypercivilization, whose features are in deep contrast with the medieval world view, and where a materialist philosophy would define the world of nature as a dead other to be exploited for human benefit, a thoroughly anthropocentric view. The new society would depend on the once-only extraction of materials and would at the same time be wasteful in the extreme, and they would create an inorganic, toxic technology. Finally, their numbers would explode. Suddenly we need to criticize the techniques and attitudes which were vital in our long historic climb, but how did such a revolution come to be?

# 8

# Materialism, Capitalism and Conquest

In the early modern age, dominator society vastly increased male power over nature, women, and indigenous cultures. Nussbaum writes that the pioneers of modern science were creating "a new heaven and a new earth ... in 1660 Europe was in revolution. At no time in its brief history as a society had any generation stood to the future with an orientation so distinct from that of its ancestors."[1] Their view of the cosmos changed radically from a great mystery penetrated everywhere by God, angels, devils and other spirits, to a perfectly understandable machine in which God was no more than a distant craftsman, the original maker. Modern science was a genuine "break in consciousness," a "collapse of the mental world of the ancient and medieval eras."[2] But oddly enough, the age had begun with alchemy and was marred throughout by a massacre of European women. The attack on alchemy and the development of an alternative, scientific view of nature, and the killing of women, also coincided with the rise of capitalism and Europe's outburst into the rest of the world as conqueror, exploiter, dominator, and transformer of ecosystems across the globe.

## The Rise of Scientific Materialism and the Defeat of the Alchemists

Materialism was not the only possibility at the opening of the modern age. In spite of official Christianity, ancient pagan animism had never been completely erased and was reviving as the philosophical orientation known as alchemy. From Plato alchemists believed the world had a soul called *Natura* (note the Latin feminine ending). For alchemists, investigating nature was a sacred occupation surrounded by taboos and rituals for the purpose of gaining a mystical knowledge of the cosmos. They believed in the *Aurea Catena*, the Golden Chain of the Magi, "that a deeply secret connection pervades all of nature, that one thing relates to the next and things depend upon each other."[3] The I–Thou relationship between the alchemists and the *Natura* was widespread and acted as a normative restraint on what they were allowed to do to the natural world. But not for long. Merchant writes, "The metaphor of the earth as a nurturing mother was gradually to vanish as a dominant image as the Scientific Revolution proceeded to mechanize and to rationalize the world view."[4] The materialistic revolution rang in "the death knell of animism in the West."[5] People today view alchemy as some primitive pseudo-science whose only goal was to make its charlatan practitioners rich, or more charitably, what chemists did before they got it right, exactly what the Church, Descartes, Bacon, Newton, Daniel Defoe, Adam Smith and the other creators of the modern mind wanted people to think.

Ancient and medieval philosophers had objected to mining on the grounds that it despoiled the earth, excited human cupidity, and provided the materials for weapons of war. The chief apologist for the newly active mining industry was Georg Agricola, whose *De Re Metalica* subtly turned the arguments based on an I–Thou concept of humanity and Earth in favor of mining. In 1556, he argued that Earth wanted humans to find her treasures, indeed had stored them up for us. And if the mining industry destroys the forests, then pleasant fields take their place and produce food. If the waters and their fish are destroyed, the profits from the mining can be used to buy fish from elsewhere. He argued further, if humans do not mine metals from the earth to make tools, they will have to go back to eating roots and berries, to primitive hunting and gathering, perhaps the earliest argument from the Doctrine of Progress. Finally, as to the bad uses to which metals are put in war, these should not be blamed on the technology, but on failed human character. Here we have one of the earliest arguments that technology and science are neutral and, by implication, that all technologies should be pursued regardless of whether humans used them for good or evil, an argument that would lead ultimately in the twentieth century to the hydrogen bomb and genetic engineering.[6]

The role of humanity shifted from central player in a sacred drama to mere observer of universal law, at best as a scientist. But even though God was more and more shut out of the explanatory structure, the old Judeo-Christian dominion teaching was left intact. Scientific apologists and philosophers such as Francis Bacon and Rene Descartes moved humans into God's place as Masters of Earth and Lords of Nature, itself now mere matter and viewed mechanistically. The revolution in thought added up to what Max Weber called a "disenchantment" of the world.[7] This new view of nature suited the capitalists admirably. The ideology of science and the productive behavior of capitalism matched perfectly and, when combined, would come to provide a degree of power to dismantle nature never possessed by any previous culture.

The new way of perceiving reality was to be mechanistic, experimental, atomistic, rational, abstractive and analytical (fragmenting nature into discrete parts). It was mathematical and quantifying, looking for universal laws "behind the phenomena" as they said. It aimed at prediction and control and was a reductive materialism, a thoroughly utilitarian approach in which nature was de-spiritualized. Berman writes that for the new scientists: "the universe is a vast machine, wound up by God to tick forever, and consisting of two basic entities: matter and motion. Spirit, in the form of God, hovers on the outside of this billiard-ball universe, but plays no direct part in it."[8] Berman sums up the implications of the scientific revolution, writing: "The holistic view of man as being at home in the cosmos, is so much romantic claptrap. Not holism, but domination of nature; not the ageless rhythm of ecology, but the conscious management of the world, not ... 'the magic of personality, [but] the fetishism of commodities.'"[9]

The leading philosophers of materialism were Francis Bacon (1561–1626), and René Descartes (1596–1650). Bacon set out to do nothing less than reconstruct all of philosophy and science. His *Institaurio Magna* (The Great Renewal) had as its aim "the total reconstruction of sciences, arts, and all human knowledge...."[10] The purpose of all this was practical. In *The New Atlantis* he announced: "The end of our Foundation is the knowledge of causes and secret motions of things, and the enlarging of the bounds of Human empire, to the effecting of all things possible."[11] And he wrote: "I stake all on the victory of art [read artifice] over Nature...."[12] No hunter-gather, no medieval peasant, no alchemist could have written such a line. With Bacon we are in the modern world. "I am laboring," he wrote, "not to lay the foundation of any sect or doctrine, but of human utility and power."[13] It was a doctrine, of course, but a thoroughly secular one. In *The New Atlantis* he "portrayed a scientific utopia in which men enjoyed a perfect society through their knowledge and command of nature."[14]

Berman writes: "Bacon is convinced that knowledge is power and truth, utility."[15] His approach to knowledge was consciously technological. To know nature, it must be treated mechanically. Will Durant was right when he wrote of Bacon that he "prepared England for the Industrial Revolution."[16] The new scientific observer was not to be passive. To the contrary, he was to annoy nature and vex it ("*natura vexata*" in Bacon's Latin)— poke and prod it, subject it to varying conditions to see what happened.[17] This was the manipulation of the alchemists but without the cautionary surround of sacred mystery. The elevation of technology to the level of a philosophy had its "concrete embodiment in the concept of the experiment, an artificial situation in which nature's secrets are extracted under duress."[18]

Descartes also revolutionized humanity's relationship to nature. In his famous *Discourse on Method*, he wrote that his discoveries:

> ...have satisfied me that it is possible to reach knowledge that will be of much utility in this life; and that instead of the speculative philosophy now taught in the schools we can find a practical one, by which, knowing the nature and behavior of fire, water, air, stars, the heavens, and all the other bodies which surround us ... we can employ these entities for all the purposes for which they are suited, *and so make ourselves masters and possessors of nature*[19] [emphasis added].

All that existed for Descartes was space and matter in motion, but if he could get control of that, his shocking boast was, "I will construct a universe."[20] Here the sin of pride, as medieval people would have said, is elevated to a mountainous arrogance. Here we are light years away from St. Francis, living literally in a different mental universe.

"I think; therefore I am," said Descartes.[21] Humans are essentially mind. "Mind is in possession of a certain method. It confronts the world as a separate object. It applies this method to the object, again and again and again, and eventually it will know all there is to know...."[22] The so-called objective universe is just the sum of its material parts, which can be taken apart and recombined endlessly. Cartesian dualism differed from Christian dualism only in that it was materialistic, leaving only two entities of note with regard to the environment—humans and nature.

Galileo (1564–1642), was also a great vexer of nature, at once a scholar and technologist and consciously interested in changing nature. This was a radical new combination in Western history. Berman observes that "Once technology and the economy became linked in the human mind, the mind started to think in mechanical terms, to see mechanism in nature...."[23] Galileo adopted an "engineering approach ... distancing himself from nature in order to grasp it more carefully...."[24] For these men, reality was to consist of the universal laws lying behind the actual phenomena themselves and these were to be expressed in equations. Nature had become a giant abstraction. Newton was to prove it so.

## Newton's Synthesis

It was Newton who most typically personifies in his psyche the cultural struggle of European civilization, the struggle by which the mystical and magical were to be banished from nature and from personality.

He united the work of Copernicus, Brahe, Kepler and Galileo and others and wove them together into a brilliant synthesis that united reason and the empirical method. His work sent a shock wave that continues to roll through Western Culture and, the planet, as that culture now dominates the world.

> Newton's precise mathematical description of a heliocentric solar system … not only summed up the universe in four simple algebraic formulas, but he also accounted for hitherto unexplained phenomena, made accurate predictions, clarified the relation between theory and experiment, and even sorted out the role of God in the whole system. Above all, Newton's system was atomistic: the earth and sun, being composed of atoms themselves, behaved in the same way that any two atoms did, and vice versa.[25]

The foregoing was written from a perspective that is still enthralled by the peculiar scientific revolution of the seventeenth century which gave humanity so much technical power. There is, however, another far less adulatory view which sees in the Newtonian synthesis the intellectual foundations of a polluted and devastated planet. E.A. Burtt describes Newton's world machine in different intonations. It was:

> the vast mathematical system whose regular motions according to mechanical principles constituted the world of nature…. The world that people had thought themselves living in—a world rich with colour and sound, redolent with fragrance, filled with gladness, love and beauty, speaking everywhere of purposive harmony and creative ideals—was crowded now into minute corners in the brains of scattered organic beings. The really important world outside was a world hard, cold, colourless, silent and dead….[26]

Isaac Newton (1643–1727) synthesized the materialistic explanation of the world developed by previous scientists and philosophers, providing the intellectual basis for Hypercivilization's understanding of nature and our concept of mastery over it which paved the way for the industrial revolution, but in fact there was a hidden, mystical side to his work which did not become apparent until the 1930s (Georgios Kollidas, Shutterstock).

As K. Sale put it: "The ideas of the scientific paradigm transformed completely the attitudes of Western society towards nature and the cosmos. Nature was no longer either beautiful or scary but merely there, not to be worshipped or celebrated, but more often than not to be used...."[27] But in fact this Newton is a deception, not the real Newton at all. The real Newton was, *horribile dictu*, an astrologer and an alchemist, a great magician and mystic, a man who wanted to explain nature in the old way, to explain both why and what, but whose efforts fell far short of his dream. After all, Newton did not even explain what gravity is; only how it works. The same was true of his theory of light.

Why did Newton, who in fact believed he had been singled out by God as the one man in his generation to whom would be revealed the ancient wisdom, who believed that he was one of the links in the *Aurea Catena*, who had in fact been immersed in the occult, particularly alchemy, publicly withhold this other, mystic and sensuous side of this thought? His thousands of pages of occult manuscripts and alchemical treatises were suppressed, first by himself, and thereafter by his descendants for centuries. It was only when the family was bankrupted that these were offered up at auction at Sotheby's in London in 1936 that the "other" Newton came to light. In one of his alchemical notebooks he had rearranged the letters of his own name to read "Jehova *sanctus unus*," One sacred Jehova—or in other words, God.

Upon discovering his full range of thought and writing, the economist John Maynard Keynes said of him that he was not "the first of the age of reason but the last of the magicians...."[28] In sum, his own terrible psychosocial struggle, arising out of a bizarre childhood, was between chaos and order. To preserve his sanity he finally chose a total identification with order. It was that theme which permeated his "scientific" writings in contrast to the rich, mystical disorder of his other corpus. It was that imbalance in favor of order that he bequeathed to the modern world's landscape of the mind. Nevertheless, Newton's synthesis of the seventeenth century "scientific revolution" was so powerful that it caused a culture to abort its long-held views of nature and adopt a whole new, mechanistic and objectified understanding of the phenomenal world. Berman writes that "Europe went collectively out of its mind."[29]

Newton's teaching was accepted for a complex variety of reasons, the most compelling of which was that this new viewpoint met the needs of capitalism. Only a disenchanted and mechanistically conceived nature can be exploited for profit without remorse or even thought of consequences. "This was an age of practical men and practical methods, in deliberate retreat from all that was subjective and intangible, or passionate and personal,"[30] and from all that was erotic and natural. More than a retreat, in fact, this sixteenth and seventeenth century movement became an attack, in particular a slaughter of women in Europe and of indigenous peoples around the world.

## The Slaughter of the Women

The attack on the feminine was not waged just against philosophical world views; it was waged against women's bodies. By the thirteenth century, Scholastic philosophers had discovered Aristotle, a blatant sexist who believed women were just incomplete or mutilated men. His thought replaced the Augustinian/neo–Platonic substrate of Christian theology. In Plato's original conception, the soul of the world was female. In the medieval Neoplatonic conception, this female world soul was divided in two: the higher portion created human souls with divine ideas and the lower, which created matter, was *Natura*, a female, subordinate to God of course, but mother to us all. Caroline Merchant writes: "Not only was nature in a generalized sense seen as female, but also the Earth, or geocosm, was universally viewed as a nurturing mother, sensitive, alive, and responsive to human action."[31] But Aristotle asserted that the animating principle of all things was male and properly ruled over the female which he viewed as totally passive. These views were projected onto the cosmos, leaving the earth still female, but totally passive—ripe for plunder as we shall shortly see in the work of Daniel Defoe.

This was the era of the great witch hunts which had for a long time been dismissed by historians as an inexplicable aberration, a curious remnant of the medieval consciousness that somehow flared up at the beginning of the modern age. In fact, they were integral to the development of the modern age and were part and parcel of the attack on the old views of nature and on the states of consciousness that would be considered dysfunctional in the newly emerging, materialist civilization. The killing of the so-called witches was an act of violence necessary to pave the way for the modern world because no power was to be left unregulated, not that of nature, or of women, or indigenous peoples in other lands. The attack was successful and purged Western society of nature mysticism for several centuries, having terrorized women into subordination and discredited their traditional healing powers. Only recently have women begun to rediscover the ancient Wiccan traditions of healing and of the feminist and nature-based spirituality that underlies it.

The slaughter was a clash of cultures. The newly rationalistic state and the newly re-assertive and anti-pagan Church, whether Catholic or Protestant, set out to destroy the matriarchal folk culture, that "vast stratum of folk religion...."[32] It is no coincidence that the witch hunts took place at the same time as Europe's overseas expansion and the rise of the Atlantic slave trade and the rise of capitalism. There would be a train of such acts as the Europeans moved into the "new" world and exterminated native nature-based peoples while kidnapping and enslaving millions of Africans to work the new plantations.

It was a period in which Western men "gained control over much of the world's natural resources and wealth, not the least part of which came in the form of human beings."[33] And the three groups of humans over whom total control was now exercised by means of terroristic violence were Africans, "Indians," and women. The Spanish philosopher, Sepulveda, argued that all three were without souls, uncivilized, barbarians incapable of self-government, identified with savage evil, an extremely convenient conquerors' rationalization.

How did the views of a handful of geniuses, Copernicus, Bacon, Descartes, Galileo and Newton and others, men who were responding to their own particular situations and to their own psychological imperatives, become the views upon which an entire biosocial system was to be built? First, it must be admitted, they do correspond *in some measure* to reality. The earth does in fact circle the sun. And they allowed for the unlocking of tremendous technical power over nature. The revival of commerce, based on so many other changes, was creating a continent of eager capitalists who were to become the driving force of the modern world. Capitalists found the new science deeply resonant and extremely useful, a paradigm based on control and even on violence. "If it moves" [without the permission of the patriarchs], "kill it," might well have been the motto of the age.

## Capitalism

Hypercivilization has produced an assault on the environment like no other before it, both in terms of scale and kind. What drives this engine of destruction lies in the fundamental contradiction between our social systems—and especially our capitalist economic system—and the natural systems of planet earth. Nature, now demystified and desacralized, became no more than a seemingly endless supply of *resources*. Environmental destruction is inherent in unregulated, free market capitalism. It can do no other. So the issue is not that capitalism can't produce material wealth, but that it is immensely destructive and therefore unsustainable because it is cannibalizing the biosphere. It is turning on itself, depleting the very resources it needs to exist. Capitalism assumes an infinite planet where one can extract resources forever and this is its greatest delusion and our greatest danger.

## How Capitalism Works

All societies develop a set of rules for their economic activity of turning nature into themselves. These rules flow from their answers to the following questions: what is nature, what is human society, and what is the good life? The rules mandate who determines who does what to nature, using what technologies, producing what goods and what bads, and how these are distributed. Capitalism is that set of rules for nearly all the world now.

In a capitalist economy, nature is reduced to resources and is divided up into parcels of legally protected private property. This amounts to the appropriation and privatization of the commons by a class of owners who become society's decision makers regarding the interchange with nature. Small owners, consumers of homes and cars, for example, shape the environment as a result of their aggregate actions. But it is the large, corporate owners of the means of production who are the predominant shapers of nature by making policy that allows them to farm, mine, drill, manufacture, and distribute with little regard for the environment. These means of production—the farms, factories, mines, railroads and machinery—all become the property of the owners, as does the final product into which nature is transformed, that is until it is sold on the market to some other owner, either an individual or another corporation. Another feature of capitalism is that its property boundaries are blind to ecosystem boundaries; for example, watersheds are divided into thousands of private as well as government-owned parcels. Nevertheless, emissions from a coal plant on a parcel of corporate private property literally roam the whole world's atmosphere. Pesticides dumped into the Rhine River from the property of an agricultural chemical plant end up in the world's oceans.

In a capitalist society everything is for sale. The market becomes an almost mythical entity. It comes to stand between humanity and nature and mediates our access to the natural world. Quite literally, people are moved off the land and out of direct contact with the Earth and its natural processes. Money becomes the means of access to goods, and since it is infinitely convertible to them, more is always better. Profit is the sole criterion of good for the owners. Economic growth is the sole criterion of progress. Even if growth degrades the environment on which civilization rests, apologists for capitalism argue that to restore the environment we need more money and only growth provides that. As Herman Daly writes in *Economics Unmasked*: "Growth must not be questioned because it is by definition the solution to all problems—even those that it causes."[34] In capitalism there is no recognition of uneconomic growth, growth that impoverishes the present or the future. Development is always viewed as progress.

The nineteenth century social philosopher John Ruskin coined the term "illth" to mean the opposite of wealth. As an example, the pumping out of an oil well ought not to be counted as income but as capital drawdown, or illth, as Daly points out in a critique of "growthism":

> To elaborate, illth is a joint product with wealth. At the current margin, it is likely that the GDP flow component of "bads" adds to the stock of "illth" faster than the GDP flow of goods adds to the stock of wealth. We fail to measure bads and illth because there is no demand for them, consequently no market and no price, so there is no easy measure of negative value. However, what is unmeasured does not for that reason become unreal. It continues to exist, and even grow.[35]

In the capitalist ideology, human nature is redefined in purely economic terms. *Homo sapiens* becomes "economic man," *Homo* the producer and consumer of processed nature. Most humans become consumers and hence need more money. They are willing to do almost anything for a job, even if they know it is destructive to the environment and imperils the well-being of their grandchildren. Most people are trapped in this way since the owners of capital and the means of production control the jobs. Society is seen as no more than the aggregate sum of its individuals, a social philosophy of atomism. Individualism is rampant and community declines. There is an almost mythic belief that the common good is achieved by everyone competitively pursuing their own private good, defined as accumulating wealth or nature transformed into goods. On the one hand, freedom, in a capitalist society, is defined as independence from government restrictions on how property owners interact with nature, as economic liberty for the owners to do with their money or material capital whatsoever they want, a policy of *laissez faire*. On the other hand, the government is welcome to aid business in their activities, such as turning over public lands to oil drilling, building transportation routes, and providing subsidies.

Competition defines the race for resources and customers in which the competitive edge is often achieved by externalizing costs to society and the biosphere, ignoring true cost accounting. Powerful psychological techniques are employed to stimulate the creation of wants by convincing consumers that these are in fact needs. Consumers then respond by exercising demand and capitalists say that the demand brings forth the supply, but in fact, the demand has been created by the marketing. Happiness is defined as having more and more and having the very latest consumer goods.

In a capitalist society the state is in the service of the owners who set the rules governing production, trade, consumption and taxation and see that government aids their endeavors with public resources. Even in a democracy, the government can be controlled for the owners by financing election campaigns and, in some cases, by outright bribery. Also the corporate ownership of the opinion-making media influences elections, and well-funded conservative think tanks and even university departments can generate public opinion favorable to the owning class. Furthermore, elected officials retain office by campaigning on economic growth that will bring jobs to their districts, since their constituents for the most part are desperate for a job in order to live. Under capitalism, corporations came to rule the world.[36] They are also able to buy up governing elites in developing countries.

The resulting condition of the environment under a capitalist system is the result of chaotic actions rather than intelligent design by society. No one is in charge. There can be no concerted and coordinated action to achieve the common good. In fact, the common good disappears as a policy consideration, or is subsumed under the unquestioned assumption that economic growth will produce it.

The nature of capitalism is to always turn more and more of the natural world into processed commodities in order to achieve wealth and power for the owners of capital and those who serve them. The financial sector also drives this destructive growth. The reason for a bank's existence is to sell the use of money, that is, to lend it at a profitable rate of interest. This means it must promote "development" such as new mines in Australia or new dams in India. And once the money is repaid, it needs to be lent out again unless the financial institution is to voluntarily go out of business. Capitalism requires permanent debt. It must grow or die, which means it needs to produce more and more in a world where there is less and less and where there are fewer sinks for its toxics.

In order to be a capitalist, one must somehow first accumulate capital. In Great Britain this was done through the enclosures at the end of the Middle Ages, a bald appropriation of the commons which was commented upon by St. Thomas More in his iconic book, *Utopia*: "your sheep, that used to be so meek and tame, and such small eaters, now, as I hear say, have become such great devourers and so wild that they eat up and swallow up the very men themselves."[37] The fourteenth-century enclosure movement began a long process not quite finished today in which the majority of people were forced off the land so it could be turned to growing crops profitable for the big landowners. Capitalism was a violent transformation of society. Anup Shah writes: "For this to work, social traditions had to be transformed. Free markets were not inevitable, naturally occurring processes. They had to be forced upon people."[38]

During the Renaissance the burghers from Bruges to Florence were embracing the new world view. Economic development went hand in hand with the new, secular humanism of the scholars, artists and their wealthy capitalist patrons like the Medicis. They had no thought of replacing Christianity, but nevertheless they introduced a wholly new perspective into its very midst, one that the nineteenth-century historian Jacob Burck-hardt called the "discovery of the world and of man."[39] The new humanism was illustrated by Renaissance artists who looked at nature in new ways. Michelangelo's David is a perfect statement of the new humanism, with its focus on the beauty of man and on the importance of accurately portraying this world. The painters were fond of portraying urban scenes, frequently inside buildings where nature is often seen only through a window, perfectly framed and under control. They almost never painted scenes in which wild nature was the focus or in which it was venerated. The place of humans in this new age was on top and was so stated unambiguously by the Renaissance philosopher Marsilio Ficino: "Man not only makes use of the elements, but also adorns them ... man who provides generally for all things, both living and lifeless, is a kind of God."[40] They were fond of quoting the ancient Greek philosopher, Protagoras, who said: "Man is the measure of all things."[41] More typical were the ideas of Pico Della Mirandola, who pronounced: "There is nothing to be seen more wonderful than man."[42] Scientific materialism, capitalism, and humanism laid the foundations for Hypercivilization.

## Daniel Defoe: The Cheerleader of Capitalism

By the seventeenth century capitalism was flourishing in England where Daniel Defoe (1660–1731) became its champion. Known today primarily as author of the novel *Robinson Crusoe*, no one's writings better show the raw nature of capitalism than his. He wrote hundreds of tracts, books, pamphlets and poems in favor of the new biosocial system, believing fervently that it would save humanity as well as make the bourgeois class rich. He was a champion of the enclosures, writing: "What an infinite number of people do these [sheep] employ. What millions of acres of land they improve, and how do they create and propagate trade even in the remotest corners of the land."[43] By land improvement, he meant an increase in the rental value of the land, not the health of soil, rural communities, or ecosystem diversity. He was a strong advocate of government intervention in the interests of humanizing the landscape for the promotion of trade. In "An Essay on Projects" he advocated "That an act of Parliament be made, with liberty for the undertakers to dig and trench, to cut down hedges and trees, or whatever is needful for ditching, draining, and carrying off water, cleaning, enlarging, and leveling the roads, with power to lay open or enclose lands, to encroach into lands, dig, raise, and level fences, plant and pull up hedges of trees ... with power to turn either the roads or the watercourses, rivers, and brooks...."[44] The landscape was a mechanical complex to be rearranged in the interest of production and profit.

He chastised the landed gentry in a long poem about economic development in Scotland, believing that capitalism would create unimaginable wealth and well-being for all members of society if only people would wake up to the new attitudes and get to work.

> Wake, Scotland, from thy long lethargic dream,
> Seem what thou art, and be what thou shalt seem;
> Shake off the poverty, the sloth will die,
> Success alone can quicken thy industry.
> To land improvement, and to trade apply,
> They'll plentifully repay thine industry.
> The barren muirs shall weighty sheaves bestow,
> The uncultivated vales, rich pastures show,
> The mountains rich flocks and herds; instead of snow.[45]

Nature, for Defoe was not a sacred Goddess, although he did like to think of nature as a female whose role was made clear in his poem, *Caledonia*:

> Nature's a virgin very chaste and coy,
> To court her's nonsense; if he will enjoy
> She must be ravished;
> When she's forced she's free,
> A perfect prostitute to industry;
> Freely she opens to the industrious hand
> And pays them all the tribute of the land.[46]

With Defoe we are a long, long way from ancient Crete where the land was a Goddess to be reverenced and worshiped.

For nature to be subservient to man came easily to Defoe. What excited him on his extensive travels throughout Britain was "that vast ocean of business, the British commerce, which gives employment and wealth to the active and numerous people we have been treating of, and is at the same time the foundation of the riches and grandeur of this noble kingdom...."[47] He looked at nature through the eyes of a bourgeois speculator. Looking at a particular district, he gives us a table of natural resources susceptible to exploitation such as lead, iron, copper, alum, coal, quicksilver, limestone and earths, including their locations, the processing techniques and the products into which they are turned. His proposals are always to increase extraction, always in terms of a growth economy, and he turned out a prodigious literature on the subject. His four volume work, *A Tour Through the Whole Island of Great Britain*, is such a tract. It is filled with such information as: "The parish of Barking is very large; and, by the improvement of lands recovered from out of the Thames, and out of the river which runs by the town, the great and small tythes as the townsmen assured me, are worth above 600 pounds per annum."[48] What interested Defoe in these drained wetlands was the amount of money they yielded when worked for profit in market-oriented agriculture. The *Tour* is an eleven-hundred-page economic geography that was highly useful for the businessmen of that day.

Defoe applauded the increase of trade and manufacture because it brought great numbers of people together in limited geographic areas and kept them from "wandering," where they procreated rapidly, thus increasing the market demand. The result of the increase in consumption is that "the rate of value will rise at market, and as the rate of provisions rises, the rents of land rise."[49] But concentrating people, who were now to be thought of as consumers, into cities where they were divorced from direct interaction with nature and were thus forced to buy their provisions from the new "owners" of nature did not stimulate "trade" fast enough. It was soon discovered and championed that demand could be stimulated even further by the creation of artificial needs. Defoe was delighted with this discovery and commented in particular upon the fashions in clothing and housing: "This must needs give a new turn to the trade, and that of course gives new methods and new measures to the manufacturers; obliges them to a continual study of novelty; introduces new customs and even gives a turn to trade itself."[50]

The side of *Natura* that the capitalists saw was the wild and impulsive side, and it made for bad business. It would simply have to be controlled. Defoe wrote in *Caledonia*:

> Thus vanishes the horrid and the wild,
> And nature's now with pleasant eyes beheld.
> When Boreas, with northern vapours raves,
> We smile, and with contempt, survey the waves.[51]

And

> See how mankind
> By her experience taught
> Has all to rule and method brought.[52]

All to rule and method—the engineering mentality.

Defoe believed that wilderness must succumb to an agriculture so intensive that it could produce no more. He wrote with pride that "there is hardly an inch of ground lost in the island of Barbados that can produce one ounce of anything more than it does."[53] And in *The Complete English Tradesman* he said, "no land is fully improved till it is made to yield its utmost increase."[54] And if more was produced than the domestic market and foreign trade can consume, Defoe's logic dictated that they increase the number of people in order to consume the maximum produce of the land, arguing economic growth for its own sake.

As we have seen, this revolution was well under way in Defoe's time—begun by Copernicus and carried on by Bacon, Descartes, Galileo and Newton himself, men who believed they were bringing hard-headed realism to society. These "practical men" were into repression: capitalists repressed workers in the new mills, the state was engaged in repression of regional autonomies and was subjecting nationalities, the Church was repressing women, and colonialists were repressing and exterminating indigenous peoples.

## Conquest

> "We can, if need be, ransack the whole globe, penetrate into the bowels of the earth, descend to the bottom of the deep, travel to the farthest regions of this world, to acquire wealth."
>
> —William Derham, *Physico-Theology*, 1713[55]

The revolution of capitalism, born in Europe, was carried around the globe by the armed merchant ship and waves of colonists who transformed the "New" World wildernesses and the "old world's" biosocial systems into what we see today. They went out and conquered Gaia. In *An Essay Upon the South Sea Trade,* penned in 1712, Defoe wrote:

> We shall, under the protection, in the name, and by the power of her Majesty, <u>seize,</u> <u>take</u> and <u>possess</u> such port or place, or places, land, territory, country or dominion, call it what you please, as we see fit in America, and keep it for our own, <u>keeping</u> it implies planting, settling, inhabiting, spreading, and all that is usual in such cases; and <u>when this is done</u>, what are we to do with it? Why, we are to trade <u>to</u> it and <u>from</u> it.[56]

The Europeans took their ideas of humanity's proper relationship to nature and went out to conquer and reshape the rest of the world. Armed with their materialist natural philosophy, capitalism, and with the newly revitalized, militant Christianity that came out of the Reformation and Counter Reformation, and with a belief in the innate superiority of their culture and race, they began the transformation of the world into Hypercivilization.

To do this they had to get there and with enough coercive power to force their will upon the newly discovered peoples and lands. The invention of the modern sailing ship made it all possible. It was a technological revolution, the first, mobile, prime converter of energy on a grand scale.

> The surplus energy derived from the sails is potentially enormous, as compared with the cost of producing the sail and hoisting it. Thus for the first time in man's history, men, using the sailing ship, came into control of very large amounts of power largely independent of plant life or the number of persons using it ... the sailing ship delivered energy at a rate previously unimagined.[57]

With the innovations of a stern mounted rudder and moveable sails, they could sail against the wind, and with the new navigational equipment they could sail out of sight of land. Its mobile energy field was vast, extending over the entire world ocean, eventually bringing European presence and power to every coast and up every major river. This new maritime technology was combined with advances in mining, metallurgy and chemistry; the ships were armed with bronze and then iron cannon, becoming the most formidable device for intimidation in history up to that time. No coastal region could withstand the awesome unleashing of their energy. "The sailing ship made it possible to bring into the valley, in the shape of men and guns, force that did not originate there and was not dependent upon the surpluses produced there."[58] In Carlo Cipolla's apt characterization, the

**Europeans Conquer the World:** The armed merchant ship was a formidable engine of conquest by means of which Europeans brought their new, materialistic, dominating culture to the rest of the world, altering ecosystems and bringing them into a global trade system (Babich Alexander, Shutterstock).

ship "was essentially a compact device that allowed a relatively small crew to master unparalleled masses of inanimate energy for movement and destruction."[59]

He writes further: "there was no power that could offer any resistance ... [and] within fifteen years after their first arrival in Indian waters, the Portuguese had completely destroyed Arab navigation."[60] Add in the European diseases to which the peoples of the Americas had no immunity and it was no contest. Thus were they able to rearrange the world.

This is not the place for a long discussion of the era of conquest. That belongs in another book. In sum, with the complicity of African slave traders, the Europeans stole millions of people from Africa and brought them to the "new world" where the survivors were forced to turn wilderness into sugar cane, gold mines, cattle ranches, and cotton fields, toiling out their lives rearranging the energy and nutrient flows of the new world biomes so that ever greater surpluses came to their European masters. The extractive economy began even prior to settlement as fishermen from England, Spain, Portugal and France had appeared off the Grand Banks about 1540. In 1578, according to Hakluyt, four hundred vessels were engaged in the trade. Each ship could expect six weeks of fishing to produce twenty to twenty-five thousand fish (approximately eight million in a season for a fleet this size). They also caught large numbers of petrels, auks, and other birds for sport and to make commercially valuable oil from their livers. They established shore bases and salted the catch before returning. Salt was obtained by evaporating sea water and the resultant deforestation to fuel this process was noticed by contemporaries. The raid on resources began even before the Europeans planted colonies on the new soil. In the seventeenth and eighteenth centuries, capital accumulation was achieved in large measure by drawing the New World biomes into the Old World system of production.

In some places like the Southwest and Southeast in what would become the U.S., and in "Mexico," they encountered civilizations as sophisticated as their own. Roxanne Dunbar-Ortiz writes an indictment of the process in *An Indigenous Peoples' History of the United States:*

> It should not have happened that the great civilizations of the Western Hemisphere ... were wantonly destroyed, the gradual progress of humanity interrupted and set upon a path of greed and destruction. Choices were made that forged that path toward destruction of life itself—the moment in which we now live and die as our planet shrivels, overheated.[61]

We learn only slowly.

The Europeans did not invent conquest, slavery, guns, technological innovation or human-induced ecological changes. They were only the most recent in a long line of changers. But unlike the others they introduced these things on a scale that, both in intensity and extent, was unprecedented in world history. Theirs was a driven civilization. They destroyed the most productive and sustainable food production in history when they took over the great Aztec city of Tlatelolco, and rampaged across the American frontier where, as late as in the twentieth century American farmers ginned up the myth that "rain follows the plow" and initiated the great Dust Bowl when clouds of Oklahoma soil darkened the skies as far east as Washington, D.C.

The Europeans now had the power to open up the earth and extract not only the New World's gold and silver that eventually funded their industrial revolution, but more important even than this, the planet's one-time reserves of mineral- and fossil-based energy that carried them and the Earth into a new era. Much of what the former European peasantry, once enclosed and dispossessed, would work at in their factory cities was quite literally the New World. It was the most efficient and effective dominator society in world history. It was brutal and it was based on the philosophy of materialism and capitalism whose ideology we have looked at already and which is illuminated again for us by the works of Defoe.

## Robinson Crusoe, or World Conquest for Beginners

The Spanish had stolen a march on the English in the race for colonies but that did not deter Defoe, whose advocacy of violence in the race for plunder was typical of the era. There was always war. "We should all wish for such a war," he wrote, "that the English might by their superiority at sea, get and maintain a firm footing, as well on the continent as other islands of America: there the Spaniards, like the fable of the dog in the manger, neither improve it themselves, nor will admit others to improve."[62] By "improve" he meant adding to nature inputs of capital and labor to produce goods for the market, and while he could not command navies, he could command ideas, which he personified in his most enduring fictional character, Robinson Crusoe. Ian Watt has written that "profit is Crusoe's only vocation, and the whole world is his territory."[63]

Wilderness, for Defoe and other Europeans, was waste land if it was not being made to maximize production for a market. What Defoe set up in his famous novel was a situation in which a rational, Christian European is shipwrecked on an undeveloped island located somewhere in the Caribbean. He has a full complement of eighteenth century technology— and this is noteworthy—the island is empty of people. Crusoe's island has become such a celebrated symbol of nature in the literature of the era that much can be learned of the guiding vision of nature that the Europeans brought to the New World, a vision that was at best schizophrenic. Crusoe's first reaction to the island was fear.

Even after a year's residence, having seen nothing more dangerous than a goat, he was still terrified. Thirteen years elapsed before he summoned enough courage to reconnoiter the island. Maximilian Novak observes that "much of his time was spent in hiding from imaginary enemies."[64] He calls his island "this unhappy island," and "this dreadful place, this horrid island."[65] And yet in other places in the book Defoe gives evidence that the island was pleasant and bountiful. He could expect two seed times and two harvests every year and he never went hunting without coming back with game for his table. He found materials necessary to make tools he had not been able to bring from the wreck. He found the woods full of pigeons and he was able to catch ample supplies of fish, and turtle eggs and turtle meat. He found sugar cane, grapes, and melons. In short, by the evidence he gives the country was idyllic. He even says so: "the country appeared so fresh, so green, so flourishing, everything being in a constant verdure, or flourish of spring, that it looked like a planted garden."[66]

It was especially useful to be able to argue that man has been wicked since the Fall because certain peoples were then identified as living more in a state of sin than others. For Defoe, such peoples tended to be found in Africa and in the New World. There is an instructive and demeaning episode in *Robinson Crusoe* in which a black woman is converted to Christianity, the whole affair being carried out in Pidgin English ("Sir, why you say God mekee all?"[67]). Defoe would advocate the redemption of such peoples by means of large doses of agriculture, commerce and Christianity. The world had to be conquered by the white man who would deserve the name of Christian. In winning the world, that is, in transforming wilderness into commercially productive land, he wins the salvation of his own soul. This is exactly what Crusoe did on his desolate island. But for that he needed a work force. He discovered and rescued an indigenous man about to be eaten by cannibals who have brought him to the island. He is first given a name, but not a human name. He is called after a day of the week, "Friday"; and then, "I likewise taught him to say Master; and then let him know that this was to be my name...."[68] Millions of Indians and Africans were to learn the same name in the New World.

How bitterly ironic that just at the dawn of the Enlightenment, a brutal form of slavery based on race was inaugurated by the English, Spanish, Dutch, French and Portuguese, the great maritime powers of Western Europe. The contradictions would run through history to the present day, enshrined the U.S. Constitution and the current wealth of the white colonial settler society which dominated and still dominates the New World and which would forever rest on the backs of people of color forced into servitude. The trafficking in humans became known as the "triangular trade," as if it were just another dimension of economics, which for them it was. Ships laden with liquor, firearms and trinkets set out from an English port (or Dutch, etc.), and made for the "Slave Coast" of Africa. There they traded their goods for the wretched humans brought to the "factories" by Arab or African slave traders (who were every bit as complicit as the Europeans), and these unfortunate people were chained together below decks for the journey to the West Indies or the American or Spanish colonies. One historian has written of this voyage that "it represented a true torture."[69] On arrival the survivors were sold onto the plantations, and the ships returned to Europe with the products of the now artifactual biomes of the New World, with tobacco, molasses from the sugar plantations from which the rum would be made, and cotton. Then it would begin again— Europe to Africa to the new world and back to Europe. They also carried naval stores such as timber for masts, and pitch back to Europe, necessities for the sailing ships of the merchant fleet and of the Royal Navy that protected them. And from all of this the merchants made vast fortunes. It was done for profit. This is what Defoe had in mind for Robinson Crusoe.

Defoe has Crusoe set about to build an empire as his island became peopled with other castaways and natives. When he finally gained additional companions, his first act was to divide the island into parcels and set up his men in farming. He makes of the island a plantation with himself at its head, and gives voice to every European's dream: "…to think that this was all my own; that I was king and lord of all this country indefeasibly and had a right of possession."[70] In the end they are at last discovered by a passing ship and Crusoe prepares to embark for home, but there is one final piece of business he must settle.

> One thing I must not omit, and that is, that being now settled in a commonwealth among themselves, and having much business in hand, it was but odd to have seven-and-thirty Indians live in a nook of the island, independent, and indeed, unemployed: for excepting providing themselves with food, which they had difficulty enough to do sometimes, they had no manner of business or property to manage.[71]

Managing business and property would become the chief activity in the New World as millions of colonists set about expanding the human footprint in countless regions.

## The Rules of the Game

The year 1776 saw not only the Declaration of Independence but also the publication of Adam Smith's *Wealth of Nations*, the bible of capitalism. By 1789 the "Americans" had drawn up a constitution that codified private property (including property in humans), limited government and give-aways of the commonwealth. The rules of the game were to be twofold: one, the interchange with nature was to be in private hands and guided by market profits, and two, the role of government would be limited to creating a safe climate for private resource decision making and for economic growth. There was to be no community planning or direction for the overall impact on nature or for the kind of habitat the corporations would create as the unplanned sum of their private acts. Anyone who could amass sufficient capital could become a decision-maker, inadvertently designing the habitat as an unintended consequence of profit-seeking, and could likely as not become very wealthy. That is why there followed a mad scramble for the nation's resources. These were practically given away to private developers by the government—given to the land speculators, timber men, and eventually the railroad and the oil and gas men, the miners and the cattle men. It continues to this day with the sale of oil and gas leases at practically no cost.

The Europeans spread their control and their occupation throughout the world. What was done in the Americas was done in India, Australia, New Zealand, and elsewhere with enormous biosocial consequences. The way was prepared for the Industrial Revolution.

# 9

# Industrialization

The new thinking of Bacon, Descartes, Defoe and the others was made a material reality in the industrial revolution, completing the transition to Hypercivilization. It was a systems-level change. It is impossible to overestimate the magnitude of the industrial revolution's impact on the natural world or the degree of discontinuity it represented from previous biosocial systems. For the first time in planetary history the machine was loose in the garden. F. Roy Willis writes that "Industrialization involved changes greater than Western society had ever undergone...."[1] Carl Cipolla writes in his work, *Before the Industrial Revolution*, that "A basic and fundamental continuity characterized the pre-industrial world even through grandiose changes such as the rise and fall of Rome, the triumph and decline of Islam.... This continuity was broken between 1780 and 1850."[2] The linkage between materialism, in its birth form of capitalism, and industrialization is clear in the minds of historians: "Capitalism signaled a radical break with all previous energy systems known to humanity. With it the primacy of biological energies ended and that of fossil energies was established."[3] Kevin Reilly, in *The West and the World*, simply pairs them: "What is responsible for the disregard of nature that has led to our environmental and energy crisis? Here we focus on the most obvious and immediate causes of the modern problem: industry and capitalism."[4]

A tremendously powerful synergy was created between fossil fuel production, metallurgy, complex machines, urbanization, capital accumulation, the mobilization of labor into the factory system, and many other interconnected changes that were growing out of the new, scientific and capitalist mode, including an industrially transformed agriculture. This synergy was recognized by contemporaries. Andrew Ure, writing in 1835, said, "Steam engines furnish not only the means of their own support but of their multiplication. They create a vast demand for fuel; and, while they lend their powerful arms to drain the pits and raise the coals, they call into employment multitudes of miners, engineers, ship-builders, and sailors, and cause the construction of canals and railways...."[5] The power of this whole new cultural complex was far greater than the sum of the parts and it made humanity the most potent ecological force on the planet.

The industrial revolution has been the most radical revolution in the entire history of humanity's relationship with the earth—its transformations the most extensive and intensive, and its final destructive impacts as yet unknown. It has been an ecological and evolutionary bomb exploding over the face of the planet.

## *The Steam Ship*

If the sailing ship was the essential technology embodying the commercial revolution of the seventeenth century which had begun the realignment of humanity and nature into new and unequal arrangements of world trade, the steamship was its equivalent for the industrial revolution. It accelerated and intensified that realignment. It truly globalized the industrial assault on planet earth. In just a single century the world's shipping, which had relied on sail power for thousands of years, went from one hundred percent sail to less than 25 percent wind power. What little commercial sail was left did not survive long into the twentieth century.

The ships were the results of advances in mining, metallurgy, and chemistry, the three industries with the heaviest environmental impacts. The steel boiler made possible higher pressures and bigger ships so that by the late 1860s, a steamer could bring three times as much cargo from China to London in half the time taken by a sailing ship. The total tonnage of steam powered vessels increased from 32,000 in 1831 to three million by 1871.[6] These new ships relied heavily on the mining and burning of coal. Up to the middle of the century, a steamship required 60 tons of coal per day, requiring not only a huge onboard supply, but coaling stations scattered around the globe.

Coal was necessary to make the coke to fire the smelters which produced the steel for their hulls, and it was the fuel for the steam engines themselves, as well as the engines that kept the mines operational by pumping out the inrushing groundwater. These ships made possible the transfer of mineral wealth, food and fiber from the periphery of the world to its industrial core in Europe and North America, with all the consequent impacts on the natural world. At the beginning of the nineteenth century, London received a total of 800,000 tons of goods from the rest of the world. It became the first "world city" with six million inhabitants and by 1900 the steam ships were bringing it eight million tons of goods each year.[7]

## *The Scientific Revolution in Agriculture*

England was the first country to go from a situation in which its productive land had produced whole communities with their physical, social and spiritual needs being met on the land, to a situation in which the land produced principally food, and of course, the profits that came from selling it on the market to the consumers who were now removed to the growing cities. The great landowners' interest in raising their income from the land led them into "scientific farming." "The result," according to Palmer and Colton, "was a thorough transformation of farming, an Agricultural Revolution without which the Industrial Revolution could not have occurred."[8] Very large landowners, who now controlled nearly all agricultural land in England as a result of a long historic process of enclosure, let it out to a small class of substantial farmers. They experimented with new technologies and techniques including new methods of cultivation, crop rotation, manuring, new tools such as seed drills and horse drawn weeders, and scientific stock breeding. They brought in new crops. This intensification of farming resulted in both higher yields and in depopulating the rural areas. By enclosing many small farms into fewer big ones, efficiencies of scale were achieved that reduced the need for as many worker-families per acre. The food supply increased and was produced with far fewer workers and the rest were now available for work in the coal towns of industrializing England.

The remaining few who worked land did not fare well. Karl Marx wrote about the child-labor gangs used for agricultural work on the scientific farms of Lincolnshire, pointing out that the "cleanly weeded land, and uncleanly human weeds of Lincolnshire are pole and counterpoles of capitalist production."[9] A Mrs. Burrows tells of her childhood in the fens in the 1850s:

> On the day that I was eight years of age, I left school and began to work fourteen hours a day in the fields, with forty to fifty children of whom even at that early age I was the eldest. We were followed all day long by an old man carrying a long whip in his hand, which he did not forget to use…. We had to walk a very long way to our work … very often five miles each way. The large farms all lay a good distance from the town, and it was on those farms that we worked.[10]

Scientific agriculture was an application of the reductionist methods of rational calculation to breeding and crop production. The single goal was ever increasing yield. Social and ecological systems were ignored and in being ignored, were often destroyed. New lands were being brought into production by draining biodiverse wetlands, a process that has continued right up to the present.

The industrial revolution was not merely the charming story of inventive genius on the part of a handful of bright tool makers. Eric Williams has shown that "the wealth from slavery and the slave plantation had been the crucial factor in the amassing of capital for the Industrial Revolution in England."[11] Europeans were able to affect this enslavement, and rearranged the ecosystems of the world so that production increased and surplus flowed to Europe because a new class, the bourgeoisie, had taken over. J.M. Blaut writes: "This provided the bourgeoisie with the legal and political power to rip apart the fabric of society in its quest for accumulation."[12]

Part of this revolution was the minute division of labor that so astonished Adam Smith in his visit to a pin manufactory. What astonished him was the productivity of this system. Where a pin maker creating a pin from start to finish could make about twenty pins a day, the new system of the division of labor, where one worker concentrated on only one of the eighteen tasks, could produce 50 times as many per day. He believed this would lead to universal prosperity. Said Smith: "It is the great multiplication of the productions of all the different arts, in consequence of the division of labour, which occasions, in a well-governed society, that universal opulence which extends itself to the lowest ranks of people."[13] Such was the optimism during the early stages of the industrial revolution.

In America, Whitney's cotton gin made possible a hundredfold increase in the output of the slaves. In fact, the first great outburst of productivity was in the cotton industry, where mechanization preceded the application of fossil fuels and created a wood shortage as the new mills consumed huge amounts of wood. By the end of the eighteenth century, metallurgy alone was consuming twelve million tons annually in France and peasants were rioting against the furnaces. An edict of 1723 reveals how desperate the situation had become—it banned the cutting of trees under ten years of age.[14]

Mass production would have to rely on coal, iron and steam engines. Kevin Reilly calls fossil fuels "the central fact" about industrialization.[15] Their use had a precedent-setting environmental impact that was enormous. The mining of coal releases methane gases and the processing of coal often results in the acidification of surface waters. Burning coal releases carbon dioxide, carbon, soot, nitrous oxide and nitrous dioxide and sulphur dioxide. Debeir points out in *Energy and Civilization Throughout the Ages*, "The clock, the windmill, the water-mill used the forces of an environment left unchanged, whereas the 'fire machine' consumed the matter from which it drew its energy."[16] Early on it was transforming the environment. Karl Marx pointed out that the steam engine was not just a single machine, but was a unique machine powering others to which it was connected and that together, this coal-fired, steam-driven complex made the modern factory city imperative. Steam powered mechanization was a realignment of all technological and social forces—turning the whole society into a true megamachine for processing nature.

The amount of coal produced in the nineteenth century was staggering. By the beginning of the century it had risen to 10 million tons per year globally and then increased one hundred fold by 1910, to a total of a billion tons.[17] By 2012 total world coal production hit 7.83 billion tons.[18] From 1870 to 1914, British coal production increased 150 percent, French 300 percent, German 800 percent, and American 1,700 percent. New coal mines were opened in Russia and in Silesia.[19] "The central ecological fact about the full-scale industrialization that began in the West in the 1800s," according to Kevin Reilly, "is that irreplaceable energy resources were used. Instead of increasing the efficiency of wind and water … industrialists turned to the fossil fuels of the earth … which could never be replaced because of the time nature took to duplicate them."[20]

Coal mining led to greatly increased iron mining, which had to keep pace in order to supply the pumps that drained the mines, as well as the steam engines, iron bridges and rails to transport the coal. And it was abundant coal that made the increases in iron production possible. Using charcoal for iron production had been limited by the amount of available wood and by the fact that the charcoal is fragile and shatters under the weight of the ore to be smelted. But coke made from coal can support far more weight and made possible ever larger iron-smelting retorts. Switching to coke made possible an increase in production at Richard Crawshay's factory in Wales from 10 tons per week to 200. The industrial revolution was built on mining and no other human activity has a heavier impact on nature. In light of such impacts, historian Keven Reilly was forced to conclude: "The civilization based on mining—the industrial West of the nineteenth century—was basically at war with the environment."[21]

Available power skyrocketed. By 1900 the average engine was thirty times more powerful than those built at the beginning of the century and they had become legion. The steam engine was, as Gustav Smil put it in *Energy in World History*, "the prime mover of nineteenth century industrialization."[22] They were very quickly applied to rail transport. "Like the steam engine," writes Debeir, "the railway was born in the coal mines…."[23] Already by 1850 a train of fourteen coaches could carry ninety tons of goods, replacing eighteen stage coaches and drivers and 144 horses.[24]

The old technologies had been organic and biodegradable and most people's lives were lived out in rural settings. Ever since agriculture, humanity's transformative power over nature was very real but it stood at a relatively low level compared to what was coming. The industrial revolution changed all of this, set humanity and nature upon a new course of development toward an immense magnification of our powers over the natural world, bringing new inorganic, invasive, and persistent technologies. And the impact of the new power and technologies, along with rising numbers, was further raised by concentrating them into the new landscape, the industrial city. Changes in kind, magnification, and concentration were the hallmarks of this new era in environmental history. "Perhaps," writes Kevin Reilly, "it was the gigantic scale of the new industry which had the most serious ecological effect."[25]

Prior to this, most people lived in rural settings, in a bio-prolific environment surrounded by plants, and working daily with animals in a rhythm set by sun, moon, and seasons. They were forced to exchange this environment for a bio-impoverished and even biocidal setting in the new industrial cities. This is not to argue that life in the country had been wholly idyllic, but coal town was something else entirely.

## Coal Town

In 1835, Alexis de Tocqueville visited the first and archetypical coal town, Manchester, England and caught the paradox of the industrial

Belching Smokestacks: Hypercivilization remains largely powered by burning fossil fuels which give off dangerous particulate matter, carbon dioxide, methane and sulfur dioxide, and can also be contaminated with mercury and other heavy metals, and result in thousands of premature deaths each year (Seroma72, Shutterstock).

revolution in his summation: "From this foul drain the greatest stream of human industry flows out to fertilize the whole world. From this filthy sewer pure gold flows. Here humanity attains its most complete development and its most brutish, here civilization works its miracles and civilized man is turned almost into a savage."[26] Manchester, whose population had trebled in the thirty years after 1773, was built on a collection of small hills and we can follow de Tocqueville as he walks about and records further observations.

> Thirty or forty factories rise on the tops of the hills.... Their six stories tower up; their huge enclosures give notice from afar of the centralization of industry. The wretched dwellings of the poor are scattered haphazard around them. Round them stretches land uncultivated but without the charm of rustic nature, and still without the amenities of a town. The soil has been taken away, scratched and torn up in a thousand places.... Heaps of dung, rubble from buildings, putrid stagnant pools are found here and there among the houses.... But who could describe the interiors of these quarters set apart, home of vice and poverty, which surround the huge palaces of industry and clasp them in their hideous folds. On ground below the level of the river and overshadowed on every side by immense workshops, stretches marshy land which widely spaced ditches can neither drain nor cleanse. Narrow twisting roads lead down to it. They are lined with one-story houses whose ill-fitting planks and broken windows show them up ... as the last refuge a man might find between poverty and death [and] below some of their miserable dwellings is a row of cellars to which a sunken corridor leads. Twelve to fifteen human beings are crowded pell-mell into each of these damp, repulsive holes.[27]

Karl Marx made similar observations in 1844:

> Even the need for fresh air ceases for the worker. Man returns to a cave dwelling which is now, however, contaminated with the pestilential breath of civilization.... For this mortuary he has to pay. A dwelling in the light, which Prometheus in Aeschylus designated as one of the greatest boons, by means of which he made the savage into a human being, ceases to exist for the worker. Light, air, etc.—the simplest animal cleanliness, ceases to be a need for man. Filth, this stagnation and putrefaction of man—the sewage of civilization (speaking quite literally)—comes to be the element of life for him. Utter unnatural neglect, putrefied nature, comes to be his life element.[28]

Walking again with de Tocqueville.

> The fetid muddy waters, stained with a thousand colours by the factories they pass ... wander slowly around this refuge of poverty. It is the Styx of this new Hades. Look up and all around this place and you will see the huge palaces of industry. You will hear the noise of furnaces, the whistle of steam. These vast structures keep air and light out of the human habitations which they dominate; they envelop them in perpetual fog ... a sort of black smoke covers the city. The sun seen through it is a disc without rays.[29]

Thousands, and then millions of humans, found themselves uprooted from the rural landscape with its trees, fields, open vistas, streams, and endless skies, uprooted from their ancestral communities surrounded by small-scale fields and wooded copses, and thrown instead into anomic cities without community support, surviving, but dwarfed by huge engines in a landscape of man-made structures, polluted with smoke, noise and toxic chemicals. The wonder is that the human body and soul were able to survive at all. The old ways were gone. The English population was half urban by 1850, the first such in the world. The workers were immiserated both in the city and in the remaining countryside.

This appalling new habitat existed because it made money for the factory owners. Even Adam Smith, the champion of "free enterprise," was willing to concede that the impetus to build these cities "comes from an order of men, whose interest is never exactly the same with that of the public, who have generally an interest to deceive and even to oppress the public, and who accordingly have, upon many occasions, both deceived and oppressed."[30] Friedreich Engels once chided one of these men, a Manchester factory owner, and received a revealing reply. "I spoke to him about the disgraceful unhealthy slums and drew his attention to the disgusting condition of that part of town…. He listened patiently and at the corner of the street at which we parted company, he remarked: 'And yet there is a great deal of money made here. Good morning, Sir!'"[31]

Competition for jobs was fierce and frequently children and women were hired over men because they were more docile and easily subdued by the foremen. In England's "first modern factory," Debeir points out, this pitiful work force was "led by overseers armed with sticks."[32] Young women worked naked as beasts of burden in the coal mines, and children as young as six were also found there. Rape and sodomy were endemic in the early industrial mine and mill. The new mills had also destroyed the handwork previously found in rural areas in the weaving and spinning industry, an employment that was often done by women and which often provided the family's margin of existence.

Coal town was inextricably linked to far flung frontiers. Industrial capitalism had a global reach. Coal became a commodity in the world wide export trade. As early as 1828, Americans living on the eastern seaboard were heating their homes with coal brought on sailing ships from England.[33] Britain was exporting one third of its production by 1900, with the Germans and the Americans not far behind.[34] And, out in the frontier districts, the movers and shakers knew what had to be done. As a delegate to the Ohio constitutional convention put it in 1850: "The earth must be subdued."[35]

By 1900, huge amounts of coal were being extracted and moved around the world as a result of the industrial revolution's ubiquitous fire machine. The coal-fired railroads linked up with the steamships. The first rail line into India was built to plunder the coal reserves of Ranigani, delivering them to the coast where the British steamers could devour them. And the French built a similar line to tap the coal fields of Yunnan in China. Coal was the fuel of imperialism. Gadgill and Gutkind point out in their environmental history of India that

> Not only did such interventions virtually reshape the social, ecological and demographic characteristics of the habitats they intruded upon, they also ensured that the ensuing changes would primarily benefit Europe. Colonialism's most tangible outcome … related to its global control of resources. The conquest of new areas meant that the twenty-four acres of land available to each European at the time of Columbus' voyage soon increased to 120 acres per European. Large-scale settlement was only one way in which Europeans augmented their "ghost acreage," for the world-wide control they exercised over mineral, plant and animal resources also contributed to industrial growth in the metropolis.[36]

Industrialism was not confined to the English Midlands where cotton was being manufactured into cloth. In fact, the Midlands were tied to the far-flung and brutal, biosocial transformation of America, India, Egypt, and Africa.

> Vast quantities of the raw material [cotton] were needed. In 1785, ten million pounds of cotton were imported; in 1850, five hundred eighty-eight million…. By 1830, three-quarters of the cotton imported by Britain was grown on the slave plantations of the southern United States.[37]

In many regions of the world, traditional peoples had developed satisfying and sustainable ways of life, interacting with nature to produce their basic needs and satisfy some material wants without jeopardizing the ability of nature to provide the same for future generations. Many of these sustainable biosocial systems were shattered by the rise of industrial civilization in distant Europe because of its global reach and its dependence on the rest of the world for labor and resources. Unable to sell their cotton products in Europe because of trade barriers, the British exported them into countries whose militaries were not sufficiently developed to pose any plausible threat to her economic penetration. Cotton established "a form of economic imperialism in which native industries of unprotected countries were severely restricted and concentration on production of raw materials for the industry of the developed countries encouraged,"[38] a pattern of global reach that has persisted into the present day. As Blaut says, "Development began in Europe and underdevelopment began elsewhere" as Europe's industrialization was continually fueled by the wealth from colonialism.[39] The old relations between people and land, which had sustained them for centuries if not for thousands of years, were being shattered on a planetary scale. India's forests were one of the first casualties; they went down before the axe in order to supply wood for the British Empire and to clear land for increased food production for export. Huge amounts of timber were also needed for railroad ties for the transport network the British were building to expedite the export of saw timber, grain, and minerals to feed the factories and factory workers of the English Midlands.

The story was the same the world over. Peoples and their ecosystems were to be rearranged to serve the needs of European industrial society. Those who would not succumb willingly would be forced to allow their resources to enter the world market. More pressure was put on the land to produce. Industrialization resulted in the intensification of agriculture world-wide. It also resulted in its spreading into previously wild lands. Simmons writes about "the breaking of the grasslands after 1870, in which many of the world's natural grasslands were converted to grain production to feed the burgeoning populations of core regions."[40] Some 281 million acres of mid-latitude temperate grasslands were plowed for the first time and made to support wheat. These grasslands had been, for the most part, extremely complex ecosystems with hundreds of species per hectare. The work of plowing that complex ecosystem under and replacing it with a monoculture was best summed up by a Plains Indian of the U.S., who said: "Grass upside down no good."[41] In ways that were both obvious and brutal, on the one hand, and complex and subtle on the other, industrialization accelerated and intensified its impacts on the biosocial system, radically altering all the interconnected elements.

The nature of work itself changed and productivity increased as owners sought to make the most out of their capital investment in the machines. The pace of work was consciously speeded up, using overseers and foremen and the labor of women and children who were more easily intimidated by violence. Factories were lit so that the work could go on through the hours of darkness. The old rhythms of nature, including night and day, were circumvented. Humans were being integrated into the megamachine and their bodily energies were in effect mechanized to produce a steady flow of nature through the factory. But the workers were hard to discipline, frequently came late, quit early, stayed in the tavern getting drunk and then raised hell. It is no accident that police forces developed in tandem with industrialization, disciplining the work force outside the factory while the foremen did so within.

The land was ripped up to provide the necessary minerals to keep the upward curves of production going. British iron output rose from a mere 17,350 tons in 1740 to 125,000 tons in 1796, a seven-fold increase, and then increased to 2.5 million tons in 1850, a 144-fold increase over the figure just 110 years earlier.[42] Ten years later, in 1860, it stood at 4 million tons.[43] French iron production increased six-fold between 1820 and 1870 and coal production went up thirteen-fold.[44] The story was the same in Germany.[45] All-in-all, industrial production rose 37 percent in each decade of the nineteenth century.[46]

## *The Industrialist's Mindset*

In 1851, Prince Albert, the royal consort of Queen Victoria, hit upon the idea of celebrating industrialization. He proposed a great industrial fair. His rationale reveals his beliefs about the nature of progress. He wrote that the exhibition was designed "to give us a true test and a living picture of the point of development at which the whole of mankind has arrived in this great task of applied science and a new starting point from which all nations will be able to direct their further exertions."[47] The Crystal Palace pavilion, where it was held, was enormous and was made out of mass-produced, standardized parts including one million square feet of glass. A typical poem of the day, in rapturous and fulsome style, gushed:

> Gather ye Nations, gather! From forge, and mine, and mill!
> Come, Science and Invention; Come Industry and Skill!
> Come with your woven wonders, the blossoms of the loom,
> That rival Nature's fairest flowers in all but their perfume;
> Come with your brass and iron, Your silver and your gold,
> And arts that change the face of earth, unknown to men of old.
> Gather, ye Nations, gather! From every clime and soil,
> The new Confederation, the Jubilee of toil![48]

The Prince went on to say that "The products of all quarters of the globe are placed at our disposal, and we have only to choose which is the best and cheapest for our purposes, and the powers of production are entrusted to the stimulus of competition and capital."[49]

One certainly wonders if the black miners in South Africa, who supplied some of Britain's minerals, toiled jubilantly to enrich the English middle and upper classes. But such self-deceits were part of the entire racist mentality that went with the European industrialization of the world, as Blaut points out, in *The Colonizer's Model of the World*: Europeans believed that the "normal, natural way that the non–European part of the world progresses, changes for the better, modernizes, and so on, is by the diffusion … of innovative, progressive ideas from Europe…. Nothing can fully compensate the Europeans for their gift of civilization to the colonies, so the exploitation of colonies and colonial peoples is morally justified."[50] We can see this mindset in Kipling's "The White Man's Burden."

> Take up the White Man's burden—
> Send forth the best ye breed—
> Go bind your sons to exile
> To serve your captives' need;
> To wait in heavy harness
> On fluttered folk and wild—
> Your new-caught, sullen peoples,
> Half devil and half child.

Take up the White Man's burden—
In patience to abide,
To veil the threat of terror
And check the show of pride;
By open speech and simple,
An hundred times made plain.
To seek another's profit,
And work another's gain.

Take up the White Man's burden—
The savage wars of peace—
Fill full the mouth of famine
And bid the sickness cease;
And when your goal's nearest
The end for others sought,
Watch Sloth and heathen Folly
Bring all your hope to naught.[51]

Incredibly this self-deceit goes on for four more verses, providing the clearest proof of the unshakable belief which Europeans in the industrial age had in their own virtuous superiority and the rightness of their new biosocial system. The racism and the exploitation of the non–European world to feed the mills of England, America, France and Germany was also buttressed by a new social philosophy masquerading as science, Social Darwinism, which was accepted uncritically although it rested on premises that were exactly the opposite of those on which the White Man's burden was founded. It rested on the premise that the non–Whites had lost out in the competition for survival and therefore did not deserve the rewards of modernism. The winners, since they had won, did. It was the law of nature and from that there could be no appeal. It echoes today.

## The Second Industrial Revolution

As the revolution gathered steam, still more significant increases in humanity's technological reach were yet to come in a second phase. These were chemistry, electricity and oil, and combined with these and exacerbating their effect on the land, air and water, was the alarming rise of the human population, beginning around 1750. These four developments pointed the way toward the flowering of Hypercivilization in the twentieth century.

### The Rise of the Chemical Industry

In the early years of the nineteenth century a chemical industry was being established to meet the demands of the Industrial Revolution.

Previously, humans had been largely confined to pushing combinations of molecules around. Chemistry enabled humans to intervene in nature at the level of the electron ring. Now they would be able to recombine the very molecules themselves and to create substances never before known on the planet. This they did with abandon, resulting in the injection of these chemical compounds into natural systems and cycles in ways that proved hugely detrimental. Historian F. Roy Willis writes: "The practical uses of chemistry won it instant acceptance in industry."[52] This was a major qualitative change.

> In 1800 industrial chemical arts differed little from those of 1700 except in the manufacture of sulfuric acid, alkalis, and bleaching chemicals.... Demand was often small.... By the end of the nineteenth century ... there was a highly developed chemical industry which had the proportions of big business. Acids and alkalis were being produced on an unanticipated scale. The fertilizer industry was making inroads.... Synthetic dyes and drugs had eliminated certain natural ones from the market. New explosives were about to supplant black gunpowder.[53]

A need developed for a new way of making cheap alkali for a variety of uses including soap making, glass works and bleaching of textiles. French chemist Nicholas Leblanc found a solution and his process made commercially feasible and soon it was in widespread production. Complaints about pollution quickly followed. Nearby landowners were irate because of the damage done by sulfurous and hydrochloric acid fumes. The pattern for pollution from the chemical industry was set early on—manufacture useful synthetic compounds and at the same time highly polluting by-products that were dumped into the commons. Closely related to the textile industry was the growing demand for sulfuric acid. Later the acid was obtained by roasting pyrites in the open air. The dyestuffs industry was born out of experiments with coal tars. Francis Bacon's dreams were coming true.

Coal tar was a nearly useless residue of coal gas manufacture, which had been introduced on a wide scale in the first two decades of the century. In 1834, Ferdinand Runge distilled an array of chemicals from it including aniline, quinolone, pyrrole and phenol. Others continued the work. The young chemistry student, William Henry Perkin, isolated aniline from coal tar and, after 1845, benzene, now a known carcinogen. The new industry also demanded heavy inputs of chlorine, sodium hydroxide and the ubiquitous sulfuric acid.

Chemistry also began to play a role in metallurgy. In 1825 aluminum was isolated but the cost was too high for commercial production until 1854 when Deville found he could prepare aluminum metal from Bauxite ore with a sodium process. In 1877, gold and silver were extracted by means of a cyanide process so effective that even trace amounts could be recovered from old tailings piles, but the cyanide polluted rivers as it does to this day in South America. In 1890, the electroplating of nickel became possible in the Mond process. The Mond Nickel Co. was formed and a huge smelter was built at Sudbury, Canada. Downwind pollution resulted in a toxic and barren environment for miles.

Electrochemistry developed with the effort to generate electricity from cells powered by various compounds. In 1857, Gaston Plante used lead sheets in a solution of sulfuric acid. While this was a commercial failure others went on to use lead oxides on lead plates. In 1867, Georges Leclanche developed a cell having a zinc electrode in an ammonium chloride solution and a carbon electrode immersed in manganese dioxide. Heavy metals were being introduced into the biosphere on a new scale. Mountains of copper would be needed for the next invention.

## Electricity

Electrical generation was a qualitative change extending the energy revolution far beyond what steam engines could previously do. The stationary steam engine reached only as far as its belts could extend. Around that space the factory was built. But the power of the steam engine attached to an electric generator reached as far as the copper wires could be made to go, extending industrial processing and, of course, lighting, into distant regions. It set off a rash of copper mining, especially in Michigan's Keweenaw Peninsula, and later in Montana and Arizona, leaving mine tailings, water pollution and, in the West, great open pits. The mining of copper spread worldwide. Electrical generation greatly magnified the demand for coal and then oil. In 1871, the ring-wound dynamo was invented. In 1879, the carbon filament electric light bulb, now extending hours of production around the clock. Edison's first bulbs were ten times brighter than gas mantles and one hundred times brighter than candles. In 1882, Edison built his first steam generating plant and in 1884 the steam turbine was developed. Charles Parsons's utility installed a 75,000-watt turbine in 1888 and within twelve years had installed a one-million-watt unit.[54] The race toward gigantism was on.

## Oil and the Internal Combustion Engine

The pace of change was now completely out of control, which was just the way nearly everyone wanted it. Invention and discovery followed quickly upon invention and discovery. Nothing could stop "Progress," a vague concept applied to almost anything new. As early as August 15, 1859, Drake had successfully drilled through rock for oil in William Penn's forests in Pennsylvania. Oil's energy density is fifty percent higher than standard coal, making it an ideal fuel for transportation, but that was not its first use. Lighting was the first big market for oil. Kerosene had been distilled only six years earlier by Abraham Gesner in London, and it soon replaced Whale oil. But the great boost to the oil industry was the development of the internal combustion engine. A German, Gottlieb Daimler, built the first light, high-speed, gasoline powered engine in 1885, the same year in which Karl Benz built the first automobile. Smil writes: "Daimler's engine, Benz's electrical ignition, and William Maybach's float-feed carburetor launched the automobile industry."[55] In 1892 Rudolph Diesel perfected his sparkless engine. Over the next hundred years, autos and airplanes would multiply like rabbits, altering the landscape and fouling the skies.

Before the turn of the century, the oil industry had sprung up and "forests of derricks stood above the oilfields around Ploesti in Romania, in Baku on the Caspian Sea, in California, in Texas after 1887, and in Sumatra after 1893."[56] Mexico would come on line in 1901, Iran in 1908. Oil would eventually power not only automobiles, trucks and trains but also the engines of war—the battleship and the bomber. These would wreak their destruction by means of the new science of chemistry which was the source of their high explosive bombs and shells, also products of the new chemical industry.

## The Demographic Explosion

The conquest of nature by steam, steal, electricity and chemistry did not exhaust the revolutionary changes of the era. Associated with these in ways not fully understood was a vast and rapid increase in human numbers, a virtual demographic explosion. There is no comparable event in all of the several million years of hominid history. By 1600 humans had attained a population of only six hundred million. By 1750, that number grew to a population of around seven hundred million. And by 1800 the population had reached 957 million. The one billion mark was reached by 1825, having nearly doubled in just two centuries. Then, having taken a million plus years to reach one billion, the population of humans doubled again in just a century, reaching two billion by 1925. Growing exponentially from such a large base resulted in the teeming numbers on the planet only one and a half centuries later when the mark hit five billion. As of this writing we humans number seven and a half billion. Areas that had been occupied by gather-hunters were overtaken by farmers. The agricultural revolution, with its ability to support many more people per hectare, was spreading into new lands that had not been previously cultivated. The indigenous peoples who were killed off or rounded up for reservations in this process were very quickly replaced with settlers and their children. On the American frontier, birth rates reached a world record high, never since repeated. New food crop species were introduced into ecosystems including wilderness and well-established farming ecosystems and occupied previously empty niches.

In the latter third of the nineteenth century pesticides were introduced as a result of the chemical revolution. Lands that were too dry to farm were irrigated. The amount of irrigated agriculture in the world went from twenty million acres in 1800 to one hundred million acres in 1900, doubling more than twice in a century.[57] Also, a drop in death rate appears to have occurred in the latter half of the nineteenth century as European cities began to improve water supplies and sewage control.

In previous doublings, which had occurred at low numbers and over immensely long periods of time, Earth had to provide twice as much food, shelter, clothing and other goods to satisfy the needs of the new humans. But in a doubling of this nature, with immense numbers and over a very brief period of time, it had to do so quickly and on a vast scale. Much more of nature had to be transformed much more rapidly. But that was not all, because the revolution of materialism included a new definition of the good life as a rising number of goods per capita, each of these new humans expected, and to some extent got, more of nature transformed into goods than had the previous generation of humans, creating exponential rates of growth in all sectors. And the new industrial mode of production had far heavier impacts on nature than had the previous, eotechnic mode, including the spewing of toxics into the environment, substances which depressed nature's natural productivity. Nothing like this had occurred on planet Earth before.

## The Industrialized Frontier in America

It was in this context of an industrializing world that Americans played out the history of their frontier. The power of industrialism to transform whole continents was evident even before the twentieth century. The sweep of the frontier across North America was evidence that the new industrial biosocial system would remake the earth and do it in but a moment of historical time. The eighteenth century had closed with Americans still confined to the eastern seaboard and the Old Northwest Territory. The nineteenth ended with them on the Pacific. It had begun with them living in a pastoral, un-mechanized, agricultural society on a continent characterized by teeming wildlife and un-humanized wilderness, especially west of the eastern mountain barrier. In almost no place did the primitive frontier last more than a generation.

The whole history of the social dynamics that would transform the wilderness is captured in this brief frontier experience. Here we see, in small, the race to mine the wilderness of resources that are to be turned into commodities in the industrial cities, each in competition with the others. Here we see the business class as the initiator and, of course, chief benefactor of the process. Here we see the government right alongside them, playing the role of promoter, assisting in the raising of capital and doing the work of surveying. But before they were able to so engage the land they had to remove the other biosocial system that had been set in place by humans and nature, the Indian–buffalo system. It was done with a swiftness and brutality made possible only by industrial technology— the railroad and the mass-produced, repeating rifle. It is a story of unparalleled violence. In 1865 there were over fifteen million buffalo living on the Great Plains. Twenty years later there were only 3000 left.[58] By 1890 only a thousand of the species remained, all captives in human hands.[59] The railroad had bisected the buffalos into a northern and southern herd. The four million animals in the southern herd were killed off in five years, from 1870 to 1875. One hunter, Tom Nixon, once killed in the space of forty minutes a hundred and twenty animals. The railroads made possible the transfer of the hides back to the eastern markets where there was an insatiable demand for them and for the tongues, which had become a gourmet delicacy. Later, the bones were picked up, ground up and shipped out for use as fertilizer.

But it wasn't just a buffalo war; the whole of the nineteenth century was a series of wars between the United States and the nations of the indigenous peoples. The principal tactic of the U.S. militaries was counterinsurgency—making war against civilians. Again and again unprotected villages were attacked and the non-combatants, the women, old men, and the children slaughtered. Crops were burned to starve the survivors. Half of Mexico was taken by force carried out by irregulars and by the U.S. Army. It was called "manifest destiny," as if it were some benign natural process, but it was calculated theft by violence, all captured in devastating detail in *An Indigenous Peoples' History of the United States*.[60]

A scant one hundred years after Thomas Jefferson drafted the Declaration of Independence, the nation celebrated its centennial with an industrial display held in Philadelphia's Machinery Hall. The giant hall held power looms, lathes, sewing machines, presses, pumps, axles, tool-making machines, typewriters, telegraph equipment and the first telephone. Historian Alan Trachtenberg notes that it was all powered by a single engine, the "thirty-foot-high Corliss double Walking-Beam Steam Engine … and its counterpart, a 7,000 pound electrical pendulum clock which governed, to the second, twenty-six lesser 'slave' clocks around the building. Unstinted but channeled power, and precisely regulated time: that combination seemed to hold the secret of progress."[61]

Out on the land Americans were using agricultural chemicals for synthetic fertilizers and pesticides. The worst pest outbreaks were in the newly cutover farmlands. Just as agriculture had always produced its own

Corliss Steam Engine: The giant Corliss Steam Engine, constructed of iron for the 1876 International Exhibition of Arts, Manufactures and Products of the Soil and Mine in Philadelphia, in which 37 countries participated. At 40 feet in height, weighing 600 tons and generating 1,400 horsepower, it ran the entire machinery floor and represents the epitome of fossil fuel-powered industrial revolution technology and what Leo Marx called, in his 1964 book of the same title, "the machine in the garden." A true altar to the worship of humanity's power to "tame" nature (courtesy of the Free Library of Philadelphia, Print and Picture Collection).

inquiline pest species, it was doing so again, but with two fundamental differences. The scale and speed of the transformations was unprecedented in history, and the effort to combat the insect emergency by means of heavy-handed chemical pesticides was also a first on the planet. The chemists turned to arsenic trioxides, specifically copper acetoarsenite and copper arsenite, which was also used to color paints green. It was sold as "Paris Green" beginning in the summer of 1867. Soon it was being brushed on potatoes, melons, squash, cabbage, other vegetables, cotton and most fruits. It was not long before mechanical sprayers were developed for a more broadcast application. It was highly toxic.

## *The Mining Frontier*

Mining and railroads went hand in hand on the American frontier. Mining damages ecosystems through cave-ins, slag dumps, settling ponds, release of acids and heavy metals into streams, and smelting. Shortly after the Utah Northern Railroad reached Butte, Montana, in 1881, the region was turned into a wasteland as the "copper kings" engineered the destruction of nature for use value and profit. According to historian Richard Bartlett: "Sterile yellow and gray slag dumps stretched out across gulches like lava from the mines, and noxious fumes from the great smelters killed all the vegetation in Butte and for miles around...."[62] And, Bartlett writes, "There were hundreds of lesser Buttes in the new country."[63]

A unique form of mining took place in California. The original gold rush of 1848 rather quickly extracted what ore could be claimed by the primitive method of panning in the streams in the upper watershed of the Sacramento River. Within five years a new and far more powerful method was devised. It was called "Hydraulicking." As the name implies, it made use of water pressure to dislodge whole hillsides. The miners would build a series of penstocks upstream to capture the water and build up a pressure head. Then by means of increasingly more sophisticated hose and nozzle devices, they would direct a high-pressure stream at a hillside along the banks. These water cannons could shoot a high pressure stream over five hundred feet in length, dashing it against the hillsides "so forcefully that boulders two feet thick were washed and batted down like so many marbles."[64] Sometimes preliminary blasting was done to loosen the hillside. All was washed away: trees, grass, soil and gravel. Most of it turned into a fine brown slurry of mud. At the bottom the gold nuggets and dust were picked out and the remaining gold extracted by an amalgamation process, combining it with the heavy metal, mercury. Wastes just washed down stream.

Not only were the hillsides lost, and river navigation destroyed, farmland was also a casualty. The silt raised the beds of downstream rivers so that any small rise in water level resulted in flooding. Soon major deluges were inundating the downstream farms and by 1880 some 33,000 acres had been destroyed and another 14,000 damaged.[65] Bartlett concludes that "The spoliation of the northern Sierras and the Sacramento Valley system is a particularly poignant example of the new country rapacity. Whole hillsides were flushed away ... while nature's own water system was twisted and reconstructed with dams, flumes, and aqueducts to the needs of the miners. A magnificent farming country was ruined and the rivers were never the same again."[66]

It was in the nineteenth century, and nowhere more apparent than on the American landscape, that not just the machine (in the sense of technology), but the megamachine—the biosocial system of industrial capitalism—was let loose in the garden. The loosing of the machine was a thoroughly conscious process. A popular lithographic print, "American Progress," published in 1873 showed the entire process in allegory and includes, for the slow-witted, a narrative text on the back side. Historian Alan Trachtenberg describes it as follows.

> The picture shows a chase. On the left, a herd of buffalo, a bear and a coyote, and a family of Indians and their horses flee before an array of Americans in various "stages" of "progress": guide, hunter, trapper, prospector, pony-express rider, covered wagon followed by stagecoach, and a farmer in a field already under plow and oxen. Three railroad lines, representing the transcontinentals, join the flow which originates from the city and its factories, schools, and churches. On the left, the text explains, we find "darkness, waste and confusion." And in the center of the scene, its presiding image, looms a white, diaphanous figure, "a beautiful and charming Female"— we are told—"floating Westward through the air," bearing on her forehead the "Star of Empire." Her knee raised through her gown as if striding purposefully, she bears in one hand a book representing "Common Schools," and with the other "she unfolds and stretches the slender wires of the telegraph...." The Indians look back at her, the "wondrous presence" from which they flee. "The Star is too much for them."[67]

## Conclusion

During the industrial revolution, from 1750 to 1900, there was an incredible synergy between the ideology of scientific materialism, the social institutions of capitalism, the development of fossil based energy sources, the introduction of new metals, the revolutions in chemistry and electricity, and the burgeoning masses of the human population that was now more and more clustered into cities. Together these revolutions represented the most far-reaching and impactful changes in the human-nature relationship since hominids had emerged some five million years previously. These revolutions made possible the gigantism of the next century, the physical construction of buildings and other steel and concrete systems that dwarfed the human, and the alteration of natural landscapes on a scale of extension never before seen in human history. All of this was devastating to the natural world. It prepared the way for the catastrophic changes of the twentieth century. It should be no surprise that some people began to react to the despoliation of nature and to the crass commercial ethic which drove its industrialized onslaught. Gradually, other voices were raised, first among the Romantics and then the Conservationists and Environmentalists; a resistance had movement developed almost from the beginning.

# 10

# Resisting Hypercivilization
## Poets, Painters and Darwin

And was Jerusalem builded here
Among these dark Satanic Mills?
—William Blake, 1804

As the nineteenth century opened the world view of the industrial capitalists was dominant and was marked by their overwhelming optimism that
they would create a materialistic cornucopia on Earth. The ancient ideologies
linking humanity and nature in a mystical, spiritual bond had been almost
everywhere discredited and natural ecosystems almost everywhere attacked.
"Nature-based peoples"[1] were regarded as dirty savages whose way of life was
hopelessly outmoded; they had nothing to teach the Bringers of Progress
about humanity's proper relationship to nature. But the rapid rise and spread
of Hypercivilization did not go un-resisted. Intellectual, artistic, and spiritual
traditions germinated and took root and, along with Darwin, would provide
the basis for the emergence of environmentalism in the twentieth century.

Thomas Malthus (1766–1834) was one dissenter. David Pepper writes
in *The Roots of Modern Environmentalism*, "Against a background of buoy-
ant optimism ... Malthus proposed the idea that the earth sets limits to population growth and human well-being."[2] He argued that population grows
geometrically and will always outstrip the food supply, which grows only
arithmetically. Famine, disease, and war all provide "natural" checks. Economic growth would simply lead to more population growth, ending again
in misery at a higher level. After all marginal land had been brought on
line, the final limits would be reached as would the final, maximized level
of misery. His arguments have been debunked time and again based on the
fact that huge amounts of land were not under cultivation at the time and
that the combination of fertilizers, pesticides, irrigation, and now genetic
engineering, would make possible a vast increase in food production per

acre. But perhaps the last chapter in this debate is not yet written as Hyper-civilization begins more and more to depress the biotic systems on which it rests and runs up against global shortages of water while the population keeps growing toward unprecedented numbers. The concern about over-population had a revival over a hundred years later in the neo–Mal-thusian works of environmentalists Garrett Hardin and Paul Ehrlich in the 1970s.[3] Overpopulation, linked with consumption per capita, remains an ongoing concern of many environmentalists.

## Pastoralists and Nature Mystics

The pastoralist poets and the nature mystics were rebelling against cold reason; they stressed feeling and intuition as the way to know nature. They include the Lake Poets in England, Rousseau in France, the Transcendentalists in America, the American painters of the Hudson River School and the English painters, Constable and Turner. To some extent, they were struggling to re-experience and affirm the participating consciousness. They preferred the old rural simplicity of a bygone era. Even the classicist, Alexander Pope wrote: "Observe 'how system into system runs,'"[4] and, "All are parts of one stupendous whole/ Whose body nature is, and God the soul."[5] Pope even elegized the self-sufficient, rural way of life.

> Happy the man whose wish and care
> A few paternal acres bound,
> Content to breathe his native air
> In his own ground.
> Whose herds with milk, whose fields with bread,
> Whose flocks supply him with attire,
> Whose trees in summer yield him shade,
> In winter, fire.[6]

Pastoral traditions were carried on into the preservationist movement by the English activists who, in 1865, founded the Commons Preservation Society, and by John Muir, father of America's national parks, and are found in later twentieth century philosophies of topophilia, or love of place, as in Yi-Fu Tuan's book, *Topophilia*,[7] and E.O. Wilson's *Biophilia*.[8] Other descendants are the deep ecology movement of Arnie Naess,[9] Green Buddhism[10] and the ecofeminism of Susan Griffin[11] and the works of Wendell Berry. As the horror of the industrial cities was beginning to manifest itself, the pastoral ideal was vigorously asserted by Oliver Goldsmith (1728–1774), whose elegy, "The Deserted Village," lamented the destructive impacts of the enclosures: "Sweet Auburn, loveliest village of the plain," he began, and then went on to enumerate all its rustic charms, ending the first section with the lines:

These were thy charms—but all these charms are fled.
Sweet smiling village, loveliest of the lawn,
Thy sports are fled, and all they charms withdrawn;
Amidst thy bowers the tyrant's hand is seen,
And desolation saddens all thy green;
One only master grasps the whole domain,
And half a tillage stints thy smiling plain.
Ill fares the land, to hastening ills a prey,
Where wealth accumulates and men decay,

[and]

A time there was, ere England's griefs began,
When every rood of ground maintained its man;
For him light labor spread her wholesome store,
Just gave what life required, but gave no more;
His best companions, innocence and health,
And his best riches, ignorance of wealth.
But times are altered; trade's unfeeling train
Usurps the land and dispossess the swain.[12]

Here was a moralist strain, a complaint about a biosocial system that had lost both decency and simplicity, about the society that was developing in the cities like Manchester, based on greed, selfishness, competition, ambition, and all the other ills to which the land was hastening. Thomas Gray's (1716–1771), *Elegy Written in a Country Churchyard* also lauded the simple life of the villager.[13]

## Denise Diderot (1713–1784)

Diderot turned around the view of indigenous peoples and, like Rousseau created the opposite stereotype of the noble savage as a critique of European society. In his *Supplement to the Voyage of Bougainville*, the Tahitians chide the Europeans.

We possess already all that is good or necessary for our existence. Do we merit your scorn because we have not been able to create superfluous wants for ourselves? When we are hungry, we have something to eat; when we are cold, we have clothing to put on.... You are welcome to drive yourselves as hard as you please in pursuit of what you call the comforts of life, but allow sensible people to stop when they see they have nothing to gain but imaginary benefits from the continuation of their painful labors.... Go and bestir yourselves in your own country; there you may torment yourselves as much as you like; but leave us in peace, and do not fill our heads with a hankering after your false needs and imaginary virtues.[14]

Diderot's critique would be revived in the late twentieth century environmentalist critique of consumerism and in their defense of the world's remaining indigenous peoples.

## Jean Jacques Rousseau (1712–1778)

Rousseau rejected the idea of civilization altogether and elevated the importance of nature. He provided a fountainhead of thought for the Romantic Movement in Germany and England and for the Transcendentalists in America and also contributed to the development of the socialist critique of capitalism. Will and Ariel Durant call him nature's "most fervent and effective apostle; half the nature poetry since Rousseau is part of his lineage."[15] It was in nature that Rousseau found God. According to the Durants:

> An almost passionate pantheism replaced the God of the Bible. There was a God, yes … but he was not the external, vengeful deity conceived by cruel and fearful men, he was the soul of Nature, and Nature was fundamentally beautiful, and human nature was basically good.[16]

In his *Confessions* he wrote about his early love affair with the pastoral: "The country was so charming … that I conceived a passion for rural life which time has not been able to erase."[17] The city was the center of and the impetus for all that Rousseau found unnatural and in so being, he believed, it denatured humans. In the city there could be no harmony and hence no authentic humanity. He stunned the intellectual world with his essay, the "Discourse on the Arts and Sciences." The Dijon Academy had offered a prize for the best ideas on this topic, assuming that all the entrants would laud the progress of reason and science and award their own age the garland for being the best and most enlightened in the history of civilization. Rousseau used the occasion to challenge the last ten thousand years of historical progress, condemning "fatal ingenuity of civilized man."[18]

He was a critic of "the established, commercial society.…"[19] And what bothered him most was its inherent selfishness, its denial of and blindness to the common good, hoping "that each should see in the good of all the greatest good he can hope to achieve for himself."[20] But this situation was only realizable in a rural habitat, in the pre-industrial landscape where people still lived in close contact with nature. Rousseau began standing morality back on its feet, defining "economy" in *The Social Contract* as "the wise and legitimate government of the house for the common good of the whole family,"[21] a sentiment echoed in the current generation by Wendell Berry.

## *Poets*

"Perhaps the most significant direct influence of Rousseau," writes LaFreniere, "was upon William Wordsworth, the paradigm of nature worship in the English Romantic tradition and a major tutor of Emerson."[22]

Wordsworth adopted Rousseau's approach that the natural landscape is what produces a decent human nature and a just social system. He says in the *Prelude* that the love of nature leads to the love of mankind. But this generosity of spirit only occurs in a biosocial system in which free people are working in harmony with nature rather than acting as its dominators. Wordsworth rejected the world view of the Newtonian-Cartesian paradigm and he broke entirely with the reductionist views of scientific materialism. In 1798, he wrote in "The Tables Turned": "Up! Up! My Friend, and quit your books," [and] "Let Nature

**Resistance: Resistance to the dominator ideology of Hypercivilization began with the poets and painters of the Romantic Movement in 19th century England including William Wordsworth (Georgios Kallidas, Shutterstock).**

be your teacher" and "One impulse from a vernal wood / May teach you more of man, / Of moral evil and of good, / than all the sages can."[23] In this poem he explicitly rejected Bacon's approach to nature. The final two stanzas read: "Sweet is the lore that Nature brings;/ Our meddling intellect / Misshapes the beauteous forms of things— / We murder to dissect." And further, "Enough of Science and of Art; / Close up those barren leaves; / Come forth, and bring with you a heart / That watches and receives."

In "Lines Composed a Few Miles Above Tintern Abbey" he set forth his nature mysticism.

> And I have felt a presence
> That disturbs me with the joy
> Of elevated thoughts; a sense sublime
> Of Something far more deeply interfused,
> Whose dwelling is the light of setting suns,
> And the round ocean and the living air,
> And in the mind of man,
> A motion and a spirit, that impels

> All thinking things, all objects of all thought,
> And rolls through all things.[24]

Here we have returned to the participating consciousness of the Old Way that hears the singing of the universe.

Wordsworth was a heathen philosopher, a fact openly stated in the poem "The World Is Too Much with Us."

> The world is too much with us; late and soon,
> Getting and spending, we lay waste our powers:
> Little we seen in Nature that is ours;
> We have given our hearts away, a sordid boon!
> The Sea that bares her bosom to the moon;
> The winds that will be howling at all hours,
> And are up-gathered now like sleeping flowers;
> For this, for everything, we are out of tune;
> It moves us not—Great God! I'd rather be
> A Pagan suckled in a creed outworn;
> So might I, standing on this pleasant lea,
> Have glimpses that would make me less forlorn;
> Have sight of Proteus rising from the sea;
> Or hear old Triton blow his wreathed horn.[25]

Jonathan Bate has written that the romantics were the "first ecologists,"[26] and, "The romantic ecology reverences the green earth because it recognizes that neither physically nor psychologically can we live without green things; it proclaims that there is one life within us and abroad, that the earth is a single vast ecosystem which we destabilize at our peril."[27]

Wordsworth was well aware what industrialization was doing to England's green and pleasant land. In *The Excursion* he decries the changes brought about by the "manufacturing spirit" as the manufacturing towns sprawl out into the countryside.

His approach to nature is ideographic. He knows nature not as universal laws but as specific places. Many of his poems are titled by place and in most of them one can sense the landscape. He makes us feel the breeze, see particular flowers and hear the stream.

> Embrace me then, ye Hills, and close me in;
> Now in the clear and open day I feel
> Your guardianship; I take it to my heart;
> 'Tis like the solemn shelter of the night.[28]

In this view of the relationship, nature conserves humans.

Wordsworth spent a great deal of his time out of doors of rural England. Already, he was beginning to see the Lake District as a "sort of national property, in which every man has a right and interest who has an eye to perceive and a heart to enjoy."[29] It would take the British more than a hundred years to begin establishing their national park system, but it was Wordsworth who first had the idea.

If it was Rousseau who gave birth to the Romantic Movement, it was Wordsworth who taught it to speak English and in so doing laid the foundation of Anglo-Saxon environmentalism on both sides of the Atlantic. It was no wonder that the American Transcendentalist, Ralph Waldo Emerson crossed the Atlantic to visit Wordsworth and John Muir came later to visit his grave. But the new interest in nature was not only happening in poetry; it was also occurring in painting.

## The Painters

It was in England where landscape painting was born in the works of John Constable (1776–1837) and J.M.W. Turner (1775–1851). Historian James Heffernan writes:

> Wordsworth, Coleridge, Turner, and Constable were all born between 1770 and 1776, and each of them grew up to practice the art of landscape: to recreate the life of natural objects in pictures or in words…. All British, all coming of age about 1800, they all also wrought significant changes in our perception of landscape.[30]

The Romantics were creating a new way of seeing nature as Edenic, as Adam might have seen it. They wanted to reject, or at least, minimize history and start afresh with the pristine. It is in this sense that Constable said: "When I sit down to make a sketch from nature, the first thing I try to do is, to forget that I have ever seen a picture."[31] In what art historian Heffernan calls the "displacement of history," the landscape painters wanted to rediscover "landscape as a prehistoric paradise unscarred by battle and unmarked by monument, a pristine spectacle never even represented before in any of the arts." While Constable frequently had people in the picture they were unheroic; ordinary peasants who leave no tracks on the land. The famous painting; *The Hay Wain* crossing the Stour shows a dog, a small mill, and the wagon with its driver and domesticated oxen. Trees and clouds dominate the scene, but it is not wilderness. It is a rural scene. The scale of human impact is small in nearly all of his paintings. Even his famous paintings of Salisbury Cathedral are principally landscapes showing the river, wild and luxuriant trees, billowing clouds and a rainbow and the old, old cathedral looking almost like a natural feature of the land. And his picture of *Old Sarum*, the original seat of the cathedral, shows the ruined town under a lowering, menacing sky, as if to say, nature will win out over civilization in the end.

Similar trends were going on in the United States and, as early as 1802,

Washington Alston (1779–1843), painted Niagara Falls, then situated upon the edge of the western wilderness. What Americans had of which they could be proud and which could compete with the architectural grandeur of Europe was a grand scenery. What first attracted the most famous school of American painters of the Hudson River School was the Hudson River Valley. William Wall's 1824 portfolio of twenty views of the Hudson River showed "an almost tender feeling for the landscape" according to art historian Wolfgang Born.[32] His "View Near Fishkill, N.Y." shows two tiny figures, one seated, in the middle distance, on the shore which is lined with high bluffs and trees. But the Hudson River School painter *par excellence* was Thomas Cole (1801–1848).

> Cole began with landscapes from the Hudson River Valley and the White Mountains which combine a loving record of details with intense expression…. Dramatic contrasts are the painter's favorite patterns. The sunrise, a motif especially dear to the German Romanticists, is not less welcome to their American colleague as a device for interpreting the mysterious energies of nature.[33]

He loved to paint gnarled, broken, and twisted trees, as in his wilderness view of "Landscape with Tree Trunks." Asher Brown Durand's famous painting, "Kindred Spirits," shows two men standing on a rock outcrop looking over a wild gorge and obviously communing about the spiritual intensity of the scene. The two men are the painter, Thomas Cole, and a transcendentalist poet, the pantheist William Cullen Bryant (1794–1878).

## *The American Transcendentalists*

Ralph Waldo Emerson (1800–1882), was the Dean of the American Transcendentalists, La Freniere writes, "The influence of Wordsworth upon Emerson is as well established as the influence of Rousseau upon Wordsworth…."[34] Emerson's seminal work was his essay, *Nature*, written in 1836 and was "strongly influenced by Wordsworth's poetry" and in turn was a "seminal book in Thoreau's life."[35] In *Nature* we see that Emerson has grasped the fundamental concept of the old cosmology.

> The rounded world is fair to see,
> Nine times folded in mystery:
> Though baffled seers cannot impart
> The secret of its laboring heart,
> Throb thine with Nature's throbbing breast,
> And all is clear from east to west.
> Spirit that lurks each form within
> Beckons to spirit of its kin;
> Self-kindled every atom glows,
> And hints the future which it owes.[36]

His epistemology was the same as Wordsworth's—one learns nature by sympathy, not by mere science. "To the intelligent," he wrote, "nature converts itself into a vast promise, and will not be rashly explained."[37] We are dealing with the mystery of the spirit world. In 1836 he said, "We are escorted on every hand through life by spiritual agents, and a beneficent purpose lies in wait for us."[38] This could have been said by an ancient, nature-based tribal person thousands of years ago.

While Emerson was a careful, empirical observer, even scientific, his epistemology was not that of the dualist, Descartes. In *Man and Nature in America*, Ekirch writes: "All nature, Emerson urged, was a unity in which man as an observer played his part—observer being fused with the observed."[39] Sounding like Wordsworth. He wrote: "For she [nature] yields no answer to petulance, or dogmatism, or affectation; only to patient, docile observation. Whosoever would gain anything of her, must submit to the essential condition of all learning, must go in the spirit of a little child. The naturalist commands nature by obeying her."[40] A far cry from the Baconian *natura vexata*, from the industrialist, from the scientists. "There is more beauty in the morning cloud," he wrote, "than the prism can render account of...,"[41] a direct reference to Newton's treatise on light. Like Rousseau and Wordsworth and modern environmentalists, going into nature heals what is sick in the spirits of modern urban people.

Toward the end of the essay he reflects on universal truths of nature. "These, while they exist in the mind as ideas, stand around us in nature forever embodied, a present sanity to expose and cure the insanity of men"[42] and "Cities," he wrote, "give not the human senses room enough."[43] Writing in his *Journals* on this point, Emerson said: "This invasion of Nature by trade with its money, its credit, its Steam, its Railroads, threatens the balance of Man, and establishes a new, universal monarchy more tyrannical than that of Babylon or Rome."[44]

It seems that Emerson knew full well the mystic beauty of the garden and equally well the dangers when the machine was let loose in it. How, then, do we account for his love of railroads and the remark in his *Journals*, that "Machinery and Transcendentalism agree well."[45] Emerson's confusion was the same fundamental error into which Jefferson had fallen and into which the entire culture in America would fall. In fact, Emerson is more complex than we have yet seen. He was a "youthful admirer of Bacon and Franklin," and in 1837 he recorded in his journal his relief to find that New England is destined to become the "manufacturing country of America."[46] After following Emerson's nature mysticism it is hard to believe that he also wrote the following lines:

The exercise of the Will, or the lesson of power, is taught in every event.... Nature is thoroughly mediate. It is made to serve.... It offers all its kingdoms to man as the raw material which he may mold into what is useful. Man is never weary of working it up…. One after another his victorious thought comes up with and reduces all things, until the world becomes at last only a realized will,—the double of the man.[47]

Bacon could not have said it any better. Here nature is to be made over entirely into our image, a rationalizing doctrine for developers. "Railroad iron is a magician's rod in its power to evoke the sleeping energies of land and water," wrote the Faustian sage of Concord.[48] But somehow or other, Blake's "dark Satanic Mills" were not going to be the result in America.

Most Americans thought they could have it all, both the garden and the technological golden age, and that somehow this virgin land would purify the machine so that here the squalid cities and impure character of industrial capitalism would not taint humanity or nature. This was the enormous blind spot in their culture and it would define their relationship to nature for the next two hundred years, explaining how they could set aside magnificent natural areas, and then surround them with the most virulently polluting culture in human history. And yet he realized that "Things are in the saddle and ride mankind,"[49] Thoreau built a whole philosophy on this.

## Thoreau

It was Thoreau who made "the first real attempt by an American to work out a philosophy of man and nature."[50]

Historian Rod Nash said that "…he came to grips with issues others only faintly discerned" and "cut the channels in which a large portion of thought about wilderness subsequently flowed."[51] Benjamin Klein calls him the "founder of the environmental movement" and says that Walden was "an attempt to warn his contemporaries."[52] Joseph Petulla wrote in *American Environmental History* that *Walden* is "an essay against materialistic commercialism, written as an apology for the simple life."[53]

It was his emphasis on simplicity and self-reliance that made him such a radical in his time and in our own. He constantly preached simplicity: "Simplify, simplify," and "Let your affairs be as two or three, and not a hundred or a thousand; instead of a million, count half a dozen, and keep your accounts on your thumbnail."[54] He said that "We work too much" and that "We fritter our lives away in detail."[55] The way to a life of simplicity is to reduce one's desires. "My greatest skill," he wrote in *Walden*, "has been to want but little," and further, "A man is rich in proportion to the number of things he can afford to leave alone."[56] It was his penetrating insight that allowed him to turn fundamental definitions on their heads, as do all true revolutionary thinkers. A society based on Thoreau's simplicity would use less than a tenth of what modern industrial society under the prodding of materialism has come to use, and have an equally reduced impact on the global biosphere.

Thoreau saw more clearly than most observers the true cost of a thing which he said was "the amount of life which must be exchanged for it."[57] But most people, wanting too much, ended up working too much, both out of overweening desire and out what was by now a deep-seated cultural fear rooted in our fundamental distrust of nature's ability to provide. "Men say that a stitch in time saves nine and so they take a thousand stitches today to save nine tomorrow," Thoreau wrote in *Walden*.[58] According to him, laying up possessions on this earth produced not wealth, but its opposite. Writing about a cart load of furniture belonging to one Spaulding, he said: "I could never tell from inspecting such a load whether it belonged to a so-called rich man or a poor one; the owner always seemed to be poverty stricken. Indeed, the more you have of such things the poorer you are."[59] And by poorer he meant what we today would call spiritually poorer and poverty-stricken in character. To wit: "Most of the luxuries and many of the so-called comforts of life are not only not indispensable but positive hindrances to the elevation of mankind."[60]

He set out to prove this at Walden Pond. By keeping his own house small, he had little need of extra furnishings. "A lady once offered me a mat, but as I had no room to spare within the house, nor time to spare within or without to shake it, I declined it.... It is best to avoid the beginnings of evil."[61] And he once made this reply to a friend who, on observing

An American eco-radical: Henry David Thoreau (1817–1862), author of *Walden* (1854) provided a more radical critique of modern civilization than even Karl Marx in that he rejected all forms of materialism and said that "In wildness was the salvation of the world." He is now considered the founding hero of environmentalism (Everett Historical, Shutterstock).

that Thoreau liked to travel, suggested that he lay up some money and take the train to Fitchburg to see the country. "But I am wiser than that." he said. "I have learned that the swiftest traveler is he who goes afoot. I say to my friend, suppose we try who will get there first. The distance is thirty miles; the fare ninety cents. That is almost a day's wages.... Well, I start now on foot and get there before night; I have travelled at that rate by the week together. You will in the meanwhile have earned your fare, and arrive there sometime tomorrow.... Instead of going to Fitchburg, you will be working here the greater part of the day. And so, if the railroad reached around the world, I think that I should keep ahead of you; and as for seeing the country and getting experience of that kind, I should have to cut your acquaintance altogether."[62] Knowing his love for being outside in the natural world, it's not surprising that he would rather walk, it is his economic defense that is enlightening. In modern materialist society, most people work fifty weeks a year in order to enjoy two weeks of leisure time at the shore.

Thoreau argued for simplicity because a biosocial system based on a highly complex division of labor both enslaved humans and ruined their character. In a trenchant line, he said: "But lo, men have become the tools of their tools."[63] And he wrote in *Walden*, "We do not ride upon the railroad, it rides upon us."[64] He worried about that "Luxury which enervates and destroys."[65] In his most famous line, he said "I went to the woods because I wished to live deliberately, to front only the essential facts of life...."[66] And further, "To affect the quality of the day, that is the highest of arts. Every man is tasked to make his life ... worthy of the contemplation of his most elevated and critical hour." But most he said, fail abysmally at this: "The mass of men lead lives of quiet desperation."[67]

It was to the woods that Thoreau went, not to the farm or to the city. For these he had little use. Indeed, he pitied men of property.

> I see young men, my townsmen, whose misfortune it is to have inherited farms, houses, barns, cattle and farming tools ... [and who, as he said, are] pushing all these things before them.... How many a poor immortal soul have I met well-nigh crushed and smothered under its load, creeping down the road of life, pushing before it a barn seventy—five feet by forty, its Augean stables never cleansed, and one hundred acres of land, tillage, mowing, pasture, and wood-lot![68]

He took out all of his frustrations attendant upon gross, commercial agriculture on the owner of the farm which gave Flint's Pond its name. It is a long excoriation but worth looking at because it reveals much about the man whose ideas have become so important to the preservationist stream of environmentalism.

> Flint's Pond! Such is the poverty of our nomenclature. What right had the unclean and stupid farmer, whose farm abutted on this sky water, whose shores he has ruthlessly

laid bare, to give his name to it? Some skin-flint, who loved better the reflecting surface of a dollar ... who regarded even the wild ducks which settled in it as trespassers.... I go not there to see nor to hear of him; who never <u>saw</u> it, who never bathed in it, who never loved it, who never protected it, who never spoke a good word for it, nor thanked God that He had made it ... who could show no title to it but the deed which a like-minded neighbor or legislature gave him,—him who thought only of its money value; whose presence perchance cursed all the shores; who exhausted the land around it, and would fain have exhausted the waters within it; who regretted only that it was not English hay or cranberry meadow ... and would have drained it and sold it for the mud at its bottom. It did not turn his mill, and it was no privilege to him to behold it. I respect not his labors, his farm where everything has its price, who would carry the landscape, who would carry his God, to market, if he could get anything for him; who goes to market for his God as it is; on whose farm nothing grows free, whose fruits are not ripe for him 'til they are turned to dollars.[69]

Here he rendered plain his judgment on capitalism as in *Walden*: "Trade curses everything it handles."[70] He wondered "Why should we be in such desperate haste to succeed and in such desperate enterprises?" a question of which most people in modern civilization are not even consciously aware.[71] He excoriated commercial civilization because it sees only money value, leaves human nature stunted and, moreover, it inevitably ruins the wild. And for Thoreau, the wild was the ultimate value: "In Wildness is the preservation of the World."[72]

In his *Journals* he wrote: "Let us keep the New World new."[73] He did not want New England to become old England, to become a factory district littered with cities.

I cannot believe that our factory system is the best mode by which men may get clothing. The condition of the operatives is becoming every day more like that of the English; and it cannot be wondered at, since, as far as I have heard or observed, the principle object is, not that mankind may be well and honestly clad, but, unquestionably, that the corporations may be enriched. In the long run men hit only what they aim at.[74]

And aiming at corporate riches left the "operatives" dehumanized automatons. "I walked through New York [City] yesterday," he wrote, "and saw no living person."[75]

He had achieved a consciousness of oneness, reminiscent of the Old Way. "Shall I not have intelligence with the earth? Am I not partly leaves and vegetable mold myself?"[76] Walking along the stony shore of Walden Pond one day at dusk, he sensed the identity of himself and Nature, writing of it afterwards: "This is a delicious evening, when the whole body is one sense and imbibes delight through every pore. I go and come with a strange liberty in Nature, a part of herself."[77] He was aware that the great and beautiful creatures of wild America were threatened and facing demise. He wrote in his *Journals*:

When I consider that the nobler animals have been exterminated here,—the cougar, panther, lynx, wolf, bear, moose, deer, the beaver, the turkey, etc., etc.,—I cannot but feel as if I lived in a tamed and, as it were, emasculated country....[78]

As a result, he foresaw the need for setting aside wild lands as parks, for nature was truly his guide. "Instead of calling on some scholar," he wrote, "I paid many a visit to particular trees…. These were the shrines I visited both summer and winter."[79]

## George Perkins Marsh

Shortly after Thoreau, a very different kind of man who was not a Romantic or a nature mystic, a practical man of affairs added the long view to environmentalism. He was a businessman and a diplomat whose tour of duty in the Mediterranean brought him into contact with lands he realized had been degraded through human overuse. He did not feel the pull of the mystical, but nevertheless in 1864 he argued in *Man and Nature,* for wise and prudential use. Marsh was aware that man "was everywhere a disturbing agent."[80] He pointed to the disastrous results of deforestation the Mediterranean areas, chronicling the decay of civilizations which had degraded their land base. He chided humanity for its carelessness: "Man has too long forgotten that the earth was given to him for usufruct alone, not for consumption, still less for profligate waste."[81] His focus on forest conservation was followed by the development of professional forestry in the United States, influencing Gifford Pinchot, first head of the U.S. Forest Service under Theodore Roosevelt.[82]

## Charles Darwin and the Rise of Modern Biology

The final nineteenth century development on the path to where we are now was the rise of modern biology under the influence of Charles Darwin (1809–1882).

Darwin's 1859 book, *On the Origin of the Species*, was the fountainhead of modern biology. And since ecology is an outgrowth of biology, much of modern environmentalism traces its world picture back to Darwin. He was an empirical scientist and no nature mystic but what he began was a process of discovery that would eventually recapture an understanding wholeness of nature, the great chain of being including ourselves, including the doctrine of plenitude, but resulting from the empirical observations of science. What Darwin saw when he began fitting his facts together was the theory of evolution, or more properly, natural selection. Organisms are what they are because they have become so over immense periods of time and have done so in response to their environment which consists of both limiting and enabling factors such as climate, soil, and the influences of other organisms. Each species is successfully adapted to its environment (at least for the moment), and has built into its biotic program, its physiology and behavior, both the mechanisms and techniques that allow it to capture energy and hence to reproduce successfully. Within any given species there is a range of differences, for example, for heat tolerance. If the climate warms appreciably, only a small portion of the species—those with high heat tolerance—will be adaptive, survive and reproduce. Their heat tolerant bioprogram will be passed on to their descendants. Less heat tolerant ones will die out and hence not pass on their less tolerant program. The species will tend toward greater heat tolerance or even to new speciation. Spontaneous mutation may also result in selecting for or against, depending on how adaptive the mutation is.

Darwin saw nature as a competitive arena in which organisms within a species and species within an environment competed for existence. It was this "struggle" that attracted the most popular attention to Darwin's theory and which stimulated the greatest misunderstanding of his work. Darwin did highlight that organisms live in a competitive environment. Not all offspring can survive. It was this single aspect which the Social Darwinist philosophers lifted out of the theory of natural

Charles Darwin (1809–1882): Darwin's theory of Natural Selection both placed humans indissolubly into the natural world and indicated our evolutionary past places constraints on what we and other creatures can and can't tolerate as in substances such as some chemicals not around when our lungs developed. (Everett Historical, Shutterstock).

selection. They applied it as a single explanatory principle for human society and so provided the capitalist captains of industry with an apology for their ruthless struggles against each other and their exploitation of labor, third world societies and nature, a rationale seemingly grounded in natural science. Allowing capitalism to work "naturally," free of social regulation, was supposed to result in the healthiest society as the weak were eliminated. This ideologically-freighted inference from Darwin became a central tenet of nineteenth and twentieth century capitalism and as such played a role antithetical to modern environmentalism. But competition was only a part of Darwin's contribution to modern environmental thought. Essential to Darwin's conception of nature was the revelation of it as a highly interactive system characterized by multiple connections, or what would be called feedback mechanisms once the science of ecology had developed out of his discoveries. He wrote of the "mutual relations of all beings" and suggested the staggering complexity of the web.

The effects of Darwin's work were to greatly complexify our view of nature and, as the quotation demonstrates, to locate humans in the landscape as players in the evolving balance of species. Pepper writes: "Darwin's particular contribution to the development of the web of life was that he included <u>man</u> in it."[83] His later book, *The Descent of Man*, set off an earthquake among the traditionally religious whose Judeo-Christian philosophy had placed humanity on earth as mere visitors, to be tested in this vale of tears for entry into heaven which was, they preached, our true home. One could say that Darwin, along with the nature mystics, returned humanity to Earth after an absence of several thousand years. This return would cut two ways as far as the implications for environmentalism are concerned. It could be argued, and it was, that if Earth and this life is all there is, then one should make the most it; enjoy its material wealth to the fullest. The flames of economic development and consumerism, so antithetical to environmentalism, were fanned by such a reading of Darwin. His web was, after all, "just" a mechanism—complex, incredibly intricate, but a variant of the Newtonian world machine nonetheless. By contrast, the ecological reading of Darwin is a very modern reading.

Darwin had opened up another avenue of knowing and down this road went the German physician Ernst Haeckel and some of his successors. It was Haeckel who, after studying Darwin's ideas, invented term "ecology" in 1869, to mean the study of organisms in relation to each other and to their environment. Bowler writes: "The concept of ecology, as Haeckel developed it, turned the life sciences away from studying particular organisms in isolation and toward analyzing organisms holistically, in the contest of other organisms and natural elements."[84] However, "Caution is also necessary when evaluating the emergence of ecology at the turn of the century.

In modern times we have come to associate the very word 'ecology' with a concern for the environment—yet ... there was no immediate link with the environmentalist movement. Some biologists saw ecology as a science that would help to support the exploitation of the environment by showing how to minimize the damage caused."[85]

It would be decades before the ecological side of Darwin would be wedded to environmentalism in the work of Aldo Leopold, coming only at the mid-point of the next century. It would take over a hundred years before the ecological Darwin would triumph over the Darwin of "survival of the fittest." During that time the brute facts of industry-caused pollution, extinction and climate modification would become frightfully obvious to all but the most obtuse ideologues. And in the coming era, the roots of genuine environmentalism which had developed in the soil of poetry, painting, landscape gardening and biology in the nineteenth century, would grow progressively into a fully-developed counter-cultural world view that at the end of the twentieth century would begin to threaten the four-hundred year reign of modern materialistic Hypercivilization.

# 11

## Resisting Hypercivilization
### Conservationists, Environmentalists and Blockadia

"A thing is right only when it tends to preserve the integrity, stability, and beauty of the biotic community. It is wrong when it tends otherwise."[1]—Aldo Leopold

From the latter part of the nineteenth century to the present, a tidal wave of environmental organizations stood up to counter Hypercivilization and many individuals put their bodies on the line. They were able to compel limited government regulatory action, although governments continued to promote economic growth, and the human population kept burgeoning. The new attitudes and willingness to organize resistance based on new knowledge provided by modern science promised hope for the future.

"Environmentalism" was a many faceted social movement with room within for serious conflicts among its adherents. Nevertheless, they were swimming against the main stream of unrestrained and unregulated development. Differing radically on the solution, some saw it as a matter of tinkering at the edges by enacting modest reforms. Others, saw it as a crisis of civilization and sought to build a new relationship to nature based on a different set of cultural values. As the twentieth century wore on, the movement gained a large popular following and widespread media coverage. Activist organizations formed all over the world from Australia to Russia, America to India, and at all levels from neighborhood to national to global. The movement had many successes both in educating the public and passing laws that regulated industry. Political action was used to restrict and even ban certain practices. It blunted Hypercivilization's onslaught on the natural world but it did not stop it. The apologists for unrestrained materialism and economic growth continued to be powerful voices in the media, corporate boardrooms, government, and in the international development organizations. Late in the twentieth century, they organized a backlash. Large corporations funded front groups such as the self-styled "Wise Use Movement" and reactionary think tanks such as the Club for Growth, the American Legislative and Exchange Council, and the Heartland Institute which worked to undo previous gains. The election of Donald Trump as U.S. President in 2016 represented their almost complete triumph. Environmentalists fought to stave off what they saw as a looming environmental disaster and worked to create a new civilization that would have a more sustainable relationship to the natural world.

## Beginnings: The Conservation Movement

The change was first apparent in rising alarm at the depletion of wild game, fisheries and forests. Self-discipline did not work because of what Garret Hardin would later call the "tragedy of the commons."[2] Only government regulation could prevent the nation's users of these resources from over-harvesting to the point at which the land could no longer provide. In 1871, Congress authorized the first of the National Parks, Yellowstone, whose 2.25 million acres became a haven for the few hundred remaining buffalo. In 1890 they created Yosemite National Park, General Grant National Park, and Sequoia National Park. By the end of the nineteenth century it had become clear that the free-for-all raid on the virgin continent of North America had produced a stunning decline in renewable resources and was in danger of permanently depleting these to the disadvantage of the nation. *Laissez-faire* had not worked. It had turned out that in the matter of preserving the flow of natural capital to society, that government which governed least did not govern best. A sea change in political philosophy occurred, though not without bitter opposition.

The first genuine "conservation" movement in the United States came as part of the Progressive Era and saw the creation of what would become the National Forests and then wilderness areas as National Parks and the creation of the U.S. Department of Forestry and executive action that set aside millions of acres. It was nonetheless a conservative approach—government intervened to preserve the market by preventing it from destroying its resource base. The approach was essentially economic and was based squarely and consciously in utilitarian philosophy. It did not challenge the fundamental Western notions of dominion, capitalism, and consumerism, but only sought to make man's dominion efficient and democratic. A bitter controversy over the best use of these lands split the young movement into preservationists and conservationists like John Muir, who wanted to keep the areas forever wild, and Gifford Pinchot, who wanted to use them as resources but in the most efficient and sustainable manner which worked for forestry, fisheries and water conservation practices but made little sense in terms of the extraction of nonrenewable resources.

The conservationists advocated efficiency as a path to greater and more sustained material prosperity. Others were concerned about what they saw as negative spiritual and social aspects of an urban life divorced from direct contact with nature where, they believed, people could find brief respite from the norms of industrial civilization and so recharge their spirits. However, by the 1940s in the U.S. the movement had become little more than an elite group of middle and upper class men who had no thought of challenging the status quo other than making certain that some wild areas were left. "By 1945," writes Hal Rothman in *The Greening of a Nation*, "conservation as a political force was moribund."[3] Its revival began in the 1950s when the Echo Park dam controversy and then Rachael Carson's 1962 book, *Silent Spring*, began to change the thinking from wild lands preservation to much broader problems of pollution.

Old organizations became revitalized and refocused, numerous new organizations were founded, membership increased dramatically and the movement was institutionalized by setting up lobbying groups in Washington, D.C., and in the state capitals. Environmental education was institutionalized in the 1970s with the American Association for Environmental Education and in many environmental studies programs at colleges and universities. These years also saw the emergence of more radical environmental groups which believed that the old line organizations, such as the Sierra Club, were too familiar with the polluters and far too timid in their efforts to blunt the onslaught of Hypercivilization. These included Greenpeace and EarthFirst! and many others including emerging Green Parties, ecofeminists, and green Buddhists, all of whom began to see the problem not as reforming a culture that was basically sound, but as replacing an inherently destructive culture. In the 1990s, a movement for "sustainability" developed which grew to include those such as Paul Hawkin, who wanted to reform capitalism so that it took accounting of "natural capital," and bioregionalists and others who were truly radical. They saw neoliberal globalization as in NAFTA and the WTO as the engine of planetary level environmental destruction.

The last third of the century also saw the globalization of the environmental movement, with the 1972 United Nations Conference on the Human Environment and the 1992 UN Conference at Rio, the so-called "Earth Summit," or more properly, the "Conference on Environment and Development." As of this writing the final outcome of the struggle against Hypercivilization is not at all clear and some parts of the environmental movement have adopted direct action, taking the battle to the streets, mines, oil patches, pipelines, corporate offices and the seas in what Naomi Klein has called "Blockadia."[4]

## Aldo Leopold

Aldo Leopold (1887–1948) was one of Pinchot's early foresters who over his lifetime came to be one of the most profound thinkers of the environmental movement, combining the best of Pinchot and Muir. Leopold's thought evolved throughout his lifetime. He started out as one of Pinchot's foresters and ended up as a major champion of wilderness and as the founder of environmental ethics. He combined a scientific approach with a value-based, philosophical passion for nature and he remains one of the eminent voices in the history of American conservation and environmentalism.

While conducting predator control as a young forester he shot a wolf. Some forty years later he wrote in *A Sand County Almanac*: "We reached the old wolf in time to watch a fierce green fire dying in her eyes. I realized then, and have known ever since, that there was something new to me in those eyes—something known only to her and to the mountain."[5] In short, there was a wisdom in the wild that humans had better appreciate. He became one of the country's leading advocates of wilderness and of scientific game management. Lorbiecki records that "Aldo Leopold was the first to put the wilderness debate on the national agenda…," and "Leopold is considered the father of the national forest wilderness system…."[6] His plea was for preserving what would later be called biodiversity, although at this time he still held with the control of predators, a view he would later completely abandon. He established the New Mexico Game Protective Association and supported novel and revolutionary idea of creating federal wildlife refuges. He also came to believe at this time that, for conservation to become a reality, it would require more that protective legislation, namely a change in the moral and ethical outlook of the general public. He wrote:

> The sad truth is, that in spite of all religion and philosophy, mankind has never acquired any real respect for the one thing in the Universe that is worth most to Mankind—namely Life. He has not even respect for himself, as witness the thousand wars ‚ in which he has jovially slain the earth's best. Still less has he any respect for other species of animals…. The trouble is that man's intellect has developed much faster than his morals.[7]

Leopold had come to see wilderness not simply as a place for hunting and recreation, but "as an area for studying healthy biotic communities," including predators such as wolves and mountain lions.[8] Inspired by Leopold's writings, four men who were angered by the ongoing destruction of wild lands, founded the Wilderness Society in 1934. They were Bob Marshal, Benton MacKaye, Harvey Broome and Bernard Frank.

In 1934 Leopold was hired as a professor at the University of Wisconsin where he established the first university course in game management. He was pioneering the pedagogies of outdoor and experiential education which would become prominent toward the end of the century. Over the next fourteen years he revolutionized America's thinking about nature. He was the prime founder of the ecological way of thinking. While his ideas had been evolving all along, it was a trip in 1935 to the Spessart Forest in Germany that proved to be a major turning point. On returning, he described the effects of what he called "slash cutting" or what we would today call clear cutting.

> I know a hardwood forest called the Spessart, covering a Mountain on the north flank of the Alps. Half of it has sustained cuttings since 1605, but was never slashed. The other half was slashed during the 1600s ... the [replanted] old Slashing now produces only mediocre pine, while the unslashed portion grows the finest cabinet oak in the world. One of those oaks fetches a higher price than a whole acre of the old slashings.[9]

Foreshadowing the ideas on which he would eventually build the concept of environmental ethics, he went on to explain that clear cutting destroyed the soil community of bacteria, molds fungi, insects and burrowing animals.

Leopold was also a pioneer in creating the ecological restoration movement. Along with others he was responsible for establishing the University of Wisconsin Arboretum and its prairie, the first of what would follow as many efforts to save and to reproduce the original Midwestern tall grass prairie. Encouraged by that project, he bought a worn-out farm in the sand country along the Wisconsin River and worked with his family restoring it as a healthy habitat for wildlife. Today it is a place of pilgrimage.

Studies of the interrelationships of the animals with the rest of the creatures and with the landforms of their habitat, in other words ecology, had to become the scientific basis on which management plans could be built. In *Round River* he wrote, "Harmony with the land is like harmony with a friend; you cannot cherish his right hand and chop off his left ... you cannot love game and hate predators; you cannot conserve the waters and waste the ranges; you cannot build the forest and mine the farm. The land is one organism."[10] He began to discuss conservation as "nonviolent land use."[11]

Leopold had begun using the term "ecology" in the early twenties. While it had been used ever since it was invented by Haeckel, its scant use had been confined to a few scientists who used it narrowly. "Leopold used the word ... in a far more expansive way, to refer to the relationships between all inhabitants and their habitat: animal, plant, and soil ecologies studied together."[12] But there was more than just "nature." As one of his students said later, "When most biologists were thinking of individuals, he was thinking in terms of populations. When it was usual to think of populations, he was thinking of ecosystems, and of humans as components of ecosystems. He was thinking beyond the preservation of nature apart—toward the integration of human and natural worlds."[13]

As Leopold grew older, his thinking began to achieve the highest stage of maturity, a blend of the disciplined, empirical science and the wisdom of philosophy. He became a critic of Hypercivilization, including both its materialistic capitalism and its big government. He wrote:

> A system of conservation based solely on economic self-interest is hopelessly lopsided. It tends to ignore, and thus eventually to eliminate, many elements in the land community that lack commercial value, but that are (as far as we know) essential to its healthy functioning. It assumes, falsely, I think, that the economic parts of the biotic clock will function without the uneconomic parts. It tends to relegate to government many functions eventually too large, too complex, or too widely dispersed to be performed by government. An ethical obligation on the part of the private owner is the only visible remedy for these situations.[14]

Out of this maturity he crafted the foundation of environmental ethics. Even early on he knew from experience that, while laws for the protection of nature were necessary, they would never be enough, and that for conservation to become a reality, peoples' hearts and minds would have to be changed. He set about to do that. "Civilization," he said, "is not … the enslavement of a stable and constant earth. It is a state of mutual and interdependent cooperation between human animals, other animals, plants, and soils."[15] He was not a primitivist who argued that humanity should go back to hunting and gathering. We would inevitably modify the land, but that must be done knowingly. In one of his most famous statements, he said that

> The last word in ignorance is the man who says of an animal or a plant: "what good is it?" If the land mechanism as a whole is good, then every part is good, whether we understand it or not. If the biota, in the course of eons, has built something we like but do not understand, then who but a fool would discard seemingly useless parts? To keep every cog and wheel is the first precaution of intelligent tinkering.[16]

And he went on to say that human's needed to adopt an "intelligent humility" toward our place in nature.[17]

His collection of essays, *A Sand County Almanac*, was published the following year and eventually sold over a million copies, becoming "the bible of the environmental movement of the 1960s and 1970s."[18] It remains widely translated and read today. Leopold was a paragon among the many men and women resisting the onslaught of Hypercivilization in the first half of the twentieth century.

After World War II, a new wave of activity developed that continued the movement to save natural areas but expanded beyond it to concerns about air and water pollution. In 1958 the Mauna Loa Observatory began to monitor $CO_2$ levels in the atmosphere. And in 1962, Rachel Carson published *Silent Spring*.

## Rachel Carson

Rachel Carson's 1962 book, *Silent Spring*,[19] ushered in the modern age of environmentalism by emphasizing not set-asides of wilderness, but the massive chemical contamination of the environment.

J.E. de Steiguer writes, "Few books published in this century could equal *Silent Spring* for its profound effect on public opinion."[20] In less than a year it sold half a million copies and was eventually published in 16 foreign countries. Carson "put before the public the chilling possibilities of a neo–Malthusian catastrophe...."[21] Biologist, gifted nature writer, and former chief editor for the U.S. Fish and Wildlife Service, her earlier book *The Sea Around Us*[22] had been a best-seller. She took an ecosystemic approach. Americans and the rest of the world had been laboring under a false sense of security about pesticides that were promoted

Rachel Carson (1907–1964) was a marine biologist and best-selling author when she published *Silent Spring* in 1962, a book credited with awakening the world to the danger of pesticides and which marks the transition from conservation to environmentalism (Linda Lear Center for Special Collections & Archives, Connecticut College).

for their apparent benefits and by propaganda that rested on little or no research from government agencies such as the Department of Agriculture or the chemical manufacturers themselves. A letter from a friend, Olga Huckins, in 1958, set off her exhaustive researches which in the end showed millions of Americans such things as massive die-offs, and introduced to them such critical concepts as ecosystem, bioconcentration, persistence, and extinctions. Douglas Strong writes, "Not until Carson launched her campaign did the public suddenly realize the magnitude of the problem."[23] And what they realized was that pesticides were not merely a threat to robins and sparrows, but posed a significant health threat to them as well.

Carson's book received rave notices from almost all of its reviewers and set off a worldwide debate. President Kennedy quickly ordered a special panel of scientists to make an independent study of the question and, when their results were published, they vindicated Carson's position completely. The Department of Agriculture and the big chemical companies fought back, calling her work "hysterically overemphatic," and "emotional and inaccurate," and the biochemist, William Darby, claimed that Carson's philosophy would mean "the end of all human progress."[24] She was even accused being "part of a Communist plot to ruin American agriculture," and was called "a fanatic defender of the cult of the balance of nature."[25] But her science was good and, in retrospect, the hysteria was on the side of those whose jobs and profits rested on the applications of untested pesticides.

Her work changed the world. Many states and then the EPA banned or severely restricted such pesticides as DDT, Chlordane, Endrin, and others. *Silent Spring* did not end the use of hard pesticides. While the U.S. banned or restricted most of them, and so did many European countries, they could still be manufactured and sold to Third World users and with tragic consequences. But it was Carson, more than any other single individual, who aroused the public about Hypercivilization's assault on the natural world.

The 1960s and 1970s saw a great swelling of environmental protection organizations and legislation, triggered in part by an iconic photo. On Christmas Eve, 1968, a picture taken from the Apollo 8 spaceship of the earth rising over the moon let humans see for the first time this unique and precious living planet against the cold dark background of space. It was an historic watershed. The next day columnist Archibald MacLeish wrote in the *New York Times*: "To see the earth as it truly is, small and blue and beautiful in that eternal silence where it floats, is to see ourselves as riders on the earth together, brothers on that bright loveliness in the eternal cold...."[26] For the first time humans could see the whole earth as a global commons. Next year saw the passage in the U.S. Congress of the National Environmental Policy Act, or NEPA, requiring the federal government to take environmental impact into all its decisions. Republican President Richard Nixon signed it on January 1, 1970. NEPA created the Environmental Protection Agency. In the following years many states enacted their own versions of this law. A groundswell of public support for environmental protection followed with the celebration of Earth Day, organized by Senator Gaylord Nelson of Wisconsin, on April 22, 1970. The same year also saw the founding of Environmental Action, the League of Conservation Voters, and the National Resources Defense Council.

In 1970 Congress passed the Clean Air Act. In 1971 several Canadians launched Greenpeace and in 1972 the UN organized the first ever global conference on the environment, held in Stockholm, and established UNEP, the United Nations Environment Program. The U.S. enacted a national ban on DDT and the Meadows published *The Limits to Growth*.[27] Commissioned by the Club of Rome, it used the first computer models to challenge the notion that economic development could continue forever on a finite planet. The year of 1972 also witnessed the passage of the Clean Water Act. The year of 1973 saw the CITES treaty or the Convention on International Trade in Endangered Species and, in the U.S., the passage of the Endangered Species Act as the means of implementing it. That same year also saw the publication of another important book, *Small Is Beautiful: A Study of Economics as If People Mattered*,[28] by E.F. Schumacher, championing the concept of appropriate technologies to counter the prevailing notion that bigger is always better. In 1977, the Sea Shepherds was founded, an organization that takes direct action against polluters on the sea.

The decade of the 1980s saw even more activity including the establishment of the Superfund in the U.S., of EarthFirst!, the Rocky Mountain Institute, the World Charter for Nature, the first North American Bioregional Conference, and the formation of the Bruntland Commission by the UN in 1983. In 1987 they issued their report, "Our Common Future," which focused world attention on threats to the global environment and on the need for sustainable development.[29] Also in that decade, the World Watch Institute issued its first State of the World report, which then became an annual edition.[30] The year of 1987 saw the first meeting of U.S. Greens and the Montreal Protocol on the protection of the ozone layer was opened for signing at the United Nations. It became the most successful environmental treaty ever and resulted in reversing the damage that chlorofluorocarbons were doing to the upper atmosphere, which protects earth from dangerous ultraviolet rays. The Montreal Protocol became the precedent for the global climate accord of 2015.

The 1980s was also the decade when support for the environment ceased to be bipartisan in the U.S., the time of the Reagan backlash. Reagan had said as Governor of California, "you know, a tree is a tree, how many more do you need to look at?"[31] He removed the solar collectors the President Carter had placed on the White House and appointed as Secretary of the Interior the arch anti-conservationist, James Watt, a Christian fundamentalist who believed in the imminent end of the world. Reagan believed that government was the problem, not the solution to problems, a mantra that became the foundation of the ever more radical, anti-environment, know-nothing conservativism that culminated in the Tea Party movement and the election of Donald Trump as President.

Not all progress was suspended. The next decade saw the U.S. pass the Environmental Education Act, the founding of the European Environmental Agency and the Global Environment Facility to help defray the costs of making international projects environmentally friendly. The world's first Earth Summit was held in 1992, a huge gathering of citizens, journalists and world leaders in Rio de Janeiro to plan for many aspects of environmental protection. Major world leaders including the American President, Republican George H. Walker Bush, attended. At his press conference in Rio he proclaimed: "Let me be clear on one fundamental point. The United States fully intends to be the world's preeminent leader in protecting the global environment."[32]

One important result of the Rio Conference was the framework convention on climate change, which eventually paved the way for the 2015 climate accord. Another result was Agenda 21, a non-binding, voluntary plan with suggestions for making development sustainable and fair. In the U.S., right-wing property rights groups have completely misunderstood and distorted it. Presidential candidate Ted Cruz said it was a UN plot to destroy American sovereignty and eliminate golf courses in the U.S.[33] By the time of presidential candidate Donald Trump, conservation had vanished from the Republican Party where climate denial became the official stance and whose 2016 platform called coal "an abundant, clean, affordable, reliable domestic energy resource."[34] In 1994 the UN opened the Convention to combat desertification and, in 1997, the Kyoto Protocol to limit greenhouse gas emissions was negotiated. In 1994 the UN held a global conference on Population and Environment in Cairo.

The U.S. had no monopoly on the origins and development of conservation and, later, environmentalism. Similar organizations and movements emerged elsewhere as people struggled to hold back the tide of development that threatened both their livelihood and the lands they loved. Their story is too long to tell here but it happened notably in Europe, Australia, India and elsewhere. But the destruction went on and by the late 20th and early 21st centuries, people all over the world were laying their bodies on the line in defense of the Earth. Presaging a disturbing trend, in 1995, Ken Saro Wiwa, a Nigerian environmental activist, and six others protesting Shell Oil's decades of fouling the Niger Delta were arrested, tried on trumped up charges, and executed by the government, which was beholden to oil exports for almost all of its revenues. More and more, activists are facing violence. In 2015, 185 environmental activists were murdered worldwide.[35] In 2017 the figure rose to 197.[36]

## Blockadia

"Blockadia" is Naomi Klein's term for citizen-based, direct confrontation with development projects that degrade the environment and take away people's rights of common access to traditional lands and hence to their livelihoods. In her book, *This Changes Everything: Capitalism vs. the Climate*[37] she outlines the rise of people-based protest against the environmental degradation around the world as they react to neoliberal globalization, and the violent, repressive measures of governments that side with the wealthy developers. Blockadia developed in part because the big green organizations had turned into professional lobbyists headquartered in Washington, D.C. She writes: "In rapid fashion what had been a rabble of hippies became a movement of lawyers, lobbyists, and U.N. summit hoppers."[38] As anti-environmental conservatives took over the government, they suddenly found themselves on the outside looking in at the center of power. Rather than being marginalized they adopted the corporate and government language of market triumphalism, gave up confrontation and sought to collaborate with big business and in turn got financial support.

Long before the term "Blockadia" was coined, the Chipko Movement in India had pioneered the tactic in the 1970s in the hills of Uttar Pradesh, after the government allotted a plot of forest traditionally used by the village of Mandal to a sporting goods company, with the wood to be used in the manufacture of tennis racquets. Spontaneously, the women of the area went into the forest and put their arms around the trees, putting their lives on the line and thus successfully resisting the ax-wielding loggers. "Chipko" means to embrace, hence the term "tree huggers." "Dhoom Singh Negi, Bachni Devi and many other village women were the first to save trees by hugging them. The success of the Chipko movement in the hills saved thousands of trees from being felled."[39] It spread throughout India. In the 1980s, EarthFirst! also pioneered some of direct action tactics now being up-scaled world-wide by Blockadia.

Klein writes:

> What has changed in recent years … is itself a reflection of the dizzying ambitions of the extractive project at this point in history. The rise of Blockadia is in many ways

simply the flip side of the carbon boom .... The industry is going further on every front. It is extracting more, pushing into more territory, and relying on more risky methods.[40]

For example, the tar sands and fracking booms had increased the number of rail cars hauling these volatile products by 4,111 percent from 2008 to 2013. More oil spilled in rail accidents in 2013 than in the previous forty years combined.[41] Another causal factor is neoliberal free-market fundamentalism, which has led to government borrowing followed by extreme austerity programs which then force governments in places like Greece to open up their long-held natural resource areas to foreign, extractive corporations that operate under the rules of the WTO.

The WTO was established by international treaty in 1995 to replace the much more limited General Agreement on Tariffs and Trade that had been negotiated after World War II. That arrangement confined itself to trade measures and had little enforcement power. Then corporations and industry trade associations lobbied governments to create the much stronger WTO, an embodiment of the philosophy of market fundamentalism, or the belief that there should be no barriers to trade. The WTO now governs world trade and can challenge not only tariffs, but any of a nation's laws which corporations deem a barrier to trade, including laws protecting the environment. Disputes are resolved in secret by a three-member panel made up of trade bureaucrats who have no requirement to be experts in these other areas and who generally have ties to global corporations, since there is no provision for conflict of interest. The proceedings are not made public, only the decision. Only nation states can bring a case to the WTO, which can nevertheless challenge not only national laws but state and local ordinances as well. Those who lose a case can either change their laws, pay fines, or have tariffs imposed on their goods.[42] It is a direct violation of national sovereignty. The first challenge was brought by Venezuela and Brazil at the urging of their oil companies and was leveled against the United States Clean Air Act, which they claimed made it more expensive to sell their oil in the U.S. They won and the U.S. had to amend the Act.[43]

The Rev. Sharon Delgado was among the thousands who formed a broad coalition of environmentalists, unions, educators, and faith-based people who came together to challenge the WTO by blockading WTO officials at their meeting in Seattle in 1999. The demonstrations were almost entirely nonviolent, although a few renegade anarchists did make some trouble which, of course, grabbed the headlines. She was arrested and dragged off to jail.

Part of the motivation for the protests has to do with losing traditional local control and sovereignty over the resources that people need to live. As Klein puts it, "how is it possible that a big distant company can put me and my kids at risk—and never even ask my permission.... How is it possible that the state, instead of protecting me from this attack, is sending police to beat up people whose only crime is trying to protect their families?"[44]

Many examples abound. In the Skoueries Forest in Greece, riot police have attacked nonviolent protestors with rubber bullets and tear gas, set up check points, and occupied their villages in force. The Greek government, citing the need to attract international capital in order to get out of debt and encourage economic growth, had leased a large area of old growth forest to the Canadian mining company, Eldorado Gold, with permission to clear-cut the forest and re-engineer the local water system to build and operate an open-pit gold and copper mine, an underground mine and a processing plant. The villagers came out in force. The Greek prime minister at the time said the mine will go forward "at all costs" because foreign investment has to be protected.[45] The government also had plans to drill for oil in the Aegean and Ionian Sea, build new coal plants, and open up pristine beaches to large-scale development. It is a pattern typical of neoliberal globalization.

In Pungesti, Romania, Chevron planned in 2013 to open shale gas exploration. Local farmers built a protest camp. Riot police charged through it, bludgeoning farmers and prevented the media from entering the conflict zone. A similar scene occurred in Canada where SWN Resources planned to do seismic testing preparatory to fracking on Indian land in New Brunswick. When the people came out to protest, a hundred police, some carrying sniper rifles and armed with tear gas and pepper spray and dogs, attacked the crowds. In remote Tibet, herders protested an open pit coal mine and were savagely repressed. And in the U.S. the environmental movement was revived by the opposition to the Keystone pipeline, a fight led by indigenous peoples and Bill McKibben's 350.org. It seemed to work. The Obama administration rejected Keystone. The oil companies turned elsewhere, including the development of a new line from Canada to Chicago. In 2016 the Standing Rock Sioux set up an encampment in an attempt to block the North Dakota Access Pipeline which Enbridge, a major pipeline company, was attempting to push across the Missouri River, the Indians' sole source of drinking water. Four thousand people from 200 Indian tribes as well as non–Indians joined the encampment. Armed riot police attacked the unarmed demonstrators with dogs, and the governor mobilized the North Dakota National Guard. The militarized police tried to tear gas them from above with crop duster aircraft and hit them with water cannons in subfreezing temperatures. More and more, these separate movements are beginning to see themselves as part of a global movement. Indigenous people from several South American countries joined the North Dakota encampment, as did whites. President Obama ordered a review. It looked as though the Native Peoples would win after all. But

President Trump reversed Obama's decisions and ordered a go-ahead for both Keystone and the North Dakota Access Pipeline. But Blockadia continues all over the world, and it is a dangerous business.

The results of the 2016 elections in the United States may be the final stand of a corporate plutocracy that has enriched itself at the expense of people and ecosystems all over the world. Their plan is to race forward with the extractive economy while denying or simply ignoring the devastating results to the global ecumene. Numerous organizations have arisen and older ones revitalized to literally stand in the way. Blockadia is alive and well in such organizations as Indivisible and the Community Rights Movement in the United States, the Extinction Rebellion, and in protests around the globe.[46]

Commenting on the opposition to Keystone, K.C. Golden captured the rational of Blockadia:

> Keystone isn't simply a pipeline in the sand for the swelling national climate movement. It's an expression of the core principle that before we can effectively solve this crisis, we have to stop making it worse specifically and categorically, we must cease making large, long-term capital investments in fossil fuel infrastructure … step one for getting out of a hole: Stop digging.[47]

And that is exactly what the world's youth are now demanding.

## *"And a little child shall lead them" (Isaiah 11:6)*

> "It's 2019. Can we all now please stop saying 'climate change' and instead call it what it is: climate breakdown, climate crisis, climate emergency, ecological breakdown, ecological crisis and ecological emergency?"
> —Greta Thunberg, tweeting on May 4, 2019[48]

In August 2018, a diminutive, 15-year-old girl in Sweden did not show up for the first day of school but instead sat down in front of the Swedish Parliament with a hand-made sign that read *"Skolstrejk För Klimatet."* When told she should be in school, she replied: "Some people say I should study to become a climate scientist so that I can solve the climate crisis. But the climate crisis has already been solved. We already have all the facts and solutions. All we have to do is wake up and change. And why should I be studying for a future that soon will be no more when no one is doing anything whatsoever to save that future."[49] Her strike went viral on social media.

Greta Thunberg launched a world-wide uprising of youth, the "Fridays for Future" movement.

Millions have walked out of schools in over a hundred countries, striking to compel their elders to take decisive action to stem climate deterioration. They continue to do so.

Later, Thunberg was invited to speak at many rallies including Stockholm, Brussels and in London at the Extinction Rebellion Rally and at the United Nations summit, COP 24, in Poland. While she speaks calmly and is extremely knowledgeable about the science, nevertheless, her message is blunt. "You are not mature enough to tell it like is," she said at the summit, addressing the Secretary-General. "Even that burden you leave to us children. But I don't care about being popular. I care about climate justice and the living planet."[50] Fearless, she even bearded the very rich at their annual gathering in Davos, Switzerland.

Because she refuses to fly, given the immense carbon footprint of air travel, she made a two-week crossing of the North Atlantic on a sailboat, arriving in New York City on August 28, 2019. She does not mince words. There, she spoke at the UN Climate Action Summit where she castigated the delegates:

> This is all wrong. I shouldn't be up here. I should be back in school, on the other side of the ocean. Yet you all come to us young people for hope. How dare you! You have stolen

Greta Thunberg, a young girl from Sweden, proved that a single person can have a tremendous influence as her decision to leave school and sit outside the Swedish Parliament to protest the adult world's inaction on the climate emergency went viral and sparked protests by millions, young and old, throughout the world. She was selected as *Time Magazine*'s Person of the Year in 2019 (Arron-Schwartz, Shutterstock).

**World-wide Student Protest: Sparked by Greta Thunberg's lone sit-in outside the Swedish Parliament to protest inaction on climate change, millions of young people have risen up and demonstrated their demands that the adult generations stop stealing their futures. Fridays for Future and the Sunrise Movement are two representative organizations (Alexandros Michailidis, Shutterstock).**

my dreams and my childhood with your empty words. And yet I'm one of the lucky ones. People are suffering. People are dying. Entire ecosystems are collapsing. We are in the beginning of a mass extinction, and all you can talk about is money, and fairy tales of eternal economic growth. How dare you![51]

She also met with former President Barack Obama and spoke before the U.S. Congress where she told the lawmakers: "But you have to understand. This is not primarily an opportunity to create new green jobs, new businesses or green economic growth. This is above all an emergency, and not just any emergency. This is the biggest crisis humanity has ever faced."[52] On September 20 she joined the Global Climate Strike march in New York. Around the world, some 4 million people marched on that day. She toured the U.S. speaking with many groups, including indigenous occupiers from Standing Rock. "Inspired by Greta's vision, three middle and high school students started U.S. Youth Climate Strike to bring this youth led movement to the United States, in order to build power and public support for a GreenNewDeal and a just climate future."[53] Thunberg also inspired the Sunrise Movement, another youth climate action group, led by Varshini Prakash. The organization occupied the office of Nancy Pelosi, Speaker of the House; it has several hundred chapters around the U.S., and is also pushing for a Green New Deal.

Many people are deriving hope from this massive youth movement inspired by Thunberg, but she tells us that's not what she wants or what the generations of the future need now. Greta says she doesn't want hope, she wants action. She wants the world to stop the emissions, period. "I want you to panic. I want you to act as if your house is on fire."[54] Thunberg has been belittled and ridiculed by oil industry executives and conservative columnists and has been verbally bullied by the U.S. President. She responds calmly by explaining that the attacks on her as a person are meant to deflect attention from her message, which is to pay attention to the scientists.

On November 13 Thunberg left the U.S. on a 48-foot oceangoing catamaran and headed for Madrid to attend COP 25, which turned out to be yet another failed meeting of world leaders who refused to treat the climate and extinction crises as an emergency. There she addressed the delegates, telling the world's leaders again that the carbon must stay in the ground and that nothing meaningful has been done, but they put off any meaningful work yet again. Greta retweeted David Wallace-Wells: "It seems like #cop25 in Madrid is falling apart right now. The science is clear, but the science is being ignored." Greta added, "Whatever happens we will never give up. We have only just begun."[55]

Hours before her speech at Madrid, Greta was named by *Time Magazine* as their Person of the Year and put on the cover. The editor-in-chief wrote: "Thunberg has become the biggest voice on the biggest issue facing the planet. This was the year the climate crisis went from behind the curtain to center stage, from ambient political noise to squarely on the world's agenda, and no one did more to make that happen than Thunberg."[56]

There is no question that resistance to Hypercivilization and its failed way of life is growing. The only question is will it be in time? Will we stop digging this hole, build on the philosophy and accomplishments of the environmentalists who went before us and create a new form of interaction with the planet, or we will go on until global collapse? And if we are to avoid such a fate, what must we do? Those are the questions we will take up in Part III.

# PART III

## Breakthrough
### *Creating Ecocivilization*

"The dogmas of the quiet past, are inadequate to the stormy present. The occasion is piled high with difficulty, and we must rise with the occasion. As our case is new, so we must think anew, and act anew. We must disenthrall ourselves, and then we shall save our country."
—Abraham Lincoln, 2nd Inaugural, 1862

"The donkey that brought you to this door must be dismissed if you want to go through it."—Idries Shah, Sufi Master (1924–1996)

We are already beginning to change our mind in fundamental ways that change our institutions and our technologies. Ecocivilization is developing in the belly of Hypercivilization. We are exploring and developing new questions, new ways of perceiving the relationship between humanity and nature, and new values. We are beginning to differentiate mere cleverness from wisdom, learning to ask how much is enough, and to explore what makes for a truly satisfying life in community. We are developing a New Story about who we are, where we are and where we are heading, and new ways of doing education, energy, transportation, urban design and food production. We are developing a peace system within the maw of the war system.

In this we will want to maintain the obvious benefits of modern civilization: a science based culture that provides modern health care, a decent and sustainable level of prosperity for everyone which requires, among

**197**

other changes, freedom from violence. But we also want to rekindle the awe and wonder we traditionally had for the miracle of the natural world. These are not incompatible aims. In short, we are already rethinking civilization and we will need magnify these changes and scale them up to create a just, peaceful and sustainable world.

# 12

# Our Changing Mind

> Throughout history, the really fundamental changes in society have come about not from the dictates of governments and the result of battles, but through vast numbers of people changing their minds, sometimes only a little bit.
> —Willis Harmon[1]

> We need a positive vision of an abundant future: one which is energy lean, time-rich, less stressful, healthier and happier.
> —Rob Hopkins[2]

To achieve breakthrough we need to change our mind, change how we think about humanity, nature and what makes for successful and happy lives. The good news is we are changing. The development of the new mindset that will guide our revolutionary project of building Ecocivilization is already underway and can be found in numerous books, films, news stories, classrooms, blogs, and in countless conversations around the world. As we are changing our basic thinking, innovative and sustainable ways of organizing our institutions and technologies are also changing.

Hypercivilization was amazing for the creative intelligence it brought to the creation of useful goods, but it was an intelligence characterized more by cleverness than anything else. Cleverness is the ability to solve immediate problems in a linear fashion, one problem—one solution, and then moving on to another problem treated as separate from the first. Ecocivilization will be characterized by wisdom. Wisdom is the ability to see and account for the whole, for the way in which a particular issue fits together with all the rest, and how a solution to one problem can become a problem for other problems, other people, species, and places.

On the surface the world appears to be made up of innumerable discrete items each occupying its own space and bounded by its exterior, but this is an illusion. The automobile parked in your garage or a simple plastic grocery bag have a long trail behind them, over which they came to be as they exist at this moment. They came through many ecological and social systems and will have a similar, complex course after they leave you. A more realistic way of perceiving the world is to realize that "Everything that exists is really no more a physical 'thing' than it is the momentary embodiment of a web of connections, of relationships."[3] These can be environmentally and socially destructive or they can nurture biological and social life. Wisdom is perceiving that the world is made up of systems characterized by linkages and feedbacks in space and time.

Wisdom leads to the realization that you can't do just one thing because so much is interconnected, and to see how that limits us. Wisdom means acknowledging the complexity of the world and avoiding simple solutions because, as Wendell Berry points out, "Simple solutions will always lead to complex troubles, and simple minds will always be surprised."[4] Wisdom means being aware of a thing's ecological footprint, knowing where it came from, how it got here, and where it's going to be a hundred years from now. Take out your cell phone and ask those questions. Wisdom also includes knowing when not to do something just because you can, or because it seems like a brilliant solution to a particular problem without considering its multiple effects.

Wisdom is to see all the connections and save all the parts and to consult all the stake holders, and to consider the not yet born. It is often ancient and it is genuinely cautious and conservative. Wisdom involves more than the head, it involves the heart. Evolutionary biologist Stephen Jay Gould said, "we cannot win this battle to save species and environments without forging an emotional bond between ourselves and nature as well—for we will not fight to save what we do not love."[5] Similarly, Edward O. Wilson argues in *Biophilia* that we have a built-in love of life itself, "that there is an instinctive bond between human beings and other living systems" and he defines biophilia as "the urge to affiliate with other forms of life."[6] Not realizing this, Hypercivilization remains stuck in an old, dysfunctional story.

## *The Old Story*

Every civilization tells itself a foundational story that explains human nature, the nature of society, the natural world, and what the good life is. These stories are the lens through which we think we see reality and so we try to reshape it to fit the story. Hypercivilization's story is a tale of separation and domination. In countless overt and subtle ways it indoctrinated us, teaching that the world is invariably divided into nation states, classes, ethnic groups, religions, male and female, each of us an isolated individual, and that these separate categories are all arranged in hierarchies of power and are in competition with each other. In this story greed, selfishness, and the willingness to use violence define human nature. Society is a zero sum game, Hobbes's "war of all against all." It's about power over, domination. It tells us that we must fear scarcity and at the same time, paradoxically, we suffer from the delusion that resources will last forever no matter how much we extract. Nature is a mine of materials to be used to create what it calls "goods" while ignoring the "bads." It tells us we have always had and always will have war, and that endless economic growth is possible and is the pathway to the good life for all. Its most basic separation is that of humanity versus nature which must be "conquered" by our species. We know how that ends. Ursula Le Guin has described it in her dystopian novel, *The Dispossessed*:

> My world, my Earth is a ruin. A planet spoiled by the human species. We multiplied and fought and gobbled until there was nothing left, and then we died. We controlled neither appetite nor violence; we did not adapt. We destroyed ourselves. But we destroyed the world first.[7]

The old story has failed, having led us to ecological disaster and the real possibility of nuclear war. We need a way out. Richard Schiffman writes:

> We need a powerful new story that we are a part of nature and not separate from it … that properly situates humans in the world—neither above by virtue of our superior intellect nor dwarfed by the universe into cosmic insignificance. We are equal partners with all that exists, co-creators with trees and galaxies and the micro-organisms in our own gut, in a materially and spiritually evolving universe.[8]

The good news is we are changing our mind, creating the New Story, developing Ecocivilization.

We live in a revolutionary time that author and economist David Korten has called "the Great Turning."[9] Buddhist leader Joanna Macy writes:

> The Great Turning is a name for the essential adventure of our time: the shift from the industrial growth society to a life-sustaining civilization.... It is happening now ....and it is gaining momentum, through the actions of countless individuals and groups around the world.[10]

We are standing on the leading edge of the great post–Enlightenment rethink. The foundational ideas of Hypercivilization, namely materialism, individualism, competition, hierarchical organizations, centralization, violence as a way of dealing with conflict, infinite economic growth, and linear thinking have served their purpose. While they have issued in a rising material standard of living for some, they have issued in poverty for millions of others, in horrendous warfare, and in unprecedented environmental degradation. The Old Story is an ecopathology. Fortunately, many people now realize that we have overshot and are in danger of collapse. As a result, new foundational ideas are rapidly developing, sometimes out of ancient wisdom, sometimes out of modern ecological science.

I am not advocating we naively aim for some utopia, but rather that we avoid the dystopia that is already looming on the horizon. We need to prosper, but not in the sense that everyone will have more and more material wealth. That doctrine that has brought us to this planetary emergency—better that we live surrounded by abundant natural life and have simple, material well-being for all. The foundational principles that follow are the guideposts that mark the way from Hypercivilization to Ecocivilization and that will serve as the standards by which we judge what we are creating.

## Community

The first and most fundamental of the ideas that will undergird Ecocivilization is the idea of community. This ought to be so obvious as to be not worth stating, yet it is crucial to lift it up in this age of radical, narcissistic individualism. Other foundational concepts flow out of and into this, including the following eleven changes.

1. From privatization to preserving the Commons.
2. From anywhere is everywhere to restoring a deep sense of place.
3. From treating the Earth as a momentary utility to seeing the Earth as a sacred community.
4. From ignoring the limits of nature to respecting its design.
5. From mindless experimentation to the precautionary principle.
6. From design by guess to redesign by biomimicry.
7. From never enough to asking how much is enough.
8. From crude measures of economic growth to determining and measuring true happiness and well-being.
9. From rigidity to resilience.
10. From war to peace.
11. From irresponsibility to responsibility and from externalizing costs to demanding accountability.

These should not be considered linearly but systemically—all interlocked with each other. The New Story is that we are community.

We have been dividers, seeing all things from human individuals to "resources" in isolation from one another and society as no more than an aggregate of individuals. We have seen iron ore deposits apart from the life of streams that flow next to the them, and burning coal in isolation from the climate, and on and on. We have lived as if everything could be taken apart and recombined endlessly, as if the biosphere and society were machines.

We humans are but a part of the larger community of the biosphere. We live and can only live within its bounds, the upper limits of the atmosphere and roughly the outer edges of the continental shelves, a thin space where everything is interdependent. From within this narrow space we humans derive our life, as does every other living thing. Within this little space is everything we love, as astronaut Russell Schweikert realized when he orbited the earth in 1969.

> It is so small and so fragile and such a precious little spot in the universe that you can block it out with your thumb. And you realize that on that small spot, that little blue and white thing, is everything that means anything to you—all love, tears, joy, games, all of it on that little spot out there that you can cover with your thumb.[11]

All our political frames and boundaries are artificial. It is nature's frames within which we live. Another astronaut, Edgar Mitchell, who flew to the moon in Apollo 14, pressed home the point.

> I was gazing out of the window, at the Earth, moon, sun and star-studded blackness of space in turn as our capsule slowly rotated. Gradually, I was flooded with the ecstatic awareness that I was a part of what I was observing. Every molecule in my body was birthed in a star hanging in space. I became aware that everything that exists is part of one intricately interconnected whole.[12]

In other words, you in me and me in you. The earth is a shared community. We are literally all in it and of it together. We only survive by cooperating. Life on earth is a group outcome. The science of ecology reveals that we are nodes in a web, each of us is a unique, but nevertheless, an expression of the whole.

Our most ancient expressions of community are found in religion, and while this has been problematic for the environment in the past, today religion is greening and is an integral part of the transition to the New Story that underlies a nascent Ecocivilization. There is core material in all the major religions that can be and is being drawn on for this new appreciation of creation and the risks we face from Hypercivilization. Religious leaders are acutely aware of the planetary emergency. Pope Francis writes, "Never have we so hurt and mistreated our common home as we have in the last two hundred years."[13] In his *Apostolic Exhortation: Evangeli Gaudiam*, he put his finger on the root cause, writing, "whatever is fragile, like the environment, is defenseless before a deified market which becomes the only rule."[14] And further, "Rather than a problem to be solved, the world is a joyful mystery to be contemplated with gladness and praise."[15] Noted evangelical author and editor of *Sojourners* magazine, the Rev. Jim Wallis, says, "We are falling down carbon drunks, fossil fuel addicts who cannot break our collective addiction cycle. Only religion ... can save us."[16] And recently a Pentecostal environmentalist posted on the web, "I'm a tree-hugging Jesus freak."

Jewish Leaders and congregations have also been greening. Rabbi Lawrence Troster writes in "Ten Jewish Teachings on Judaism and the Environment" that "This is the most fundamental concept of Judaism. Its implications are that only God has absolute ownership over Creation" (Gen. 1–2, Psalm 24:1, I Chron. 29:10–16). Thus Judaism's worldview is theocentric, not anthropocentric. Like Christianity, Judaism also has a strong social ethic which is the foundation for Jewish belief and action to aid vulnerable communities on the front lines of climate deterioration. Furthermore, we should stand in awe and wonder before the creation, even the more so now that modern science has shown us its richness and complexity. We should be humbled by it. He writes: "Love and humility should then invoke in us a sense of reverence for Creation and modesty in our desire to use it."[17] And to use it sparingly: "In Judaism, the halakha (Jewish law) prohibits wasteful consumption. When we waste resources we are violating the mitzvah (commandment) of *Bal Tashchit*" ("Do not destroy").[18]

The same is true for Islam. The "Muslim Declaration on Nature," proclaims, "For the Muslim, mankind's role on earth is that of a Khalifah—vicegerent or trustee of Allah. We are Allah's stewards and agents on Earth. We are not masters of this Earth; it does not belong to us to do what we wish. It belongs to Allah and He has entrusted us with its safekeeping."[19] And from the *Islamic Faith Statement* we find a critique of the Hypercivilization of the modern West, arguing that we humans are committing blasphemy in the extremist materialism by which we justify the despoliation of the earth: "We see men now wherever we look, so blinded by arrogance and the worship of man as God that they are doing things no one but the insane would do ... men maddened by the belief that they are both omniscient and omnipotent, that they are indeed God."[20]

Islam has a long tradition of legal protection of the environment including the other than human inhabitants:

> Six hundred years ago the classical Muslim jurist, Izzad-Din Ibn Abdas-Salam ... formulated the bill of legal rights of animals back in the thirteenth century. Similarly, numerous other Muslim jurists and scholars have developed legislation to safeguard water resources, prevent over-grazing, conserve forests, limit the growth of cities, protect cultural property, and so on. Islam's environmental ethics then are not limited to metaphysical notions; they provide a practical guide as well .... *Surah Shari'ah* institutions as haram zones, inviolate areas within which development is prohibited to protect natural resources, and *hima*, reserves established solely for the conservation of wildlife and forests, form the core of the environmental legislation of Islam.[21]

Prior to the Paris Climate Summit of 2015, Islamic leaders at the International Islamic Climate Change Symposium in Istanbul issued a call for zero emissions.

> The Muslim leaders furthermore called on the people of all nations and their leaders to phase out greenhouse gas emissions as soon as possible in order to stabilize greenhouse gas concentrations in the atmosphere, and to commit themselves to 100 % renewable energy and/or a zero emissions strategy as early as possible. They specifically called on richer nations and oil-producing states to lead the way in phasing out their greenhouse gas emissions as early as possible, and no later than the middle of the century.[22]

These three religions of the book, as they are known, provide ample material for the faithful who are seeking divine justification for their care for the Earth.

Buddhism has long recognized the fundamental truth of deep community. A 1990 book, *Dharma Gaia: A Harvest of Recent Essays in Buddhism and Ecology*, is a garden of such concerns.[23] Ken Jones writes in one of the essays, *Getting Out of Our Own Light*: "As the crisis deepens, it will become clearer that if we are to get through in good shape, caring for all our people as well as other planetary life, we shall have to undertake the most radical social transformation since the Neolithic Revolution."[24]

The foundation of Earth care in Buddhism is the ancient teaching of dependent co-arising. I am because you are. What and who we are is dependent on others without whom we could not live. And the others include more than just our fellow humans. If we could put a large, hermetically sealed glass bell jar down over a human being, she would die within a few minutes for lack of oxygen, a product of the world's plants. Provide oxygen and she would still die of thirst within a few days. Provide water—a free gift of Earth—and she would still die of starvation in a few weeks. Without the complex soil community, sunlight and water there would be no food. Without the farmer, the trucker and many others we would not eat. And the bell jar is also a cultural barrier to humanness. Under the bell jar, without our parents, teachers and the media we would not have the ideas we have about reality. We would have no recognizable mind at all. There is no such thing as a free-standing, self-made individual. We are all essentially parts of one body. It's not that I am my brother's keeper, but that my brother and I are inescapably related parts of a single whole. The Dalai Lama writes that the word *self* "does not denote an independent object. Rather it is a label for a complex web of interrelated phenomena."[25]

One could write at great length about the greening of religion, and not just these four but Hinduism, Taoism and humanism. All of this squares with the ethics appropriate to a human-land community as voiced by Aldo Leopold in *A Sand County Almanac*. "The land ethic simply enlarges the boundaries of the community to include soils, waters, plants, and animals, or collectively: the land…. In short, a land ethic changes the role of *Homo sapiens* from conqueror of the land community to plain member and citizen of it. It implies respect for his fellow-members, and also respect for the community as such."[26]

We all live downstream of one another and the well-being of any one is ultimately dependent on the well-being of all, and vice versa. We live a common life. The first principle that flows out of the concept of community, then, is the requirement to protect that commons.

## Preserve the Commons

An ecological and social dimension of the Earth community is that it is a commons. The commons is that on which all depend. We all have a right to the public benefits of the commons. The global air-shed, the world's oceans, and global biodiversity are all a part of the global commons. Just as no one person or corporation has a right to pollute the air we breathe, or to privatize our water, no one has a right to extinct a species by careless actions or laws, however unintentional. All of us participate in the commons in various ways from breathing in pollutants that cross international borders to suffering the climate effects of burning carbon, the destabilizing impacts of deforestation, diminished and polluted water, and diminishing biodiversity. To respect the commons requires us to pay attention to both global and local matters. Even when we live far from polar bears and the big cats, we feel in our spirits that our world will be sadly diminished should we continue to drive them on toward extinction. In our interdependent global economy, access to food is also commons.

## Developing a Sense of Place

And here is paradox; to do that we must also attend closely to our local places, for it is in local places that degrading elements enter the global commons. It is in England that sulfur fumes from power plants originate that kill the forests in Scandinavia. It is in the local river that the pollutants from agricultural runoff and industry enter the world ocean. It is from a specific power plant in Beijing, or Arizona, that $CO_2$ enters the global atmosphere, and of course endangered species can only be protected where they are living. That means we must each get to know our place intimately, to know its biology and ecology, but also its culture and history if we are to truly be considered conservatives. Recognizing the commons involves us in *glocalism*, attention to our immediate locale and to the planetary whole at the same time. And to pay attention to our locale requires us to know it intimately, to restore our sense of place. Most of us who live in the over-developed world have become disassociated with any sense of place, that "combination of characteristics that makes a place special and unique."[27] Hypercivilization is characterized by rootlessness, a loss of any deep identification of with a particular place. The international style in architecture made every city look pretty much like every other, and when you have been in one international airport you have been in them all. Shopping malls—everywhere the same—became the major outside-the-home habitat for many people. The English lawn became ubiquitous, driving out native landscapes.

The industrializing and commodifying of agriculture in the late nineteenth and twentieth centuries drove millions off the land so that now only a tiny percentage of the population, using destructive technologies, produce our food and fiber. We were removed from our place and from a natural setting by the industrial and consumer culture. Few people were intimately connected to or knew much about the place they were living in or the distant places where their life came from. Nor did they care. Gloria Flora points out that "We live Here. We get our goods, like oil and cheap gadgets, from Over There, and Away is where things go when we don't want them anymore" [and] "In our madness we've failed to see that Here, Over There, and Away are simply one place with interchangeable names."[28] Today few people could name the common plants, animals, and insects where they live, much less the soil types, or know when the robins come back from the South. Few spend free time out of doors. In short, we don't know where we are. And as Wendell Berry says, "if you don't know where you are, you don't know who you are."[29]

How does one get to know a place intimately? Berry provides a set questions each of us must be able to answer in order to live successfully in a particular place.

1. What has happened here?
2. What should have happened here?
3. What is here now? What is left of the original natural endowment? What has been lost? What has been added?
4. What is the nature, or genius, of this place?
5. What will nature permit us to do here without permanent damage or loss?
6. What will nature help us to do here?
7. What can we do to mend the damages we have done?
8. What are the limits: Of the nature of this place? Of our intelligence and ability?[30]

The wisdom then gained about a place will be passed on to generations down the line. Ecocivilization will be built on a renewed sense of place that can only come from living in one place for a long time and from getting out of the house, out from in front of the LED screens and into the real world, the local biosphere. Once there, with a newly developed sense of wonder and awe, we will realize that it is sacred.

## Acknowledge That the Earth Is a Sacred Community

In 1990, twenty-four distinguished scientists, including Carl Sagan, Stephen Jay Gould and several Nobel laureates, drafted an appeal to world religious leaders to join them in protecting the global ecosystem from climate deterioration, massive extinctions, deforestation, and nuclear war. They said at the outset that these looming disasters have "a religious as well as a scientific dimension" and "that what is regarded as sacred is more likely to be treated with care and respect. Efforts to safeguard and cherish the environment need to be infused with a vision of the sacred."[31] What does it mean to say that the earth is "sacred?" One way is to see the earth as a living sacred being. Llewellyn Vaughn-Lee, a contributor to Joanna Macy and Thich Nhat Hanh's book, *Spiritual Ecology*, writes: "The world is not a problem to be solved; it is a living being to which we belong…. The deepest part of our separateness from creation lies in our forgetfulness of its sacred nature, which is also our own sacred nature…. We are all part of one, living spiritual being."[32]

Not only is it precious beyond imagining but it is partakes of another dimension of the sacred: it is mysterious, a point Wendell Berry emphasizes:

> But even more important, we must learn to acknowledge that the creation is full of mystery. We will never clearly understand it. We must abandon arrogance and stand in awe. We must recover the sense of the majesty of the creation and the ability to be worshipful in its presence. For it is only on the condition of humility and reverence before the world that our species will be able to remain in it.[33]

Such humility is not often discussed in environmental circles and almost never in the halls of university engineering departments or corporate boardrooms. Instead, what we are hearing lately is that we humans have moved geological history into a new era, the "Anthropocene," in which we humans will now control the course of planetary evolution by applying our sciences to "geoengineering." They argue that we no longer live within the earth but within the "Anthroposphere," which we will reshape more and more to meet our material wants. Unfortunately this is just the latest version of the doctrine of the conquest of nature, and as Rachel Carson pointed out years ago in *Silent Spring*: "The 'control of nature' is a phrase conceived in arrogance, born of the Neanderthal age of biology and philosophy, when it was supposed that nature exists for the convenience of man."[34]

The idea that we can manage the planet is a silly, foolish and dangerous delusion. Anyone looking honestly at the record of Hypercivilization would find it laughable were it not so tragic. What we need to manage is ourselves. We need to fit our civilization into the patterns of the biosphere, and that means, in large measure, reducing the scale of our impact including, ultimately, our numbers—even if all we want is abundant life for our own species. Our thinking has been upside down so to speak, as Berry points out: "We've lived by the assumption that what was good for us would be good for the world. We've been wrong. We must change our lives so that it will be possible to live by the contrary assumption that what is good for the world will be good for us. That requires that we make the effort to know the world and learn what is good for it. We must learn to cooperate in its processes and to yield to its limits."[35] Thinking in terms of "Anthropocene" is an obscene case of hubris and, like all over-weening and self-congratulating pride, leads inevitably to a fall.

## Respect the Design

Nature is what nature is. We can't alter the basic laws by which it functions. Whether we want to "believe" in particle physics or bioaccumulation or natural selection or global climate change is irrelevant. When people believed the Earth was flat, it wasn't. The Earth was not made to our design and while we can work within its laws, we can't do just anything that pleases us at the moment. It is true that we can build a world within the world, but we can't build just any world, and we certainly can't build the world we have been trying to build without seeing it crash and bring down far too many other species with it. Certainly, working with nature, we can alter it to a degree. We can turn a wilderness into a farm but we must accommodate ourselves to the processes which produce and protect life on the planet over the long term, to say nothing of knowing and respecting the limits and laws that will keep this farm productive a thousand years from now. There are some things we can do and some we cannot do without dangerous consequences, if not for us, then for our descendants. In our case, we were able for about two hundred and fifty years to build a thriving industrial economy on the basis of burning fossil fuels. In that process a large percentage of people were able to climb out of poverty and some to become as rich as kings, although we have to admit that, because of population explosion, more people now live in poverty that ever before. While some studies make much of the fact that today a smaller population is now living on less than a dollar a day than fifty years ago, living on two dollars a day, or five, isn't much consolation.

The consequences have caught up with us. The ironic tragedy is that while we have made great strides as a result of the fossil fuel revolution, it now threatens everyone, rich and poor. The same thinking that got us here—the conquest of nature, economic growth through further industrialization, will not get us out of here. To respect the design we need to close all the linear processes, to make them into loops as nature does. All waste must become feedstock. And what can't be reused, recycled or repaired ought no longer to be made. Furthermore, nature functions on the principle of diversity. Hyper-civilization is a radical simplifier of nature. Nature is systemic, characterized by feedback loops, interdependence, and limit. Hypercivilization acts as if it can get away with linear processes, ignore feedback, and exceed limit. Nature makes changes slowly, on an evolutionary time scale. Hypercivilization makes rapid and often ill-considered changes in the natural world, which is why Ecocivilization will be based on the precautionary principle.

## Employ the Precautionary Principle

The complexity and ultimate unpredictability of nature requires us to figure out how to interact with it, for interact we must even on the basis of partial knowledge.

> And so the question of how to act in ignorance is paramount … it is dangerous to act on the assumption that sure knowledge is complete knowledge—or on the assumption that our knowledge will increase fast enough to outrace the bad consequences of the arrogant use of incomplete knowledge.[36]

The only way to avoid the error of hubris is to enshrine the precautionary principle in our cultural foundations. As we have seen all too often, Hyper-civilization acts on the opposite basis: first transforming an aspect of nature and only later finding out if there were destructive consequences as was the case with PCBs, CFCs, DDT, nuclear waste and countless other experiments. A very simple statement of the precautionary principle is one we all learned in childhood: "Look before you leap." Whoever wants to introduce a change must prove it is safe prior to introducing it. It's just applied common sense. And as we act to create new institutions, new technologies and to redesign our cities and our food and industrial systems, we need to imitate nature, to employ biomimicry.

## Redesign by Biomimicry

In her book, *Biomimicry,* Janine Benyus opens up the possibility of an exciting new era of innovation that will lead us into Ecocivilization.[37] The term comes from the Greek, *bios* (life) and *mimesis* (imitation). She opens with a quote from Vaclav Havel, former president of the Czech Republic. "We must draw our standards from the natural world. We must honor with the humility of the wise bounds of that natural world and the mystery that lies beyond them, admitting that there is something in the order of being which evidently exceeds our competence."[38]

The three principles of biomimicry are:

1. *Nature as model.* Biomimicry is a new science that studies nature's models and then imitates or takes inspiration from these designs and processes to solve human problems, e.g., a solar cell inspired by a leaf.
2. *Nature as measure.* Biomimicry uses an ecological standard to judge the "rightness" of our innovations. After 3.8 billion years of evolution, nature has learned: What works. What is appropriate. What lasts.
3. *Nature as mentor.* Biomimicry is a new way of viewing and valuing nature. It introduces an era based not on what we can *extract* from the natural world, but on what we can *learn* from it.[39]

One of the things we learn from nature is the principle of limit. Benyus puts it this way:

"A species cannot occupy a niche that appropriates all resources—there has to be some sharing. Any species that ignores this law winds up destroying its community to support its own expansion."[40] And then it crashes: "humanity now uses eleven times as much energy, and eight times the weight of material resources every year as it did only a century ago. And most of this increase has occurred in the last fifty years."[41] Moreover, that accounting does not include the use of non-plant resources. The Global Footprint Network tries to account for the total impact of Hypercivilization on the planetary environment and they calculate that we are now taking 3/2ths of the productive capacity of the Earth.[42] If that seems impossible, it's explained by the fact that we are living off a one-time reserve of fossil fuels, minerals and natural energy sinks. But when these run out or fill up, the society has to power down, to shrink back to sustainable limits. Unfortunately, we do not know where the point of too much lies, where we are taking so much of the biomass that we deprive other organisms on which our life depends of biomass they need to flourish. However, it is a good bet, as we look at the extinction crisis, that we have passed that point. How much can and should we take? To answer that we first we need to embrace a concept very foreign to Hypercivilization—the concept of "enough."

## Understand How Much Is Enough

Building a civilization which mimics nature means building a civilization that understands how much is enough. How much is enough for how many and for whom? In an interdependent system characterized by endless feedbacks, no one can maximize for all the variables. One cannot have maximum biodiversity, stable climate, healthy soils, clean water, thriving fisheries, etc., and unlimited growth of population, extraction, consumption and waste over a long period of time. There are inevitable trade-offs. There is the obvious trade-off between population and per capita consumption. Within the limits of the biosphere we can have a smaller number consuming more or a larger number consuming less per capita. According to World Watch Institute, the biosphere can support 13.6 billion people living what it calls "low income lifestyles" or 2.1 billion living "high income lifestyles."[43] But perhaps income and happiness, or a high quality of life, are not directly proportional. Surely our goal is not to achieve the maximum number of people who can survive in misery on the planet. It is obvious that a minimum level of consumption is necessary to support life, something like the United Nations level of "basic needs" for food, clean water, shelter, clothing, basic education and health care. While more can be better if one is trying to live above a minimal level of consumption, more is not always better, especially for the already developed societies where less is better. In fact, at some point more is obviously worse. Further growth becomes uneconomic because it depresses the natural capital that supports the economy. This truth implies the necessity of shifting to a steady state economy. Taking too much, we degrade and dismantle the geobiotic structures such as the community of soil organisms or the food chain in the seas which support life. And considering how much is enough, we have to ask, "For what?" Is unlimited material consumption the road to happiness and well-being?

## Determine and Measure
## True Well-Being and Happiness

The goal of our interaction with nature is to reproduce our lives physically but also to achieve well-being and happiness. The great mistake of Hypercivilization was to associate well-being with gross domestic product, with a rising average income in a community or a nation. It's a measure that does not even always equate with the material well-being of ordinary people, to say nothing of the all-important non-material aspects such as functioning community life, psychological and spiritual contentment, creativity, and social justice. A story from the early days of the Inter-American

Development Bank just after World War II illustrates this for those of us who have been brainwashed into thinking that well-being equaled money or even modern "conveniences." Bank President Philipe Herrera was in Colombia looking to finance a huge hydro-electric dam, another example of the purely techno-materialist concept of development, when he realized there was money left over from the project. He approached the indigenous elders of the village where they were staying and said the Bank would like to make a gift of the money to the village. What did they want? They replied: "We need new musical instruments for our band." Thinking they misunderstood, he said: "What you need are improvements like electricity. Running water. Sewers. Telephone and telegraph." But the Indians pointed out that on Sundays after church they all gathered in the village square. "First we make music together. After that we can talk about problems in our community and how to resolve them. But our instruments are old and falling apart. Without music, so will we."[44] Once again, we are led back to community as the core concept for Ecocivilization.

Hypercivilization teaches us to equate well-being and wealth. Madison Avenue brainwashed whole populations into consumerism by guilting people into internalizing what Wendell Berry has called the "hysterical self-dissatisfaction of consumers."[45] This does not appear to make us happy or free. Buddhist author Sulak Sivaraska points out: "We try to overcome the emptiness of our lives by increased consuming. We are at the mercy of advertisers and, inevitably, we are exploited."[46] Once hooked into the consumer lifestyle, the goal always remains elusive as Patrick H.T. Doyle points out: "There's never enough of the stuff you can't get enough of."[47]

We have become loud and frantically busy with no time for, well, for doing nothing. For contemplation. For solitude and silence. Thomas Merton, a Trappist monk and one of the more cogent observers of the twentieth century, said: "Those who love their own noise are impatient of everything else. They constantly defile the silence of the forests and the mountains and the sea. They bore through silent nature in every direction with their machines, for fear that the calm world might accuse them of their own emptiness."[48] Sivaraska indicts the affluent, industrialized North, pointing out "Individuals lose their sense of meaning and peace."[49]

Dietz and O'Neill point out in *Enough Is Enough* that true happiness is not directly correlated with wealth. Although GDP per capita has tripled in the U.S. and the UK since 1950, surveys indicate that people have not become happier. While it is true that in industrial societies as incomes increase up to about $20,000 a year, satisfaction increases; after that it plateaus.[50] In so-called "developing" countries that number is far lower. In fact, where traditional cultures that had provided social cohesion and life satisfaction at the village level, what the World Bank promoted as "modernization" actually resulted in social disintegration and environmental destruction, even though GDP per capita rose. Sulak Sivaraska and others ask, what are the true indicators of well-being? Sivaraska lists these:

- The degree of trust, social capital, cultural continuity, and social solidarity
- The general level of spiritual development and emotional intelligence
- The degree to which basic needs are satisfied
- Access to and the ability to benefit from health care and education
- The level of environmental integrity, including species loss or gain, pollution, and environmental degradation.[51]

Ever since the King of Bhutan coined the term "gross national happiness" in 1972, many people have been working to develop indicators by which happiness and well-being can be measured. The UK government is pursuing an index of happiness to help shape national policy. In Australia they have a program designed to address the question, "Is life in Australia getting better?" The Japanese government is pursuing a similar set of happiness indicators. In the U.S., surveys indicate that a variety of factors besides income influence happiness including: "living with a partner, enjoying good health, holding a secure job, having trust in institutions, volunteering, and limiting the amount of time spent watching television…."[52] The New Economics Foundation has developed the Happy Planet Index which subtracts the damaging economic activity from that which promotes life, determined by a composite of life expectancy and life satisfaction. "HPI gauges how well we transform the limited resources available to us into long and happy lives."[53] When measured on the HPI, Costa Rica comes out on top and the U.S. in the 114th place.

What we are looking for in our interchange with nature is happiness, satisfied lives, a sense of meaning and purpose, recognition of our own worth and dignity, and a sense of leaving a better world for future generations.

Ecocivilization will be very careful in how it measures these intangible values, knowing that everything that counts can't necessarily be counted, and certainly cannot measured by something as crude as GDP. There is a profound Buddhist insight about means and ends. "When you know that what you need is not the snare or the net but the hare or fish, it is like gold separated from the dross, it is like the moon rising out of the clouds."[54] We don't need multi-ton SUVs and interstate highways, we need transportation. We don't need giant agribusiness, we need food. We don't need Walmart, we need well-being. We don't need instability, we need continuity, and so we need to build resilience into our communities.

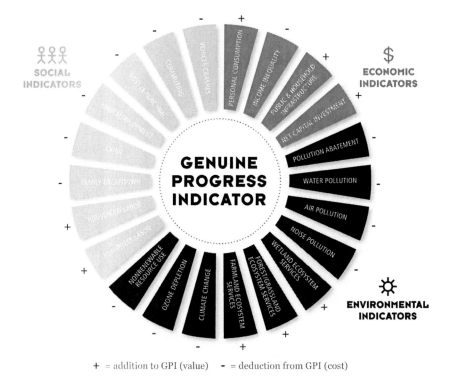

+ = addition to GPI (value)  − = deduction from GPI (cost)

Measuring Progress: GDP, a measure of the dollar value of all goods and services has traditionally been used to measure progress, but it is crude and includes profits made from polluting. A host of more sophisticated measures of well-being have been proposed including this one. (Berik, G. and E. Gaddis. 2011. "The Utah Genuine Progress Indicator (GPI), 1990 to 2007: A Report to the People of Utah," https://utahpopulation.org/wp-content/uploads/2014/11/Utah_GPI_ Report_v74_withabstract.pdf).

## Build in Resilience

Communities need to be resilient, adaptive in the face of change so as to stay balanced, avoid major degradation or worse, spiral into collapse. In systems involving humans this requires knowing the ecosystem and the social system and how they interact. Resilience comes from a combination of social, economic and environmental diversity, interdependence, self-reliance, appropriate scale, redundancy and managing change so that it is incremental. It means the ability to absorb shocks from the outside, be they ecological, financial, climatic, or political. It means that communities must be much more self-sufficient than in recent history. And there is no worse shock than warfare, which we must end, if we intend to survive and prosper on earth.

## Establish the Peace

War, peace and the environment are intimately related. Like the oceans and the climate, peace is a commons. Any war anywhere degrades the Earth for all of us everywhere. Warfare between particular societies destabilizes the global order by stimulating other societies to invest precious resources in building up defenses and preparing for war, thus making the world even more dangerous. "War is a self-perpetuating and self-fueling infection in the global community."[55] We cannot save the environment if we are constantly preparing for war or are at war with each other somewhere on the planet. This is true for four obvious reasons and one more profound and subtle reason. First, war distracts us from the true threat to our security, the rapidly deteriorating environment. Second, preparing for modern warfare involves the creation of toxins that poison the environment including spilled jet fuels, $CO_2$ emissions from military sorties, practice bombing, radioactive waste from nuclear weapons, the potential horrors of accidental biological weapons releases and so on. Third, war is frightfully expensive; trillions of dollars wasted over the years which are now desperately needed for investment in renewable energy, mitigating climate damage and ecological restoration. Fourth, the practice of warfare destroys the natural and built environment from spraying herbicides to scattering "depleted" uranium weapons and land mines to the all-out desecration of a full scale nuclear holocaust. But more subtly, the war game perpetuates the myth that we various peoples on the planet are divided from one another and in a deadly competition when, in truth, we are a single species dependent for our well-being on the well-being of countless other species. We will not achieve one people, one planet, without one peace. Amster writes that "... we will need to redefine our engagement with one another and with the habitat, reframing our master narratives and breaking out of the mindset of 'scarcity begets conflict' toward one that emphasizes the abundance of creative and cooperative capacities inherent in human beings...."[56] For all of this, we are the ones who have to take responsibility. We are the only ones who can transform Hypercivilization into to a sustainable world before it's too late.

A human society turning away from Hypercivilization and toward Ecocivilization will attend seriously to international laws and agreements that protect people and other creatures in one part of the commons from of those who live in other parts who thoughtlessly and selfishly degrade the lives of all for short-term gain. And nothing is more degrading in this sense than warfare. David Suzuki writes, in *The Sacred Balance: Rediscovering Our Place in Nature:* "War is a social, economic and ecological disaster. It is totally unsustainable and must be opposed by all who are concerned about meeting the real needs of all people and future generations."[57] Randal Amster writes in *Peace Ecology* that "...demilitarization can serve to promote biodiversity and thriving ecosystems," and further "...whereas militarism produces the inverse: desertification, resource depletion, and toxification."[58] And Anita Wendon cites a fundamental truth: "...peace will require environmental sustainability and environmental stability will require peace...."[59]

I and others have written extensively on the War System and how it can be replaced in such books as *From War to Peace: A Guide to the Next Hundred Years,* and *A Global Security System: An Alternative to War,* where one can find detailed analysis of the War System as well as detailed description of a Peace System which must replace it if we are to save planet Earth.[60] So for Ecocivilization to be born, we will have to outlaw war, on the surface a seemingly impossible task. But unknown to the casual observer, we have been moving in that direction for over a hundred years. In fact, we are living in the midst of a little commented on peace revolution. Some twenty historic trends point in that direction. They are:

- The rise of civilian-based peace organizations, beginning around 1815, and blossoming into thousands of them all over the world.
- The development of international courts to adjudicate disputes and so prevent violent resolution, including the International Court of Justice (better known as the World Court), the International Criminal Court, and regional courts.
- The development of international organizations to develop, keep and restore peace including the League of Nations, and then the improved version, the United Nations, and further efforts to improve upon it.
- The rise of neutral, armed peace keeping: e.g., the UN Blue Helmets and others.
- The development of the philosophy and techniques of non-violent struggle, by Gandhi, King, Sharp, and many others, as a demonstrably practical strategy to replace violence.
- The development of civilian-based, non-violent peace keeping, e.g., The Nonviolent Peaceforce and Peace Brigades International.
- The rise of the human rights movement as enshrined in the Universal Declaration of Human Rights.
- Decolonization, freeing millions of people for self-determination after the age of colonization.
- The rise and rapid spread of peace research, peace education and peace journalism.
- The emergence of regions of long-term peace.

- The decline of institutionalized racism.
- The world-wide movement for women's rights.
- The rise of the environmental sustainability movement.
- The spread of peace-oriented forms of religion.
- The legalization of conscientious objector status.
- The emergence of an international development regime.
- The rise of global civil society, tens of thousands of international non-government organizations knitting people together all over the world for beneficial purposes.
- The development of the international conference movement, beginning with the Rio Conference on the global environment in 1992 and with many conferences since.
- The slow but gradual rise of a sense of planetary loyalty, made more obvious by the climate crisis.

None of this has happened before in the history of the world. While we have a long way to go, we are undeniably on the way.

## Reestablish Thrift

Hypercivilization is a throwaway society. Perfectly good and useful things pile up in our landfills and oceans. The sense of thrift we once had is gone. Our fascination with material goods is in fact very shallow. We quickly use them up or tire of them. If we truly respected them, we would take care of them and, when they began to wear or break, we would fix them. In the early part of the twenty-first century in the U.S. it became common practice to knock down perfectly good houses and build mini-mansions on the same site. In Ecocivilization the old time adages will again apply to our consumer goods: reuse, repair, recycle, restore, rebuild, and renew, and this means that we abandon built-in obsolescence. These concepts also apply to degraded ecosystems including abandoned and vacant lots in our depopulating central cities such as Detroit, Michigan, and farmland degraded with chemicals and erosion. If we are to achieve Ecocivilization in time we all need to be involved, to take responsibility for our children's future.

## Take Responsibility—Demand Accountability

Responsibility and accountability are two sides of the same coin. Conservatives are correct in pointing out that as individuals we need to take responsibility for our lives, and I would emphasize, for the way we impact the commons in which we all live. What is our family's ecological footprint? Do we choose the affluent lifestyle, owning several big cars with V-8 engines, or drive a hybrid or an electric? Do we recycle? Do we live in an adequate sized house or in a mini-mansion? Do we find our satisfaction in consumption or in relationships, the beauties of nature and in spiritual practices? Do we care about environmental justice or does it not matter to us that the poor and people of color are the ones who find their neighborhoods sacrificed to coal plants, toxic dumps, and chemical factories? Do we exercise our political responsibility and privilege to attend to legislation and to the politicians behind it, taking time to learn whether it is environmentally hostile or friendly? Do we take responsibility for creating a more peaceful world?

Responsibility must especially apply to corporations whose standard practice is to externalize the environmental costs of production whenever they can get away with it. We need to pay attention to their lobbyists. We need to pay attention to our elected officials when they are tempted to take money from corporations in ways legal or illegal. We need to demand transparency from our governments and our corporations. The ancient principle of Roman law, "That which touches all is the concern of all," must be the standard by which we judge our leaders political and corporate. It is the cornerstone of democracy. We all need to be held accountable for our actions and more especially our inactions because it is all too easy in this culture to become distracted by drugs, sports, and computer games. Time is short and, in the final analysis, it is the only wealth. We need to spend it wisely. And above all we need to tell ourselves a new story.

## The New Story

How are these general concepts leading us to a New Story that guides us in changing our technology and our institutions (economy, education, governance, religion, etc.)? Like the old and ruinous story, the new story is about where we came from, who we are, and where we are heading. A New Story would tell us that each of us, and all of us together, are inextricably interrelated as the result of a long, long evolutionary adaptation to nature's laws. In a real sense, we are our past. The atoms of which we are composed are billions of years old. Our bodies are mostly sea water, our hand-eye coordination probably evolved before we walked upright in the savannahs. We are unavoidably dependent upon other creatures and the environmental conditions they create, most obviously the photo-synthesizing plants that make our breathable air, and the decomposers who keep the world from getting buried under organic waste. Whatever else we are, we are biological beings. We are also social beings. We cannot live as individuals by ourselves. Not only are we totally vulnerable in our early years, but we all depend on the achievements of countless unknowns who created the civilization in which we are embedded and which is our cultural parent. And by its very definition, civilization thrives mainly on cooperation, not on competition which is, in fact, only a minor theme. And for cooperation to exist, evolution has provided us with reason which makes our mutual dependence obvious to anyone who thinks about it for a few moments. Evolution has also provided us with huge capacities for empathy and compassion, not only for other humans, but for animals as well. Human reason also makes it possible to see ahead in a way that other animals cannot. We can read the possible futures and avoid the coming catastrophe that will be our fate if we do not abandon Hypercivilization and make the transition to Ecocivilization. And that transition is underway.

Sarah van Gelder, editor of *Yes! Magazine*, went looking for this future in 2016. She took a 12,000 mile trip around the U.S. and found what many people are already doing to create what she calls a "Culture of Connections" in the midst of the Old Story's "Economy of Extraction." Here is a summary of how she contrasts the two concepts of community in her 2017 book, *The Revolution Where You Live*.[61]

In a Culture of Connections we live in a restorative economy, that is, we harvest only what is needed and in such a way that we leave our place more alive and healthy than when we found it; soil and water are sacred. We are rooted, that is, nourished by our place and it supports us. We are attached to our homes and resist displacement. The Commons belongs to all beings and generations; it is our job to protect them. By contrast, in an Economy of Extraction, I profit as an individual by what I can take from soils, forests, and waters and these are merely material resources to be used to build monetary wealth. I can live off them until they're gone and then move on. Exiled, I can live anywhere and am at home nowhere. The market determines where people live and who can afford a home. There is no such thing as the commons—everything is best privately owned and available for dumping.

In a Culture of Connections, to be educated means to acquire deep learning about our place including those aspects of no apparent economic value to humans. By contrast, in our dominant Economy of Extraction, learning means to know how to extract wealth from humans and the earth. Love of place is dismissed as a romantic myth by the dominant culture because it can interfere with extracting resources from earth in order to build wealth for the already rich. In the one, creating sustainable, local jobs is seen as the bedrock of a vibrant community, while in the other, community has no economic value and it is better to replace workers with technology or move the jobs on to wherever in the world labor is cheapest at the moment. Real wealth is money that can be extracted and concentrated in a few hands as a reward for merit, and can then be invested in more extractive industries. In a Culture of Connections, real wealth is human, community, and ecological well-being. Money is widely distributed and circulates locally. An economy of cooperation creates a place for everyone, whereas the dominant, corporate economy creates big financial rewards for a few winners. Money earned in the community is moved out as fast as possible to big financial institutions in distant cities. Money comes first, whereas in a Culture of Connections, relationships come first—family, community, place, and environment. Investment is in infrastructure, the commons and the well-being of the whole. All costs are internalized, unlike the Economy of Extraction in which money is placed in abstract financial instruments that create private gain for a few while costs are externalized to everyone else. Oil and gas are profitable, to be got out as fast as possible and then move the enterprise on to the next global location. In the Culture of Connections the goal is sustainable prosperity of a particular place so that the full range of human and economic values are maintained.[62]

Transforming our societies into van Gelder's Cultures of Connection, or Ecocivilization to use my term, is what Thomas Berry called our "Great Work." It begins with our ideas and values as discussed here and so eloquently put in the preamble to the Earth Charter.

We stand at a critical moment in Earth's history, a time when humanity must choose its future.... To move forward we must recognize that in the midst of a magnificent diversity of cultures and life forms we are one human family and one Earth community with a common destiny.[63]

As we change our minds along these directions, we will be able to derive new institutions and patterns of living, and from those we will know what are the appropriate technologies to adopt.

# 13

## Changing Education, Food and Industry

"It's not too late at all. You just don't yet know what you are capable of."—Mahatma Gandhi

We need to redesign education, the food system, and systems of industrial manufacturing. The authors of *The Natural Step: How Cities and Towns Can Change to Sustainable Practices*, suggest four "system conditions" to guide us.[1]

- "In the sustainable society, nature is not subject to systematically increasing concentrations of substances extracted from the Earth's crust."
- "In the sustainable society, nature is not subject to systematically increasing concentrations of substances produced by society."
- "In the sustainable society, nature is not subject to systematically increasing degradation by physical means."
- "In the sustainable society, human needs are met worldwide."

As Martin Luther King, Jr., said: "We must learn to live together as brothers or perish together as fools."[2] It is as important to protect distant peoples' lands from climate deterioration as our own, wherever "we" happen to live. We are a single people on a totally interconnected planet.

### Changing How We Educate

As Wendell Berry has noted, most of the damage done to planet Earth has been done by well-educated people, experts, and that we are turning out "itinerant professional vandals," and ecologist David Orr writes that we "unleash" on the world "minds ignorant of their ignorance."[3] We need a different kind of expertise because our present education is radically dysfunctional. Currently the goal of most education is to fit graduates into narrow slots in the growth-oriented, corporate production and consumption machine of Hypercivilization. We justify this by the imagined need to compete in the "global market" and other world destroying nonsense. The world does not need nor can it tolerate any more MBAs in finance. In Ecocivilization, education will be about the whole Earth community, especially as it exists in the locale where the students reside. The goal of education will be to know the Earth intimately and to live sustainably with it and at peace with one another.

Most of us who realize something is radically wrong with Hypercivilization had very early experiences in which adults who loved the natural world and were awed by it took us outside to explore or just to enjoy unstructured play. For each of us there is a meadow, a garden, a woodland, a beach, or a streambank where we first experienced wonder. Maybe it was just a backyard where we found a fossil or first saw the beauty of a butterfly. Getting children outside is critical. I recall my mother telling me that she was near-sighted as a child, and when she got glasses she was almost stupefied looking at the night sky, and saying to herself in wonderment, "So there really are stars." It's that wonderment children need to experience.

Richard Louv, author of *Last Child in the Woods*, writes that we need "to bring about a world in which we leave no child inside."[4] Outdoor experience away from LED screens needs to be a major component of education *at all levels but especially in the early years*. Keeping children in and giving them baby laptops and baby cellphones will not aid them in growing up to be lovers of the natural world or even very smart and certainly not wise. At the Fiddlehead Forest School in Seattle children spend four hours a day, rain or shine, in cedar grove "classrooms" under the big trees at the University Botanic Gardens. The natural setting provides the basis of the curriculum. Similar preschools operate in San Diego, Midland, Michigan, Lincoln, Massachusetts, and elsewhere. The number of these schools is growing; by 2016 there were ninety-two, up from 12 in 2008.[5] In 2016, Oregon approved funds to enable all children to have some outdoor schooling.

Once in school, we should turn to a more structured, age-appropriate learning organized around the environment. Over the last forty years environmental education has been widely established in the schools and colleges, but too often as an add-on to the curriculum. We study economics here and ecology over there. Just inserting a course in ecology will not reveal the whole complex problem of the interaction between ideas, values, technology, economics, government, and nature which together make up the overall biosocial system in which we live. A holistic and systemic approach can make Earth the core around which all learning takes place.

**Outdoor Education: Hypercivilization contained its children in-doors for an education focused primarily on human civilization and on an abstract and materialistic approach to science. Today we are seeing more and more children being taught out of doors, engaging in hands-on learning about the juicy biosphere in their locale (Rawpixel.com, Shutterstock).**

Our graduates need to learn to think in terms of large, complex systems. As Paolo Lugari, founder of the groundbreaking eco-community of Gaviotas in Colombia, said: "The world has too many specialists. We need more generalists who can see all the connections and possibilities."[6] Louv is also critical of "an overly abstract science education" that focuses on such things as microbiology, or studying tropical rain forests via computer without learning anything first-hand about the woodland just outside the school.[7] A graduate of a nature-based curriculum should know about the importance of distant rain forests and coral reefs, but also be able to answer Berry's questions about her own locale.

We need to abandon the test-based education that leaves little room for hands-on experience, imagination and creativity. We need to jettison our over-reliance on educational technology such as computers because they turn nature study into an abstraction and limit the child's experience to pale imitations of the real world. A picture of a rain forest is not a rain forest, and a chart showing rising $CO_2$ levels engages only the right side of the brain. It would be wise to put off computers in the classroom until high school. Children need to be able to engage their senses—to touch as well as see, to smell and to hear the sounds of living nature. They need to know the names of things. Biologist Elaine Brooks says: "humans seldom value things they cannot name."[8] I am continually surprised at how many of my friends do not know the names of trees they have lived under for decades or the names of the local butterflies and birds. Their parents did not take them outside and teach them these things, nor did their schools.

Furthermore, it is critical to not expose children to the great damage already done to nature before they have a chance to develop a sense of wonder at nature and a love for it, lest they think it is so damaged they just give up on experiencing and preserving it. Lessons on the damage of Hypercivilization and a critique of it are necessary at some later stage in a child's education. We also need to abandon the individualized competition for grades. Learning needs to be done in small, cooperative groups that are the training ground for community participation and which give real-time lessons in the reality of interdependence.

## Re-Skilling

Hypercivilization has rendered most of us helpless. If we are to make our local communities resilient, we need to re-skill, learn to use basic hand and power tools, to garden, to preserve and cook food, to do basic carpentry, wiring and plumbing, to make clothes, care for animals, and entertain ourselves without the use of electronic media. We will need some people who know the care and working of horses. We need to learn about how others have lived in the past and are living now. We also need to study those disciplines which de-parochialize us, which liberate us from thinking that our way is the norm for all humans and that we are somehow exceptional. That requires studying history, anthropology, cultural geography, sociology, comparative religions, art, literature and languages. And in order to know how to live at peace with one another, children need to attend to the discipline broadly known as peace studies including conflict resolution, non-violent communication, group dynamics, ecological justice, peace history and peace systems. We require a total revolution in education from earliest childhood through graduate school and on into life-long education, a curriculum and methodology that will provide the foundation of Ecocivilization.

## Fixing the Food System

As late as 1922 in the U.S. "localized food systems were still intact and vibrant, and a centralized, over-industrialized food system was not necessarily inevitable."[9] We chose the wrong path and the consequences can be summed up in one word, erosion, "the erosion of topsoil, values, communities, and even democracy."[10] A healthy and successful food system will be characterized by soil health, and clean surface and ground waters. There are three paths to this outcome. One is eliminating artificial fertilizers and herbicides and returning to organic systems. The second is changing the method of tillage. Using seed drills and low tillage methods leaves some organic cover on the land after harvest and does not leave the soils exposed to the forces of erosion, but there is a problem with this. Under no till, heavy doses of herbicide must be applied. Better to plant cover crops and till in the spring if feasible. Another and still experimental technique is to switch from annual grains which require tillage and herbicides to hardier perennial grains which do not. Wes Jackson is experimenting with this at the Land Institute. Perennials are also better for the soil since their roots go far deeper and can both sequester carbon more deeply and draw up nutrients untapped by conventional crops.[11] Jackson's Sunshine Farm Project in Kansas uses renewable energies including biodiesel for the tractors, solar and wind power, and draft horses.[12]

In a successful food system most food will be grown as close to where it is consumed as possible, providing fresher food and reducing its carbon footprint. Today in the U.S. food travels on average about 1500 miles to get to one's dinner plate.[13] In an Earth-sustaining food system, the role of the middle man, which has become dominant and controlling, would be reduced and we would see more food going direct from farm to consumer and to local groceries and a greater share of the income to farmers. Farmers markets and CSAs are important, although much of their energy-saving value is wiped out if the customers drive individual cars to and from the market. It would be more efficient for the local growers to distribute food from a single truck, unless, of course, pick up by the customer were combined in a once weekly trip to get all other necessities.

In Ecocivilization we will eschew giant, mono-cropping farms. The demise of the small, mixed farm in favor of giant agri-business operations which began in the 1950s was foolish as Berry points out.

> Once plants and animals were raised together on the same farm—which therefore neither produced unmanageable surpluses of manure, to be wasted and to pollute the water supply, nor depended on such quantities of commercial fertilizer. The genius of American farm experts is very well demonstrated here: they can take a solution and divide it neatly into two problems.[14]

Instead, food will again be produced organically on small, mixed farms where a portion of the land will be left wild to gain the ecosystem services that wild places provide to farming, such as habitat for pollinators, groundwater conservation from leaving potholes and wetlands, and other forms of synergy. The ideal model for this synergy is permaculture. "Permaculture is a philosophy of working with, rather than against nature; of protracted and thoughtful observation rather than protracted and thoughtless labor; and of looking at plants and animals in all their functions, rather than treating any area as a single product system."[15] In a sustainable society Concentrated Agricultural Feeding Operations (CAFOs), where animals often stand in their own manure and require heavy doses of antibiotics, and which draw huge amounts of water, often drying up local wells and foul the air and water for miles around, would be out-lawed.

A successful food system will increase the number of grain and vegetable varieties by returning cultivation to many hardy traditional seeds, reversing the mistake of the Green Revolution which drove these locally adapted varieties out of the field. It will provide healthy, fresh food rather than the over-packaged, over-salted, over-sugared and over-preservative-laced edible *food-like products* that now dominate the center section of every supermarket. It will be democratic; a few giant industries will not dominate it by monopolizing the production of chemicals and equipment. Their practice of patenting food genes will be illegal. Growing food is a human right. It will support the revival of many once resilient, economically healthy rural communities. At the same time it will provide food security by encouraging not only small scale local production but medium scale regional production in case of local drought or flood, and it will ensure that all sectors of the population have access to healthy food.

One of the obstacles to returning to a healthy food system and healthy rural communities is the concentration of land into the hands of relatively few big farmers and agri-business corporations that are currently mono-cropping with heavy chemical inputs and carbon and soil outputs. If we are to get more people, and especially younger people back on the land, we must somehow make land available to those who want to carry out mixed, organic agriculture and to produce for local markets. Government programs to buy up the bigger farms as the aging population of owners retires could be used to make land available in smaller parcels to younger, landless families. That requires providing government assistance to newly trained young farmers so that they can buy land and appropriate technology, small-scale equipment, and livestock. Low- or no-interest financing and outright grants will be necessary. Big private foundations could help with funding. We will also end all subsidies to big corporate farms, de-incentivize them especially for ethanol production. Giant agri-businesses need to be broken up as unfair monopolies. Their rationale for existence and for their methods is that they are "feeding the world," a claim only marginally true. While it is true that we must retain some capacity for moving large quantities of food globally in case of regional famines, it is far better to and to assist local communities around the world to produce their own food. Also, university agriculture and extension programs must break the funding leverage held over them by "Big-Ag" so they can assist newcomers, linking them up in mentoring programs with farmers already practicing permaculture, agro-forestry and other sustainable methods.

The dumping of commodities by big Northern producers which has destroyed many a local food industry in the Global South must also come to an end. Food assistance programs will need to include aiding local farmers in rebuilding once thriving agriculture in their own countries. The cultivation of export crops such as flowers and palm oil in countries which are experiencing food shortages will cease and those lands return to food production for the local inhabitants. Apart from luxury items that can only be grown in certain locales, like coffee, tea, and cocoa, countries that are importing food ought not to be exporting it. We need to acknowledge that there will be losses here—only very lately in history we have come to expect fresh fruit year round. Perhaps we should adopt the maxim that seasonal is reasonable. Also, when we have achieved a truly functional food system there will be no more food deserts in our central cities. Furthermore, food production and marketing will be transparent—no one will need to wonder what is in their food or in its packaging. Workers in the food industry—especially field workers—will be protected from harmful chemicals and dangerous working conditions and will be properly paid.

Great gains can be made by changing diet. In a healthy food system, the affluent Northerners would greatly reduce the amount of meat in the diet, returning to the level considered normal in the 1950s. This would open up far more land for the production of cereal grains and vegetables, hence feeding many more people, and result in reduced rates of heart disease, colon cancer and other maladies. Over the last couple of decades we have seen important trends in reduced meat intake, including vegetarianism and veganism, which relieve the pressure on the world's poor and on the land. If everyone in the world is to eat well, some of us need to eat lower on the food chain. Also, a healthy food system would not replace traditional cuisines with fast food. Philip Ackerman-Leist's *Rebuilding the Foodshed, a Project of the Post Carbon Institute* is hopeful: "I believe the prospects for positive change are remarkably encouraging.... I see more reasons for optimism in the next half century than what I have seen and experienced in food and agriculture this last fifty years."[16]

He is encouraged by the rapid growth of the organic and local food movements where the old ways had given over to the gigantism of industrialized agriculture and to marketing that unleashed the "hounds of homogenization"[17] upon our food supply. A revolution is underway. Community Supported Agriculture and farmers' markets are the most rapidly growing segments of the food system. So, too, is the revival of urban agriculture, or: "the practice of food production within a city boundary or on the immediate periphery of a city, including the cultivation of crops, vegetables, herbs, fruit, flowers, orchards, parks, forestry, fuelwood, livestock, aquaculture, and bee-keeping."[18]

The opportunities for growth are almost limitless, especially in suburban areas where lawns can easily be replaced with gardens, and in decaying urban areas where vacant land has become available. Rohit Kumar writes: "Imagine if we grew food instead of grass. Every community is a local food economy waiting to come to life."[19] Some urban and suburban communities are rapidly changing old ordinances which inhibited urban food production, eliminating such ordinances favoring lawns only and prohibitions on raising backyard chickens.

One of the most encouraging developments is the emergence of community gardens that serve not only as sources of food but as educational and community resources. D-town Farm "is a seven acre organic farm in Detroit's Rouge Park where they grow 'more than 30 different fruits, vegetables and herbs that are sold at farmers markets and to wholesale customers. The farm features four hoop houses for extended-season growing, bee-keeping, large-scale composting, farm tours and an annual harvest festival.'"[20] It is part of the Detroit Black Community Food Security Network and was inspired by Will and Erika Allen in Milwaukee, Wisconsin, whose pioneering work has been the model for many urban agricultural projects in the U.S. Another encouraging development is the appearance of rooftop gardens. McCormick Place, a giant convention center in Chicago, boasts a 20,000 square foot garden growing produce for the convention goers. They plan to expand it to 108,000 square feet, which would make it larger than the Brooklyn Grange's 2.5-acre rooftop garden in New York City.[21] Rooftop gardening also mitigates temperature extremes in the buildings. Vegetables could also be produced in urban high rises using hydroponic methods. Even the ultra-conservative United States Department of Agriculture is getting involved, committing small sums for research and to encourage urban farming.[22]

Finally, a healthy food system will address the shockingly high levels of waste in the current system. Worldwide, as much as one third of the food (1.3 billion tons annually) produced is never eaten. It's forty percent in the U.S. The two kinds of waste are loss—food that never makes it to the plate—and food that is prepared but not eaten. Both the nutrition and the embedded energy are lost as most of this food ends up in landfills, losing also the benefit of composting. Ironically, it's the most developed and technologically advanced societies that waste the most.[23] Cities could pick up the organic wastes and turn them into compost to substitute for chemical fertilizers or into "engineered soil" that can be used on rooftop gardens. It can be used in biogas generators to produce methane, a clean burning fuel, to produce electricity, and for municipal heating. Portland, Oregon, is starting this practice. Food is also wasted in restaurants where it is hard to predict how many customers will appear for any given meal, and any unused but perfectly good food must be discarded to comply with legal sanitary requirements. Perhaps these need to be relaxed. The other great waste stream coming from the food system is packaging. Staying out of the center of the super market will help here but not entirely. Not everything needs to be in plastic. Even small changes at the individual level can make big changes at the macro-level, such as bringing cloth bags to the supermarket. Overall, the leading principle for our food system needs to become soil to soil, what is being called a regenerative food system. It is the same idea forward thinkers are advocating for manufacturing—close all the lines into loops, or cradle to a cradle production system that moves from fabrication to use, to recovery and remanufacture.

## Fixing the Industrial System: New Ideas—New Technologies

Our industrial and manufacturing systems are outdated and run on a dangerous source of energy. They rely on the one-time extraction of finite supplies and on a throughput mentality that creates waste. We will need to radically change our technology. Finally we must keep in mind the four system conditions. These will be hard standards to achieve, but they point us in the right direction.

There are several ways that manufacturing can be made sustainable. The first is for corporations to follow "the principle of the triple bottom line," a phrase coined in 1994 by British consultant John Elkington.

> His argument was that companies should be preparing three different (and quite separate) bottom lines. One is the traditional measure of corporate profit—the "bottom line" of the profit and loss account. The second is the bottom line of a company's "people account"—a measure in some shape or form of how socially responsible an organization has been throughout its operations. The third is the bottom line of the company's "planet" account—a measure of how environmentally responsible it has been.[24]

Corporations will only pay attention to what they measure, so measuring their impact on people who work for them and who live around their sources of supply and their manufacturing plants and the public in general, including the future generations, will at least enable them to consider mitigating these impacts. Without the measurement they would be ignored. The same is true for environmental impacts all along the chain from supply to manufacturing, marketing and eventual disposal. Just as governments now require corporations to exercise due diligence with their investor's money, requiring them to do everything legally possible to ensure a return on investment, they could require due diligence in pursuing the triple bottom line. To some extent in the Global North they already do by means of labor and environmental regulations, but these are often avoided by sending manufacturing to countries which ignore such constraints. One strategy to counter that is the Fair Trade Movement in which companies certify that their products were not produced in socially or environmentally harmful ways, although it's largely confined to a few commodity trades such as coffee, bananas, tea and crafts. Adopting triple bottom line business model relies entirely on the good will of corporate boards; without government intervention for the common good, it will remain a minority practice. Since corporations in the U.S. are chartered by States, one path to making corporations accountable is to withdraw the charters and reissue them with triple bottom line requirements. Of course some industries simply have to cease operating including fossil fuel extraction and the production of many toxic chemicals.

One change that adopting triple bottom line would stimulate is the necessary reformulation of much of our industrial chemistry into harmless, biodegradable substances. Too much that is manufactured relies on fossil feedstocks, in particular the ubiquitous plastics which are formulated specifically to not biodegrade. Many manufacturing processes require the use of highly toxic chemicals. Ecocivilization will accept the challenge of completely reformulating our industrial chemistry and what can't be reformulated or substituted for with less harmful materials will have to be abandoned. There is precedent in this regard; for example, most countries have banned DDT, PBCs, CFCs and many other harmful chemicals. Ecocivilization will adopt the precautionary principle regarding any new chemicals.

Ecocivilization will adopt "Cradle to Cradle" manufacturing. (The term is a registered trademark of McDonough Braungart Design Chemistry consultants.) Also known as C2C, it was articulated in 1970s by Walter R. Stahel and popularized by McDonough and Braungart in their 2002 book of the same name. C2C is an industrial model that:

> ...seeks to create production techniques that are not just efficient but are essentially waste free. In cradle to cradle production all material inputs and outputs are seen either

as technical or biological nutrients. Technical nutrients can be recycled or reused with no loss of quality and biological nutrients composted or consumed.[25]

Cradle to Cradle has implications beyond manufacturing:

> It models human industry on nature's processes viewing materials as nutrients circulating in healthy, safe metabolisms. It suggests that industry must protect and enrich ecosystems and nature's biological metabolism while also maintaining a safe, productive technical metabolism for the high-quality use and circulation of organic and technical nutrients … [and] it can be applied to many aspects of human civilization such as urban environments, buildings, economics and social systems.[26]

Some industries such as carpet manufacturing are already employing this model, saving hundreds of thousands of tons of landfilled old carpet by using it as feedstock for new rugs. Other companies recycle old plastic water bottles into carpet and other flooring. In fact:

> Companies around the world employ C2C methodology to improve their products. Some of the more well-known brands include Ecover, a Belgium-based brand making sustainable dish soap, laundry soaps and other household cleaning products. Method, another cleaning product company, also uses C2C to create their extensive line of cleaners, soaps and products. There are dozens of other companies with building materials, home design and clothing doing manufacturing better with C2C principles.[27]

The *Cradle to Cradle Products Innovation Institute* certifies that products are indeed made to C2C standards. Certification ensures the following principles are met.

- Material Health: Value materials as nutrients for safe, continuous cycling
- Material Reutilization: Maintain continuous flows of biological and technical nutrients
- Renewable Energy: Power all operations with 100% renewable energy
- Water Stewardship: Regard water as a precious resource
- Social Fairness: Celebrate all people and natural systems[28]

C2C and the other strategies noted here constitute a revolution already underway. Another revolution is underway in energy, which we take up in the next chapter.

# 14

# The Energy Revolution
# Is Underway

Our current energy system is leading to a climate catastrophe and the possible collapse of civilization. Even if it were not, the health hazards of burning coal and other fossil fuels would make it imperative to switch to clean and sustainable energy generation. The good news is that the revolution is underway. The more sobering news is that it is in its infancy and if it is not rapidly accelerated it will be too late to avoid calamity, and doing so will require a systematic, comprehensive response by governments. This is not to say that there isn't much individuals and families can do. There is, and these measures are described below, but individual solutions are not enough to address systemic problems.

A new and sustainable system must satisfy seven principals. First and foremost, they must head off the worst effects of climate deterioration. Second, upstream solutions are always preferable. Stopping emissions before they start makes sense. Third, renewables are obviously preferable as long as they are clean. Fourth, any solutions must not impair human health or the health of ecosystems including water supplies. Five, new energy generation should be dispersed, a 180 degree turn from our current fossil-based system. Six, they must not place a burden on future generations. Seven, they must come on line as fast as possible, meaning they will require government stimulation. The foremost concern now is preventing a climate catastrophe.

First we will discuss the non-starters: fossil fuels (including so-called "clean coal"), mechanical carbon capture, nuclear energy, certain forms of biomass, and hydrogen fusion. Then we will look at the off-the-shelf, ready to go alternatives including first and foremost, conservation, followed by wind, solar, and the electrification of transportation including light rail, then hydro, tidal and geothermal power and a smart grid to maximize the efficiency of all these systems. Finally, there are natural methods of carbon capture by means of forestry and agriculture whose implementation can be greatly accelerated.

## The Failure of Fossil Fuels and Mechanical Carbon Capture

Fossil fuels result in air pollution, compete with water, destroy habitat as in mountain top removal and open pit mining, and are the main source of greenhouse gas emissions. Robert Pollin lays out the challenge here in *Greening the Global Economy*: "The Intergovernmental Panel on Climate Change (IPCC) estimates that to stabilize the global average temperature at its current level of around 60.3° Fahrenheit ... emissions will need to fall 40 percent within twenty years to 27 billion tons annually and 80 percent by 2050 to 9 billion tons."[1] Pollin's is a conservative estimate, and other say we need to do far more.[2] Therefore, most of the coal and oil needs to be left in the ground, especially tar sands, the dirtiest fuel whose mining destroys vast areas of forest that function both as an energy sink and crucial wildlife habitat. The same is true of mountaintop removal for coal.

There is no such thing as clean coal. All coal mining yields huge amounts of overburden which have to be dumped on lands that could be wildlife habitat or food producing land, and these overburdens often contain toxic materials. Mountain top mining destroys streams and watersheds. All coal has to be washed, making it a competitor for scarce water and leaving a dirty residue that has to go somewhere, and when burned, coal leaves a toxic ash that pollutes groundwater. Also, deep shaft mines have to be pumped out, requiring energy which, if it is fossil fuel, decreases the net energy gain and if it's clean power is an oxymoron—i.e., using renewable energy to create non-renewable dirty energy. Coal is also responsible for unacceptable levels of respiratory and heart disease. Nevertheless, the energy industry is touting a tech fix, mechanical "carbon capture" or "sequestration," i.e., capturing the $CO_2$ either at the smoke stack or by means of giant machines that suck it out of the air and then pipe it into deep underground repositories. This fails because it is a downstream solution which allows the habitat destruction that inevitably comes with mining and drilling to continue.

Using mechanical capture as a rationale for continuing the production of oil fails because it stimulates demand as low prices encourage people to continue buying gas-guzzling vehicles. Continuing high profits for corporations with a sordid track record in delaying climate action by confusing the public about climate deterioration and who have a vested interest is in prolonging the fossil fuel age by lobbying politicians, is naive social policy.

Third, the opportunity cost of scaling up large, mechanical carbon capture machines to tens or hundreds of thousands placed around the world would deprive the economy of capital to invest in upstream solutions such as renewables and conservation. Fourth, the machines themselves use energy to power giant fans and the chemical extraction process requires "proprietary chemicals" the industry is unwilling to tell the public about. Pollin lists other objections including: the technology renders the cost of a power plant 76 percent higher to build than coal-fired plants without it; timing—there isn't a single full-scale plant on the planet and most governments and utilities have scaled back plans to build them; scale of sequestration—the amount necessary to put it into the ground would equal the amount of oil and coal now coming out of the ground and thus require an enormous infrastructure; permanence and transparency including the danger that the sequestered $CO_2$ could escape—creating a huge spike in emissions, thus requiring a huge government monitoring apparatus to make certain it was not escaping.[3] And if it did escape, it would constitute a carbon bomb that would rapidly spike the level of $CO_2$ in the atmosphere. Mechanical carbon capture is a non-starter.

In the U.S. the federal government attempted to shut down the dirtiest coal fired utilities and in 2016 President Obama moved to end new coal leases on federal lands and to increase the fees for existing ones, as a source of revenue to be used in mitigating the impact on mining communities.[4] Unfortunately the Trump Administration reversed these policies and encouraged coal production. It was also friendly to big oil interests and opened vast public lands to mining and drilling for fossil fuels. In 2017, Rex Tillerson, a former CEO of the global giant Exxon-Mobil, was chosen as Trump's first U.S. Secretary of State, Scott Pruitt, who had repeatedly sued the federal government on behalf of the fossil fuel corporations became head of the EPA and was succeeded by Andrew Wheeler, a coal company lobbyist.

Oil still has its advocates—the big corporations that have unwisely invested huge sums in reserves still in the ground. While not as dirty as coal, it's still a dirty fuel accounting for 31 percent of all energy produced and contributing a huge share of greenhouse emissions. Oil also needs to stay in the ground which, however, will make these fuels stranded assets, severely affecting the values of these outmoded energy corporations. It could lead to what some industry analysts are referring to as the "bursting of the carbon bubble."[5] Still, overall global demand for oil is still climbing, although as of 2016, there was a temporary glut caused by overproduction of shale oils in the U.S. Nevertheless, in some leading oil users consumption is actually declining. Germany, which peaked at 3.3 million barrels a day in 1979, was down to 2.4 million in 2013. Japan was at 5.8 in 1996 and fell to 4.6 in 2013 and the U.S. fell from a high of 21 million in 2005 to 19 million in 2013.[6]

## Natural Gas

Natural gas is being touted as a "bridge" to renewables but we need to recall there is no such thing as a clean fossil fuel. While natural gas emits about half the pollutants as coal, using it as a so-called "bridge fuel" involves us in emissions that will continue for at least fifty years as the fossil fuel companies and utilities will need to get their investments back on the heavy infrastructure needed to recover and distribute the gas. Furthermore, the current natural gas boom is the result primarily of hydraulic fracturing which creates three problems of its own. It pollutes the ground water and some of the waste well contaminants are carcinogens. The wells and pipelines leak methane, in some cases in the Bakken Shale fields in North Dakota, up to 10 percent. Since methane is a far more potent greenhouse gas than $CO_2$, the result "would be worse than burning coal" according to Pollin.[7] Furthermore, even switching fifty percent from coal to natural gas would provide only very modest gains, reducing $CO_2$ emissions by only 8 percent.[8] Finally, the extraction of natural gas requires the drilling of many, many wells but the supplies in each tend to be quite limited so that, as Lester Brown writes: "the new wells are depleted quite rapidly..." and therefore "it makes little sense for society to invest in expanding gas infrastructure and then have to abandon it."[9] Fracking is also associated with other problems including earthquakes, and frack sand mining pollutes air and water with tiny silicate particles that get in the lungs. The sand mines also destroy agricultural land. The whole fossil fuel business belongs to a previous age.

## Nuclear Energy

Many people, including some environmentalists, are advocating nuclear energy as the solution or at least as a bridge to clean renewables, but nuclear has its own insurmountable problems. Currently it supplies only 4.8 percent of global generation and would have to be ramped up on a colossal scale to become relevant. However, nuclear generating plants take a long time to build and are horrifically expensive, much more so than clean alternatives that are now on the shelf. Even if we started to build them now, they would not come on line in time to ameliorate the climate crisis. They also fail as a solution because they perpetuate a highly centralized, undemocratic power source. Nuclear plants are not fossil-fuel free. They depend on large expenditures of fossil fuels for fuel processing, transport and construction. They have their own large, carbon footprint. Further, they leave us with horrific toxins lasting ten times longer than civilization has yet been around and with useless, heavily irradiated structures that will need to be dismantled and buried at an astronomical cost per plant. Then there is the risk of the next Fukushima type meltdown. Perhaps the most telling argument against them is that as the world continues to warm, they will become inefficient and even unusable. This is because they rely on water for cooling to prevent melt down, and already some plants have had to be powered down or even shut down because the water they draw from rivers is too warm to safely cool the reactors.[10] Some proponents are arguing that we can avoid this problem by building liquid sodium cooled fast-breeder reactors, but liquid sodium bursts into flame on contact with air and explodes on contact with water, a highly dangerous technology to surround a nuclear reactor with. Furthermore, fast-breeder reactors produce bomb-grade plutonium. Like all nuclear reactors, they complicate the problem of nuclear weapons proliferation and are prime targets for terrorist attacks. Finally, nuclear power just doesn't make sense. It is an incredibly complex and dangerous technology just to heat water into steam. Heating water in this way is just silly. Expanding nuclear is a non-starter.

## Biomass

Wood is a renewable fuel, and while it is local, it is a high emission renewable, putting $CO_2$ and other pollutants into the air, and whether it's firewood or pellets in a home furnace or wood chips in a big utility, it's still wood. Another objection is that to burn wood on any meaningful scale would result in the loss of trees which are a natural carbon sink, a case of robbing Peter to pay Paul. Another non-starter that has unfortunately already started is corn ethanol which is an energy sink and; over a thirty-year cycle it is equivalent in terms of emissions to burning gasoline.[11] Growing corn for ethanol competes with food production, raising the price of bread in parts of the world that cannot afford to pay more for food. The uprisings of the Arab Spring were caused in part by a spike in food prices. And corn is not an energy dense fuel. Lester Brown reports that "filling the tank of a large SUV one time with ethanol takes as much grain as could feed a person for a whole year."[12] If, as some argue, we should clear forest lands, we have already seen the reason that makes no sense. Advocates say we can replant but opponents are skeptical, saying that "achieving that sort of balance is almost impossible, and carbon-absorbing forests will ultimately be destroyed to feed a voracious biomass industry fueled inappropriately by clean-energy subsidies. They also argue that, like any incinerating operation, biomass plants generate all sorts of other pollution, including particulate matter."[13] While other sources of biomass such as rice hulls can be used, it seems to be a non-starter on any meaningful scale.

Burning other forms of biomass, such as algae, is another alternative to fossil fuels. Over a hundred labs are working on perfecting algae as fuel, trying to get the most robust strains either by traditional selection or by genetic engineering. It can be grown as a crop and then sent to refineries as oils and lipids for conversion into jet fuel and biodiesel. The benefit side is highly touted. Per acre it produces up to ten times the biomass as corn, could be grown on arid land or in brackish waters, and it sequesters $CO_2$.[14] One lab has already developed an herbicide resistant strain: "Like the widely grown Roundup Ready soybeans, these algae are resistant to the herbicide Roundup. That would allow the herbicide to be sprayed on a pond to kill invading wild algae while leaving the fuel-producing strain unhurt."[15] That would allow for even more of this possibly carcinogenic and wildlife-destroying pesticide to be sprayed all around the world. But as with all technologies, there are major risks. What if a particularly robust strain escaped, a likelihood given the scale on which it would have to be produced? Our prior experience with escapees from lab, factory, field and fish pond have been devastating. Would it out-compete all the other strains of algae? Would it multiply in every ditch, stream, and lake? Large algal blooms deprive water of oxygen resulting in fish kills and creating toxins dangerous for human health. What about the ecological impacts of turning coastal wetlands and inland ponds into commercial algae farms? Innovations of this magnitude need to conform to the precautionary principle, especially when narrowly trained scientists get over excited about linear solutions.

## Hydrogen Fusion

Hydrogen fusion is currently a fantasy technology not now feasible. Nuclear fusion is what happens in the interior of the sun. It requires super-heating hydrogen and other gases. If it were ever perfected, its advocates claim is that it would provide us with unlimited power too cheap to meter, a claim we heard in the early days of nuclear power which ended up to be the most expensive form of power generation. Furthermore, do we want unlimited energy? Given our track record with the huge increase in available energy that fossil fuels gave us, we do not appear to be ready for unlimited energy. Energy turns nature into things: skyscrapers, open pit mines, automobiles, jet aircraft, highways, a billion I-phones, and so forth, while destroying habitat and degrading the biosphere. I can think of no greater danger to the stability and health of the biosphere on which all civilization rests than a species equipped with hydrogen fusion. Nevertheless, several governments and private research firms are working on it. One of the seemingly insurmountable problems are the temperatures needed to achieve ignition—in one line of research 100 million degrees, in another over 3 billion degrees.[16] Nuclear fission has to take place within a lead lined container but no physical container on earth can withstand the temperatures needed for fusion, so the notion is that they can contain it within a magnetic field. It would require huge, technologically complicated power plants. This is centralized gigantism in order to find a very complex way to boil water. If it ever were to be feasible it would be long after the climate emergency had gone critical and, since we can't let that happen and we already have other technologies in place to produce clean power, why bother pursuing fusion? It is simply another manifestation of the outmoded Faustian conquest of nature philosophy that has created our planetary emergency, as are the so-called "geoengineering solutions."

## Geoengineering

There are many proponents of schemes to alter the atmosphere so as to block out incoming solar radiation and theoretically cool the planet. These include massive cloud seeding, spraying aerosols into the upper atmosphere (supposed to act like the cooling effects of volcanic ash), and even floating a "thin, wide sheet of carbon fiber into a Lagrange point, which is a relatively stable point in the complex system of gravitational pulls between the Earth, moon, and sun."[17] Others have proposed fertilizing the ocean with iron and lime to stimulate the growth of phytoplankton, which absorb carbon and deposit it on the ocean floor, as well as combating acidification, a scheme already outlawed by the UN. The scale on which these would have to be deployed is enormous, and the results speculative, uncontrollable, probably irrevocable, and hence foolish. We don't know enough about the Earth to engage in such reckless behavior. At least one State Legislature agrees. The Rhode Island state legislature passed the *Geoengineering Act of 2017*, which claimed that "geoengineering encompasses many technologies and methods involving hazardous activities that can harm human health and safety, the environment, and the economy of the state of Rhode Island."[18] Simonson asks, "What if large-scale cloud seeding, for example, alters the jet stream and delays the monsoon season across Southeast Asia? What would this do to rice crops? Or what if dumping tons of iron into the ocean destroys the fish population along the coast of Chile?"[19] So if we can't use the fossil fuels or ethanol or hydrogen fusion or geoengineering, what can we do, since we need to avoid climate collapse and we need electricity?

## Modalities for a Practical and Sensible Energy Regime

### Conservation

Conservation involves two principles: using less and using it more efficiently. The world needs to power down if we are to survive. This raises equity questions which we will address shortly, but the first strategy is energy conservation because it is an upstream solution reducing demand and increasing efficiency. Reducing demand requires no technological innovation. For families and individuals, it means combining trips to the stores, shopping once a week instead of every day. It means turning thermostats down and air conditioning up. Most buildings in the developed world are over heated, over lit, over cooled, and under insulated. Heat escapes unnecessarily into the atmosphere. Lights glow in unused rooms; TVs and other electronics are on all the time. Changing thermostat settings by a couple of degrees would provide large gains. As President Carter once said, "Put on a sweater."[20] While the government could regulate some changes like this, and certainly could require them of federal facilities and vehicles, these are efforts for the common good that individuals can achieve without the need for government regulation, something that should appeal to political conservatives. In fact, there are ten changes individuals and families can easily adopt to aid in holding off the worst of climate deterioration. They are: (1) stop wasting food—40 percent of food produced in America is never eaten, the embedded fossil energy wasted. ( 2) Get rid of the lawn. Gas-powered lawn mowers and leaf blowers are the most inefficient gas engines on the planet and fertilizer and pesticides are derived from petroleum stock, just encouraging the drilling. (3) Drive more slowly. Cutting your speed from 75 to 65 or 65 or even 55 saves a huge amount of emissions. (4) Cut down on single-use plastics, also derived from petroleum. (5) Further seal and insulate your home to prevent wasted emissions. (6) Stop buying so much stuff; everything manufactured and transported has embedded carbon in it. While the U.S. has slipped to second place in world emissions, that is only because China is doing much of our manufacturing for us, creating what scientists call "embedded emissions" in all the stuff which would not be produced were we not importing it. Those emissions should really be charged to the U.S. The same is true for the E.U. (7) Take mass transit. (8) Don't fly, or cut down on trips. Air liners are the worst emitters of all. (9) Reuse and repair. (10) Start a garden—whatever food you grow prevents food coming 1,500 miles by air or truck.

Conservation also involves efficiency; that is, designing energy consuming machines such as appliances, power tools, and automobiles to use less energy per unit of output. Fuel efficiency standards are already resulting in some gains. In the European Union:

> The energy consumption of the transport sector has been decreasing quite rapidly since 2007. Around 40% of that reduction is due to the economic recession, with a decrease in freight traffic and the stability of passenger traffic. Almost 60% is due to improvements in energy efficiency, mostly for passenger cars. The average specific consumption of the car fleet decreased from 8.1 [liters] l/100 km in 1995 to 6.8 [liters] l/100 km in 2012 at EU level, thanks to the progress achieved with new cars.[21]

Now we need to convince people to buy these fuel-efficient vehicles.

Most buildings could achieve gains through added insulation and high R-value windows and green roofs. Ford Motor's River Rouge plant "contains state-of-the-art green building features including a 500,000-square-foot green roof, solar panels and fuel cells."[22] Energy Star appliances such as new refrigerators use far less energy than the older models. Replacing incandescent light bulbs with LEDs and many other options are available already. The trick is to greatly accelerate energy conservation through efficiency. Certainly we can expect some relief here from technological innovation. Some companies are even now experimenting with hybrid or even all electric airplanes.[23] But conservation is the quickest change we can make. Large scale investment in energy efficiency our best option available for building a global clean energy economy. The beauty of conservation is that it is an upstream solution—averting the creation of negative impacts rather than trying to clean them up after they have been generated. The technologies for conservation are off-the-shelf and ready now.

We are making some progress. Residential buildings in the U.S. have become more energy efficient and showed a nine percent gain from 1985 to 2004 as a result of efficiencies in heating, air conditioning, refrigeration, clothes washing, small appliances, and the insulation of roofs, walls, doors and windows. Tighter state building codes have played an important part in driving this trend. And there are great gains yet to be made. The U.S. Department of Energy laments that "Using today's best practices, builders have demonstrated that it is possible to design and construct new houses that are 30 to 40 percent lower in energy intensity than a typical code house, at little or no additional cost. Still, such high-performance homes hold a very small market share."[24] In spite of gains from more efficient technologies and houses, the growth in the size and number of households has increased overall energy demand. Economic growth almost always negates efficiency gains. Unfortunately the Trump Administration moved to roll back both appliance and automobile efficient standards.[25] It's a different story with commercial buildings which have shown the opposite trend over time and have become more energy intensive. But conservation is our primary defense against climate catastrophe and everyone has a role to play. However, conservation only works in a steady state economy, otherwise growth wipes out gains and the best we do is to inch up emissions more slowly, which is not good enough.

## Electrify Transportation

Hypercivilization has relied almost totally on fossil powered engines for transportation. Much of the decline in oil demand in the Global North can be explained by savings in the transportation coming from more efficient vehicles including hybrids and electrics. A recent article by Justin Gillis in the *New York Times*, titled "Oil Industry's New Threat? The Global Growth of Electric Cars," points out: "Counting plug-in hybrids, more than 20 electric-car models are on the market. Sales are growing worldwide at a rapid pace: They jumped 49 percent in the first half of 2016 compared with the year-earlier period."[26] Volkswagen projected twenty electric models in its own line by 2020. Toyota, Ford, Chevrolet, Peugeot, Volkswagen and many other manufacturers now offer hybrids, and in some cases, all-electric vehicles. Volvo has set a goal of selling one million electrified cars by 2025, in two hybrids and one electric model. Tesla reported in April of 2016 that it had 325,000 deposits for its new, high-end all-electric S Model.[27] Unfortunately they are having trouble meeting production quotas. Tesla is also building a chain of recharging stations around the U.S. Even powered by fossil fuel utilities, there are gains with electric vehicles, but to maximize gain in reducing emissions, electric vehicles need to be powered by renewable sources of energy. Gillis writes: "Norway, one of the world's leading oil producers, has become an unlikely proving ground for the proposition that electric cars are ready for prime time. Because of big tax breaks, they have come out of nowhere to seize a third of the new-car market in just five years."[28] Unsurprisingly, the big oil companies are belittling the trend because, as Gillis writes, they "are counting on political leaders failing to meet their stated goals for limiting climate change. The stock market seem[ed] to be betting with the oil companies."[29] But their strategy belongs to an age of dinosaurs. Rami Syvari, head of international sales for Fortum Charge & Drive, which is hurrying to install car chargers in Norway, says: "These technology disruptions seem like they are not going to happen at all, and then they happen so fast you can't keep up."[30] Several European countries are considering a ban of fossil fuel driven cars by 2030. Electric cars will be a major component of the transportation system of Ecocivilization.

Another positive development is the rising rates of ridership on mass transit—a 20 percent increase from 2000 to 2011, and to more people working at home. While all this is encouraging, big car manufacturers in the U.S. successfully lobbied the Trump administration to abandon stiffer fuel economy standards and are again focusing on gas-guzzling SUVs and trucks. Nevertheless, some municipalities are electrifying bus fleets. Los Angeles has committed to a fully electric fleet by 2030 and some electric busses are already on the road there. Electric school busses are now available as well. Sacramento, Elk Grove and Twin Rivers Unified School Districts in California have deployed an electric fleet.[31] Many cities, like Madison, Wisconsin, and Portland, Oregon, are also making it easier for people to move about on bikes. These kinds of real world changes indicate that we know what to do, have begun doing it, but need to vigorously accelerate these changes across the board.

## Clean Renewables: Hydro, Geothermal, Tidal, Wind and Solar

Geothermal energy, gained by tapping into heat deep in the earth, is clean. Iceland, China and the U.S. are important users of this energy for space heating, generating electricity and heating fish ponds for aquaculture. To cite a single example, Ball State University in Indiana (U.S.A.) put in geothermal and closed its two coal-fired boilers, cutting its carbon footprint in half and saving $2 million annually.[32] Ground source heat pumps can be used both for heating and cooling and use twenty-five to fifty percent less electricity than conventional sources.[33]

Hydropower is a mixed bag. On the plus side, it tends toward clean power and there is room for expansion since most of the world's smaller dams which do not produce electricity could be upgraded to do so. Also, hydropower operates 24/7 so it could fill in for peak power demands when wind and solar are at low output. Ninety-three percent of the new dams planned or under construction world-wide are big dams but these have problems of their own. They are costly, take a long time to come on line, are subject to both flood and drought which lower their output, and in some cases the decaying organic matter in the lakes behind them release methane which could make them competitive with coal in terms of greenhouse gas emissions. They also displace millions of people, create social unrest and they destroy vast habitats. The really big dams, such as the Three Gorges Dam in China, create such a weight of water in their reservoirs that they are associated with earthquakes. Dams also destroy habitat for fish, a needed source of protein. Finally, dams eventually silt up, reducing pressure head and lowering electric output. Like nuclear plants, big dams have finite life-times. Two other uses of water power include making use of tides and wave currents. These are currently in the experimental stages but hold promise. France has built 240 megawatts of tidal capacity and South Korea 254. The U.S. and Canada have a joint project at the Bay of Fundy and the Scots are planning a 386 megawatt facility in the Firth of Forth. Another experiment with water power relies on the temperature gradient between warm surface waters and colder waters in the deep to generate electricity. China is experimenting with this technology.[34]

Enough solar energy falls on the planet every hour to power the world's economy for a year.[35] Wind and solar are the best energy sources of a future which is already developing rapidly. We see the transition all around us. Unlike the fossil and uranium fired plants, these are dispersed sources of power. Centralizing the generation of electric power in gigantic facilities made some sense when the fuels—coal, oil and gas and then uranium—were located in particular deposits and then had to be transported to the plants. It was inefficient to build a small power plant every few miles. It made more sense to generate it in one central location and then disperse the electricity, although that required huge transmission networks requiring immense amounts of copper which led to large, polluting open pit copper mines that yet scar the land. Also, gigantic installations result in gigantic damage when things go wrong as they inevitably do with human technology, however sophisticated it seems. By contrast, with dispersed power, individual producers can take advantage of net metering; when they produce more than they need the surplus is fed into the grid benefiting everyone. The utility gets power so it will not need to build more expensive generating capacity and the rest of the people get a more reliable source of clean power.

Unfortunately, in an attempt to maintain profits and control, some utilities have been penalizing individuals for installing solar power. This is selfishness in action and it is short-sighted.

In Ecocivilization, power will come from many small sources rather than one big, vulnerable power station. We are not free when we rely on highly centralized corporations. Concentrating electrical power concentrated social, political and economic power, giving occasion for exploitation by the utilities and fossil fuel giants like Exxon-Mobil. What makes the coming era different is that the fuel is already dispersed by nature; wind, and especially sun, are everywhere on the globe. Small wind and/or solar farms can be located in and around villages, and it can be installed on individual dwellings keeping the political and economic power at home, bolstering local control, democracy and the resilience of local economies.

Solar energy can be used for space and water heating and electrical

An Early Solar House: Once prohibitively expense, the cost of solar heating and photo-voltaic electricity generation plummeted in the first decades of the twenty-first century and are now cost competitive in much of the world, to say nothing of their health and environmental benefits (paparazza, Shutterstock).

generation. Passive solar is the ideal solution for space heating and involves designing or retrofitting buildings so they capture the sun's energy through large expanses of south facing glass and store it in the mass of the building to be slowly released at night. Skylights let in natural light, reducing the need for much electrical lighting. Various earth-berming and planting strategies increase the efficiency of a building by conserving the heat absorbed. Just a little short of that is the use of photo-voltaic cells and wind turbines to generate electricity which can be used to power electrical resistance heating, cooling, and stationary and mobile engines. as in electric cars.

Clean renewables are already here. In 2013, Denmark generated 34 percent of its electrical needs from wind and in January of 2014, 62 percent. In 2013 Portugal and Spain were getting 20 percent. In South Australia wind was generating more power than coal. South Dakota is getting over 26 percent of its electricity from wind and Iowa 50 percent in 2018. China has developed over 91,000 megawatts of wind energy and is aiming for 200,000. Wind is providing more power than nuclear and is heating water for 170 million Chinese homes. France is rapidly developing wind power. In the U.S. there are hundreds of utility scale solar farms either built, in the building stage, or planned.[36] Lester Brown reports: "The world-wide use of solar cells to convert sunlight into electricity is expanding by over 50 percent a year."[37] This phenomenal growth is due to the fact that the cost of generating power by sun and wind is plummeting.

Photovoltaic cells were invented as long ago as 1954, but the cost of generating electricity from them was prohibitive. By 1972 it was $74 per watt but by 2014 it had fallen to 70 cents per watt.[38] The cost of solar and wind energy has been declining precipitously in the last decade. In South Australia and South Africa the cost is well below that of coal generated electricity and in Denmark it is at half the cost of coal and natural gas.

In order to maximize gain from clean, renewable electric energy we will need to build a smart grid, "an electricity supply network that uses digital communications technology to detect and react to local changes in usage."[39] The U.S. Department of Energy describes this as:

"Smart grid" technologies are made possible by two-way communication technologies, control systems, and computer processing. These advanced technologies include advanced sensors known as Phasor Measurement Units (PMUs) that allow operators to assess grid stability, advanced digital meters that give consumers better information and automatically report outages, relays that sense and recover from faults in the substation automatically, automated feeder switches that re-route power around problems, and batteries that store excess energy and make it available later to the grid to meet customer demand.[40]

**Wind and solar, once considered prohibitively expensive dreams, are now cheaper than coal and even natural gas in many places and are the fastest growing sector of the energy market (Vaclav Volrab, Shutterstock).**

The renewables revolution is coming on rapidly, as reported in the *New York Times*.

Last year, for the first time, renewables accounted for a majority of new electricity-generating capacity added around the world, according to a recent United Nations report. More than half the $286 billion invested in wind, solar and other renewables occurred in emerging markets like China, India and Brazil—also for the first time.

The average global cost of generating electricity from solar panels fell 61 percent between 2009 and 2015 and 14 percent for land-based wind turbines. In sunny parts of the world like India and Dubai, developers of solar farms have recently offered to sell electricity for less than half the global average price. In November, the accounting firm KPMG predicted that by 2020 solar energy in India could be 10 percent cheaper than electricity generated by burning coal.[41]

A criticism of solar and wind is that they are not constant. Solar does not work at night and wind is variable. Utilities correctly point out that they need to be able to assure base load. Several technologies are developing to deal with these drawbacks. One is to build taller wind generators because winds are stronger at higher elevations, thus getting more generating power from them. Another strategy is to build the turbines offshore where winds are stronger. While this does not solve the problem, it can increase the power available over a wider area as a country which is generating enough to meet its own needs can send that power to another country as Denmark recently did with Germany.

Another solution is pumped storage. During periods of low demand, such as at night or times of moderate temperature, electricity can be used to pump water up to higher levels and then during peak demand it can be run down through water driven turbines. But batteries have the best potential for solving the problem of intermittency. They can obviously power automobiles, but now we are seeing progress in putting batteries in solar homes to provide night time power and even more importantly, teaming battery arrays with grid-scale solar arrays. While this technology is in its early stages of implementation, the *Washington Post* reports that "SolarCity and the Kaua`i Island Utility Cooperative jointly announced last week that they've entered into a solar power purchase agreement in which SolarCity will provide 20 years of power from a 52-megawatt-hour battery installation that will be able to send as many as 13 megawatts of electricity to the island's grid."[42] Up to now big batteries have been expensive and not very efficient, but the same source reports that a number of investment companies are putting money into research into longer lasting, safer and cheaper battery technologies. In 2016 Tesla, a maker of electric cars and storage batteries, joined SolarCity in a project that has installed 15,000 panels and 30 of Tesla's large power pack batteries as a first step in plans to merge the two corporations.[43] And as pointed out, hydropower, tidal and wave power can provide peak time supplements to coastal and riverine cities. The transition away from fossil fuels is underway, impeded only by fossil fuel industry's efforts to hold back the tide in order to preserve the profits of a now obsolete way of life.

## *The Ethical Dimension*

Shutting down most coal and oil extraction will result in economic hardship for many people and communities. While necessary, we cannot advocate such a far-reaching policy without caring for the workers in these industries and the economies of the towns where they live, here in the Global North. Not only is it the ethically right thing to do, if we do not provide such help we will never keep the coal and oil where they can do nor further harm. Therefore, compensation, retraining, pensioning off, and relocation assistance will have to come from the larger community. As for the corporations invested in these fuels, they should not be compensated since some of them, such as Exxon-Mobil and Shell, knew long ago that they were contributing to climate deterioration and kept that knowledge secret while they spent millions on propaganda to confuse the public.[44] And they have made billions on the production and sale of these minerals while accepting public subsidies. The more difficult question is about the relative prosperity—or lack of it—between the North and the Global South where a billion people still live in poverty. As Ted Trainer points out, there are questions whether renewables are adequate.

> The crucial question is can renewables meet the future demand for energy in a society that is fiercely and blindly committed to limitless increases in "living standards" and economic output. The absurdity of this commitment is easily shown. .... If 9 billion people were to rise to the "living standards" we in rich countries will have in 2070 given 3% p.a. economic growth, then total world economic output would be 60 times as great as it is now![45]

Halting overall economic growth with its rising curve of consumerism and waste must be part of the solution, but we can't expect the poorest billion people to continue living in extreme poverty. If we are to have any chance at all to head off climate disaster and other forms of environmental deterioration, the developed world must power down while the parts of the world still experiencing a billion people in extreme poverty need to power up as efficiently as possible. Overall, humanity must cut its carbon use dramatically but the developing societies in which the poorest billion people live without access to clean water, good food, education, and health care have a right to those things and it would be the height of hypocrisy if the developed societies asked them to stay poor so that those who are responsible for the overwhelming majority of climate deterioration could continue to live in over-developed affluence. Even if we were so crass, they would not do it. As Robert Pollin points out: "…there is no reasonable standard of fairness that can justify working people and the poor sacrificing opportunities for rising living standards to achieve climate stabilization."[46] So rather than have the developing nations further develop the filthiest fuels to gain a barely reasonable standard of living, we in the developed world must cut our usage drastically while aiding them to develop alternative energy. It is both just and enlightened self-interest. It is noteworthy that the proposed 2017 U.S. Federal budget eliminated funding for that very thing. Critics will argue that this is a transfer of wealth, and it is, but it must be viewed not as a giveaway to the underserving, but as an investment in global survival for all of us.

## We Need New Policy for Conservation and Transition to Renewables

Only government can marshal resources in a way that individuals can't and only it can prevent the market from making foolish short-term choices based solely on profit for a few. That the market is unable to make the right responses is illustrated by the recent action by the U.S. Congress in removing the ban on oil exports, leaving demand to the market, a move which has encouraged the continued production of dirty fuel. Back as 1977, President Carter said: "…the United States is the only major industrial country without a comprehensive, long-range energy policy."[47] We still have no policy for energy conservation or transiting to clean renewables because many Congress persons are in thrall to the fossil energy industry, but without large scale public investment we will not achieve the goals in time to head off climate disaster.

The good news is that we can afford it. Robert Pollin argues that we can make the transition while expanding employment and reducing the worst global poverty if we "…devote between 1.5 and 2 percent per year of [global] GDP to investments in energy efficiency and clean, low emissions renewable energy sources."[48] In 2013 global GDP was $87 trillion so we would have needed to invest about $1.3 to $1.7 trillion per year, a third in efficiency and the rest in renewables. We are already investing somewhere between .4 and .6 percent so we are about a third of the way there and doing so has stimulated job creation.[49] We can have an energy system that achieves our goals of health and a sustainable economy.

What help can we expect from the business community? Can they adjust? Are capitalism and Ecocivilization compatible? Are market solutions viable? The big energy companies like Royal Dutch Shell do not appear to understand the imperatives of the present. Ben van Beurden, the chief executive of Royal Dutch Shell, Europe's largest oil company, said that while Shell was investing in biofuels and other renewable energy, there was no alternative to oil and gas that would pay the company's shareholders the large dividends they expected.[50] No one could have put the case for incompatibility more accurately. The wealthy elite demand their dividends regardless of the common good and businesses are legally obligated to satisfy them. Furthermore, so-called "market solutions," such as a carbon offset program for reducing airline emissions, seldom work.[51] But the world of business is complex. Walmart buys up wild land for conservation while at the same time it expands the footprints of its stores and promotes the consumer culture. Its products arrive on the store shelves after long, fossil powered journeys. Dell, which has hitherto moved its computers from China on daily Boeing 747 flights, now has switched to more efficient cargo ships.[52] And yet their product, like all computers, is full of toxics. Dell is also reported to be looking into cradle to grave manufacturing and using waste plastic to create its packaging. Reed reports that: "Many business leaders say they recognize climate change as a serious problem and that they are under pressure to demonstrate to their customers and employees, as well as governments and regulators, that they are doing something about it."[53] Others will try to hold back progress to protect their stranded assets. Whether large, corporate global capitalism itself survives is another matter. All in all, we are beginning to make the right decisions, but we need to rapidly scale up. In the U.S. we need a national industrial and economic plan similar to our effort in World War II.

Many are now calling for such a plan, sometimes under the heading of a Green New Deal, as proposed by Representative Alexandria Ocasio-Cortez and Senator Ed Markham. Their rollout looked like this.

The Plan for a Green New Deal (and the draft legislation) shall be developed in order to achieve the following goals, in each case in no longer than 10 years from the start of execution of the Plan:

1. Dramatically expand existing renewable power sources and deploy new production capacity with the goal of meeting 100% of national power demand through renewable sources;
2. Building a national, energy-efficient, "smart" grid;
3. Upgrading every residential and industrial building for state-of-the-art energy efficiency, comfort and safety;
4. Eliminating greenhouse gas emissions from the manufacturing, agricultural and other industries, including by investing in local-scale agriculture in communities across the country;
5. Eliminating greenhouse gas emissions by repairing and improving

transportation and other infrastructure, and upgrading water infrastructure to ensure universal access to clean water;

6. Funding massive investment in the drawdown of greenhouse gases;
7. Making "green" technology, industry, expertise, products and services a major export of the United States, with the aim of becoming the undisputed international leader in helping other countries transition to completely greenhouse gas neutral economies and bringing about a global Green New Deal.[54]

While not getting everything she wanted, Ocasio-Cortez did persuade the new House Leadership to establish a Committee on the Climate Crisis. And Paul Hawkin had already in 2017 assembled a group of experts who produced grist for this Committee in the book, *Drawdown: The Most Comprehensive Plan Ever Proposed to Reverse Global Warming*, with a hundred different techniques and practices implementable in the short range. At the bare minimum, a Green New Deal would do the following.

## A National Plan

Here is a representative, but not exhaustive, list under 4 headings:

### 1. Slow and halt the production of fossil fuels. Policies here will de-incentivize production and will include:

- Eliminate all existing incentives to production of fossil fuels, including tax breaks, direct subsidies, and selling of digging and drilling permits on federal lands.
- Eliminate subsidies to large, corporate agricultural operations including for ethanol production since it is an energy sink.
- Prohibit the export oil and coal. American coal burned in China poisons Americans' atmosphere.
- Enact fee and dividend, with sharply rising costs per unit of production.
- Direct the BLM, Army Corps of Engineers, and Environmental Protection Agency to cease permitting any new fossil fuel infrastructure such as pipelines and export terminals.

### 2. Incentivize practices which assist in speeding up the development of clean electric power including:

- Subsidize electric vehicles, rooftop solar, light rail and electric buses.

- Electrify the railroads
- Build more wind generators, tidal generators, and geothermal plants where feasible.
- Avoid any new nuclear power plants but keeping existing plants on line, and praying we don't have another Fukushima.
- Create disincentives by means of sales taxes the production and sales of vehicles that run on fossil fuels.

### 3. Implement federally run and funded programs to:

- Build a national grid of charging stations for electric vehicles (just as we built the inter-state highway system—a boon for fossil powered vehicles).
- Start up a federal program to insulate and weatherize buildings. This is the low hanging fruit and is an upstream solution and is a jobs program that does not require a college degree.
- Inaugurate a major program of reforestation to capture carbon naturally. (Another jobs program)
- Require "zero deforestation" supply chains for the food industry and manufacturing.
- Assist the exponential spread of urban food production.
- Fund innovation, especially energy storage technologies.
- Start a major program to get farmers to use conservation agriculture practices which capture carbon naturally including agro-forestry, no-till, cover crops, hedge rows, permaculture practices, etc.
- Assist young organic farmers to access capital and land to start farms that supply food to local consumers.
- Institute an educational/promotional program to cut down on food waste and meat consumption.

### 4. Retake international leadership

- Negotiate a new international climate treaty with realistic goals and enforcement powers based on trade sanctions.
- Provide aid to first world countries so that their peoples will not have to migrate as climate refugees.
- Provide a major increase in funding for the UN High Commission on Refugees, to help prevent the violent chaos of unregulated mass migrations.

This has been a discussion of technical and policy solutions to the climate crisis. Elsewhere we have discussed the problems of overpopulation and biodiversity. While we can expect some relief from technological ingenuity, by itself it will not save us. As Richard Heinberg, a fellow at the Post Carbon Institute points out, technical solutions also often bring technical problems right along with them. This raises profound challenges for us in the Global North. Heinberg is skeptical of the ability of technology alone to save us. He writes: "However, the real problem isn't just that we aren't investing enough money or effort in technological solutions. It's that we are asking technology to solve problems that demand human moral invention—ones that require ethical decisions, behavioral change, negotiation and sacrifice."[55] Keeping this in mind, the changes already underway in education, food production, manufacturing that we saw in the last chapter, and in energy and elsewhere, give us hope that whatever the rocky road ahead looks like as we experience the further environmental deterioration, we can see what a successful future might well look like. We are living in a paradoxical moment in which things are getting better as they get worse. We turn now to the ethical and social issues of rebuilding our communities for resilience, redesigning the economy to serve humanity while protecting the planet, and to rethinking governance.

# 15

## Creating Resilient,
## Place-based Communities

"In difficult times, we need each other more, not less."[1]
     —Rob Hopkins

"What we are facing isn't a financial crisis, but a crisis of the imagination."—Headline from British Website, Transition Network.org[2]

Ecocivilization is rooted in a sense of place. To save the planet and ourselves we all need to start at home because that is where it is being ruined—everywhere at once. A particular river is polluted by a global chemical company, this forest threatened by a mega cattle company, the air we breathe here is polluted by a coal-fired power plant whose owners live far away, this local butterfly habitat is under attack by real estate development, and so forth. We also need to act nationally and globally because often the forces degrading our home areas are generated in distant centers of power, in the boardrooms of global corporations and distant halls of government. We need clean air and water *here*, species diversity *here*, healthy soils *here* where we live each day.

In *The Revolution Where You Live*, Sarah van Gelder asks: "Can we build a new economy, rooted in our communities that can support us and protect the natural world? We need ways of meeting needs and providing livelihoods that don't extract wealth from most of the people of the world and from nature."[3] As we have seen, she makes a fundamental distinction between "cultures of connection" and the "economy of extraction" which is rooted in the neo-liberal globalization so characteristic of Hypercivilization.[4] We need to understand that the Old Story of conquest, command and control, and power over has failed us. The world is literally falling apart. Species are falling out of existence, polities are disintegrating, the climate is deteriorating, and soil is washed and blown off the land and into the sea bearing toxic chemicals. We know the results of the Old Story. The corporate-led globalization of the economy, world-wide wars and terrorist attacks, and a hardening of religious extremism constitute the final phase of Hypercivilization. If allowed to go on unchecked it will bring further ruin upon the world. One part of the alternative is to relocalize economies and re-empower local communities world-wide, restoring genuine community and sustainability. Doing so will allow us humans to go on for a long time in balance with the natural world on which we so obviously depend. Re-localization will make our villages, urban neighborhoods and bioregions more resilient. Re-localizing does not mean a rejection of national and international involvement, but rather a change of emphasis, especially in terms of the economy, from the larger to the more local. It means nesting just, environmentally and economically healthy local communities within larger circles—the state, the region, the nation and the world. The good news is that we see the beginnings of these changes all around us. The New Story is beginning to play out in the womb of the old. But first, why is it so necessary, desirable and urgent?

## Neo-liberal Globalization: A Wrong Turn

The Old Story that justifies the political economy of corporate driven Hypercivilization jettisons traditional values including family, neighborliness, generosity, community, the land, thrift, healthy environments, the well-being of the future and even religion. Wendell Berry observed that we don't even realize it anymore because, "The good of the whole of Creation, the world and all its creatures together, is never a consideration because it is never thought of; our culture now simply lacks the means for thinking of it."[5] Our culture reduces all value to economic value. This has not worked for the common good. Since World War II, powerful elites have been pressing forward the extreme neoliberal policy of globalization. It has left in its wake environmental devastation, lost jobs, impoverished communities with little or no control over their economies, labor rights or environment. Meanwhile, in the context of an international anarchy based on the myth of sovereignty, nation states grapple for empire. A concomitant of this global regime is permawar and terrorism as a response to foreign occupiers.

The entire edifice of neoliberalism is based on the abstract idea of the *market*, a notion that totally free trade, independent of any regulation, will result in the common good, and that the common good is simply the aggregate of all individual goods selfishly pursued. The facts on the ground demonstrate otherwise. Unfortunately, nearly all of the world's political and economic elites have bought into this myth of endless, mindless, unregulated economic growth since, for the short term at least, it benefits them. If we are going to avoid collapse brought on by continuing to undermine the natural base of civilization we will need to opt out of this corporate, smash-and-grab form of global capitalism and recapture control over our lives by re-localizing our interaction with nature. This is not to argue that we should go back to living in dirt huts without the benefits of modern science, health care, and electronics, or that there will be no international trade. Rather we are describing a process of going forward to something better. This alternate path of development flies in the face of the dominant ideology of unregulated markets, so-called "free-trade," and consumerism that is promulgated everywhere as the only path of progress. David Morris writes: "The suggestion that we choose another path is viewed, at best, as an attempt to reverse history and, at worst, as an unnatural and even sinful act."[6] However, the alternative we are talking about is not a state-owned, demand economy, even though those who benefit greatly from the current situation would like us to believe what David Korton calls "the fallacy that the only alternative to rule by Wall Street capitalists is rule by communist 'bureaucrats.'"[7] It's not. We are championing instead main street capitalism, small scale entrepreneurship, neighborhood cooperation, and protection of the commons.

The corporate-driven global economy is systemically flawed. For a century it has drawn people off the land, turning on its head the stable pyramid of food producers to food dependent people living in the cities. It has replaced farming with giant agribusinesses that do not care about the long term health of the land or the quality of the food produced, much of which it dumps on third world markets destroying their indigenous food systems and livelihoods. It has gutted our once stable rural communities. It confuses well-being with wealth, reducing all values to cash and credit. It requires endless economic growth at the expense of the planet. It requires indebtedness of individuals and developing nations and an ever increasing inequality. It has no sense of how much is enough. It allows for phantom wealth for a few who get rich by speculating on money without producing anything useful. It requires that most people must trade their time and labor to someone else, usually geographically or socially distant, creating goods that they will not own and robbing them of self-reliance. It creates a plethora of shoddy goods that are unrepairable and end up in landfills. It funnels the benefits to a few and externalizes the costs to the many. It favors giant, multi-national corporations that erode national sovereignty while putting main street businesses out of business. It swiftly depletes finite resources. It creates instability by hunting for the lowest labor costs of the moment, gutting one city or nation by temporarily locating in another where labor costs can be driven down even farther, and then after a little while, it moves on yet again leaving abandoned factories, pollution, and unemployed people.

The so-called "free market" allows a few people who make stupid or malicious decisions to throw the whole system into chaos, putting millions out of work. It has abandoned our central cities, leaving unemployment, poverty, blight, food deserts and crime in its place. It has intensified our lives, speeding us up so we are always on the go in quest of money but, in fact, working longer for less. Wherever it can, it replaces people with robots. In their new book, *People Get Ready*, John Nichols and Bob McChesney predict a jobless economy as countless corporations are investing in robots, automation and digitization.[8] It results in a race to the bottom for income and destroys laws that protect labor and the environment. It allows for a concentration of wealth in the hands of a tiny minority who then corrupt democracy. It funnels huge amounts of money to the military industrial complex which creates things that not only have no use value for citizens but are designed to destroy things which do. It results in huge inefficiencies of energy, moving materials and goods all over the globe. It pits us against each other as individuals, cities and nations through a system of ruthless competition. It creates scarcity, needless poverty and destabilizes the peace throughout the world.

Perhaps the most problematic for breakthrough is that the global corporate economy must generate overall economic growth or collapse. Because a nation's money supply is created mostly by banks loaning it into existence, it requires continual debt. David Korton writes: "a nation's money supply and the stability of its money system depend on continuously expanding debt to create enough money to pay the old debts,"[9] hence ever more throughput, ever more extraction, energy expenditure, waste and climate deterioration.

The global corporate economy violates the prime directive of Ecocivilization by eroding community. It fails Aldo Leopold's dictum: "We abuse land because we regard it as a commodity belonging to us. When we see land as a community to which we belong, we may begin to use it with love and respect."[10] It fails the people as well. The word "free" in "free market" Wendell Berry points out "has come to mean economic power for some, with the necessary consequence of economic powerlessness for others."[11] And he goes on, saying that it is a "falseness and a silliness" because it "always implies, and in fact requires, that any community must be divided into a class of winners and a class of losers," and, "the losers simply accumulate in human dumps," and, "The danger of the ideal of competition is that it neither proposes nor implies any limits. It proposes simply to lower costs at any cost, and to raise profits at any cost. It does not hesitate at the destruction of the life of a family or the life of a community."[12] Over twenty years ago Jerry Mander wrote that the result of globalization is "a new kind of corporate colonialism, visited upon poor countries and the poor in rich countries."[13] He asked "Where will the resources—the energy, the minerals, the water—come from to feed the increased growth? Where will the effluents of the process ... be dumped?"[14] A minority will get richer and richer from this planet plundering "while the rest of humanity is left groping for fewer jobs and less land, living in violent societies on a ravaged planet."[15] In short, as Kalle Lasn (founder of Adbusters and who inspired the Occupy Movement), writes: "The global economy is a doomsday machine that must be stopped and reprogrammed."[16] In contrast, Ecocivilization is a profoundly conservative, community-based biosocial system, committed to restoring health to the land and the larger biosphere and to caring for people.

## The Different Political Economy of Ecocivilization

The social norms of Ecocivilization are very different from those by which global, corporate capitalism operates and will yield a different kind of political economy. They are designed to favor community, equity, justice, peace, and above all, environmental sustainability. What I am advocating is that people recapture power over their local ecosystems, sources of materials, jobs, money supply, and governments. The good news is that the movement forward to build a new economy and a new more democratic polity is already underway, but first, what do these new economic and social norms look like?

Economist David Korton and journalist Sarah van Gelder have clarified the contrasting norms of a place-based community versus those of the global, extractive economy as discussed in a previous chapter.[17] To briefly reiterate, Ecocivilization will restore the health of ecosystems, be rooted in place, protect the commons, educate for deep learning about place, take responsibility for long term well-being for future generations, create jobs for people rather than for robots, appreciate that real wealth is community and ecological well-being widely distributed, put relationships ahead of money, invest in long-term well-being of community infrastructure while internalizing all costs, and protecting local, democratic decision-making.

The primary goal of the New Economy will be to produce not money but goods for the support of life. The current money system drains wealth out of local communities and into the coffers of big corporations including banks. Much of it is off-shored by the global elite, as the 2016 revelation of the "Panama Papers" demonstrated. It's not that we don't need money. We do. One alternative already underway is to develop local currencies because, for the most part, exchange cannot work on a large scale without money. While barter is perfectly legitimate, the two actors must each have some item the other wants. When that is not the case, they need something that can be used to obtain some other thing neither has, that is, a medium of exchange. Thus if a local craftsperson wants to start building furniture from locally sourced wood but can't get a bank loan of the national currency because there isn't enough money, or the multi-national bank won't risk it because they don't know her or the potential purchaser, nothing happens. Skills, local renewable resources and livelihood are paralyzed. While a truly locally owned bank could make a difference, local currency could also facilitate the exchange. A local currency is not supposed to replace the national currency but to supplement it. The Schumacher Center for a New Economics points out that:"Centralized currency issue serves centralized production whereas regional currencies represent a democratization of currency issue, supporting local businesses and educating consumers about how their money circulates in the local economy."[18]

Local currencies are a reality in many places. Some are issued by local banks, others by the chamber of commerce, citizen's organizations, or just a group of friends. It's perfectly legal in the U.S. and elsewhere. In the Berkshire region in Massachusetts one can get BerkShares by exchanging federal dollars. Here's how it works.

> Federal currency is exchanged for BerkShares at eight branch offices of three local banks and spent at 400 locally owned participating businesses. The circulation of BerkShares encourages capital to remain within the region, building a greater affinity between the local business community and its citizens. The members of BerkShares, Inc. envision a diverse and resilient regional economy that supports and prioritizes responsible production and consumption, wherein community members rely on the land and each other to fulfill the basic needs of food, culture, clothing, shelter, and energy.[19]

Approximately 1,500 communities around the world have issued local currencies.[20] Since the local currency won't be honored outside the community, it can't migrate to national and international banking centers. The advantage is that it keeps the money at home to facilitate exchange.

The New Economy must be powered by renewable energy and protect the soil, water, air and biodiversity. It must provide meaningful work and jobs. Saving labor is not always a good idea and is frequently a bad one. There is nothing wrong with hard work if it's meaningful. The much touted trend toward robotics and other labor saving devices and efficiencies at the expense of jobs will be discouraged because it displaces people. The economy is for people, not for profit at the expense of people. The New Economy will need to move toward a steady state, no-growth, true cost economy. This

BerkShares: Local currencies are legal in the U.S. BerkShares are a local currency for the Berkshire region of Massachusetts and encourage money to remain within the region, rather than being drained off to distant corporations and banks, thus serving as a tool for community economic empowerment and development toward regional self-reliance (courtesy Rachel Moriarty, BerkShares).

is utter heresy among most economists and politicians. However, we are coming up against the limits of the biosphere and, as Wendell Berry points out, "To hit these limits at top speed is not a rational choice. To start slowing down, with the idea of avoiding catastrophe, is a rational choice...."[21] Traditional economists and politicians argue that because we will need more money to invest in renewables and ecological restoration, etc., we will therefore need to grow out way out of these problems. Herman Daly at the Center for the Study of the Steady State Economy argues that such a path is self-defeating.

> Even if we could grow our way out of the crisis and delay the inevitable and painful reconciliation of virtual and real wealth, there is the question of whether this would be a wise thing to do. Marginal costs of additional growth in rich countries, such as global warming, biodiversity loss and roadways choked with cars, now likely exceed marginal benefits of a little extra consumption. The end result is that promoting further economic growth makes us poorer, not richer.[22]

Vandana Shiva explains the environmental contradiction in the ideology of free market growthism. "Nature shrinks as capital grows. The growth of the market cannot solve the very crisis it creates."[23] Therefore if our interchange with nature is to be sustainable, we must give serious thought to creating a steady state economy. As Eric Zencey writes: "All of economics is divided into two schools: steady state theory and infinite planet theory," and goes on to conclude: "They can't both be right,"[24] which brings to mind Kenneth Boulding's iconic formulation: "Anyone who believes exponential growth can go on forever in a finite world is either a madman or an economist."[25]

Continued quantitative growth in our finite planet is *uneconomic* growth and is damaging the well-being of people and planet. A sustainable, true cost economy is one that does not allow externalizing the impacts of resource depletion and pollution and counts in the real, albeit sometimes less tangible benefits of intact ecosystems. The goal is not to maximize production, which is impossible in a finite biosphere, but to maximize well-being. There is a profound difference between growth and development. Progressive change in science, technology, education, ecological restoration, income equity that promotes more democratic governance, in community and family stability, control of violent conflict, and in the arts can all bring greater well-being to people without requiring biosphere-destroying growth. After all, a steady state economy is inevitable, as Zencey points out:

> No matter what we do we'll eventually have an ecologically sustainable civilization with a steady state economy, one that's in dynamic balance with its host ecosystems. That's because by definition an unsustainable system doesn't last. We can make the transition haphazardly, through crisis, catastrophe, and collapse, at much cost in human pain and suffering, or we can anticipate the necessary changes and give ourselves a better, less brutal path forward. To find that path we'll have to identify and correct infinite-planet suppositions wherever they are embedded in our system.[26]

Better to achieve steady state rationally than in a catastrophic breakdown.

We cannot leave this issue without noting that if the hundreds of millions now living in poverty, most but not all of whom are in the developing world, are to have at least minimally decent lives, there has to be a redistribution of consumption. The North will have to consume less material and energy so that the South can consume more. While this appears to be a very difficult change to make, we can note that the nations of the global North have already assented to the principle in their pledge to share technology and money with the nations of the South in order to assist them to cut their carbon footprint and move toward renewable energy. Even in the global North there is currently a great deal of dissatisfaction at the gulf between the rich and the poor. Some measure of redistribution, the bugaboo of the neoliberals, is not an impossible daydream. And throttling back also brings great benefits including better health, more leisure, less traffic and congestion. Finally, in Ecocivilization the economy will generate only real wealth in goods. Speculative financial manipulation will not be allowed because it generates no useful goods or services beyond increasing the bank accounts of an elite few. Abstract dealings in abstract things like derivatives and other economic fictions will be prohibited and banks and other financial houses will be strictly regulated.

Above all, economic activity will center far more on local production and consumption, though not to the exclusion of larger scales as long as large scale economic activity is transparent and regulated for the common good and all its costs are internalized so the market is honest in setting prices. Most businesses, including especially banks, will be locally owned so that the lenders live in the region they are impacting with their loans. Local economies will reverse the now overly long trend toward colonialist economies in which a locale relies on exporting a single commodity to the national or global market, such as coal, forestry products, corn and soybeans, or cattle. Such a system results in exporting jobs and the young people follow, leaving our communities demographically truncated. Such communities are also extremely vulnerable to changes in the market and to ecological changes. Thousands of ghost towns in places like Oklahoma, Iowa, Texas and many other places already testify to that. The solution is to return to a diverse local economy based on all the things the land will produce without damaging the biosphere, thus keeping the jobs at home and strengthening the resilience of the local economy, strengthening its ability to buffer national and global downturns. Of course, once local needs are met, surplus production could be exported to the larger economy.

Localizing implies a great complex of decentralized, small scale industries. The Old Story's objection, that centralized, large scale production achieves efficiencies of scale is rendered false by the fact that these efficiencies were achieved at the expense of jobs, the environment, community, and the future because they externalized the true costs. Berry writes: "If we wish to make the best use of people, places, and things, then we are going to have to deal with a maxim that reads about like this: As the quality of the use increases, the scale of use (that is, the size of operations) will decline, the tools will become simpler, and the methods and the skills will become more complex."[27] Furthermore:

> A proper economy … would designate certain things as priceless. This would not be, as now, the pricelessness of things that are extremely rare or expensive but would refer to things of absolute value, beyond and above any price that could be set on them by any market. The things of absolute value would be fertile land, clean water and air, ecological health, and the capacity of nature to renew herself in the economic landscapes.[28]

The same is true for retail. There is no need for a global octopus like Walmart or Amazon in Ecocivilization, and the demise of such operations is accomplished by simply refusing to do business with them, by switching to local-sources where they exist and creating them where they don't.

Income inequality will need to be addressed. In Ecocivilization income will be distributed much more equitably and income above a reasonable level will be distributed for the common good to provide such things as free education, health care and eco-restoration. The world will also need to address the problem of gross inequality between the global North and South, especially if the North expects the South to eschew dirty coal and adopt green energy in its place, a transition crucial to saving the planet from climate induced ecological disaster.

While it is true by some accounting measures that the number of people in the world who are living in extreme poverty has been halved in the age of globalization, there is good reason to be skeptical of this. First is the problem with measurements of GDP that average incomes, so if the rich are getting fabulously rich, the averaging makes it look like the poor are becoming better off. Even if the poor are making more money than before, going from a dollar a day to three dollars a day still leaves people in relative poverty. It is not much progress, especially when the environmental and social costs of the changes necessary to achieve such gains outweigh the gains. For example, the so-called free trade agreements have resulted in devastating traditional rural economies all over the global South, a process exacerbated by global climate deterioration driven primarily by the global North. Hundreds of millions of people who once lived more or less successfully in natural and agricultural settings, though not buried in Western style material wealth, have been driven into heavily toxic and polluted urban slums to seek work in sweatshops. Many of these burgeoning cities are controlled by urban gangs such as we see in Mexico and Honduras, causing desperate parents to send their children on dangerous journeys of migration, hoping that on their own they will be able to cross borders into a safer country. Another example is the million farmers in Syria who suffered total crop loss for three years as a result of climate change induced drought. They were forced by necessity to migrate into cities that were unable to accommodate them, helping spur revolution and a ghastly civil war.

Hypercivilization has seen hundreds of millions leaving the farm where one could at least raise some food to feed the family, placing the new urban migrant at the mercy of the global commodity prices and dumping of inferior goods. The price of so-called "progress" is that they have lost control of their lives, a loss hardly compensated for by an increase in income of a few dollars a day. By contrast, Ecocivilization will be characterized by a true-cost economy where environmental and social degradation cannot be externalized.

Another failing of Hypercivilization is that the international financial institutions impose so-called "structural adjustments" on poor countries, forcing them to cut social services in order to pay off debts to the global North. International development institutions such as the World Bank and the International Monetary Fund will need to make a great deal of further progress in redirecting their loans to environmentally sustainable projects that benefit rural communities. At the same time they need eschew completely the so-called "austerity" programs which have destroyed safety nets, school systems, environmental protection and other public benefits. The

World Trade Organization, with its secret tribunals whose decisions can attack local communities, will cease to exist. As we are looking at fundamental changes here, we can expect that we will need to change some of our fundamental laws that privilege corporations.

## The Community Rights Movement

Our understanding of property law needs to change. In Ecocivilization, ownership will not constitute absolute sovereignty. Even now, just having a deed to the land does not confer unlimited rights to use it howsoever one sees fit. Ecocivilization will not abolish privately owned business property—to the contrary, small scale local entrepreneurs will be encouraged. However, the uses to which property is put will be governed by the rule of the common good as determined by those who are impacted by its uses. Sarah van Gelder writes: "Ultimately we will need to change the laws that favor big, transnational extractive companies over enterprises that are locally rooted and locally contributing."[29] Transitioning to Ecocivilization requires not just a revolution in concepts and values, not just new technologies, not just a place-based economy, but a thorough-going legal revolution. The Community Rights Movement, now some twenty years old, is redefining what law must look like so that global corporations will no longer have the legal right to overrule the people of a local community in order to carry out ecologically and socially destructive enterprises as they do now. Thousands of examples could be cited; one will do.

In Wisconsin, as of this writing, the Canadian oil transmission company, Enbridge, has been given the right of eminent domain by the legislature in order to add a new line to carry heavy tar sands from Alberta to Chicago and then on to the Gulf coast for processing and export. In Wisconsin the line will cross hundreds of rivers. including a Federal Wild River and a State Wild River. Under American law, local communities, such as a counties or municipalities are unable to say "No!" to this company or to any other major corporation that wants to introduce an environmental intrusion the local people do not want, be it a CAFO or a chemical plant, if the state or federal governments wish to permit it. With regard to our example, the only path open to local people of Washburn County, where the line will imperil two Wild Rivers, is negotiate with the company in the hopes that they will place the shut off-valves closer to the rivers. The landowners will have no say in the 200 foot wide pipe line corridor, even though this company has been responsible for devastating spills in the recent past. "In other words, instead of questioning why corporations can legally treat our communities as sacrifice zones, we accept their authority to do so, while trying to make their proposed project a little less harmful."[30] Why? Because the Interstate Commerce Clause of the Constitution and state constitutions give higher levels of government preemptive powers over local governments, and, unfortunately, it is a feature of politics in America that politicians are highly susceptible to campaign assistance from corporate donors.

The Community Rights Movement argues that, for local people, there is no real self-government, no real democracy, in America.

> …our system of law in the United States systematically strips communities of the power to adopt and enforce local laws, when those laws come into direct conflict with decisions made by corporations that want to use those communities for various projects. And … our system of law elevates the corporate "right" to decide, above the rights of you and me to clean air, clean water, and sustainability in general.[31]

They argue that regulation doesn't work. At most, citizens can hope to influence unelected bureaucrats in the regulatory agencies in an attempt to get corporations to moderate their plans, but as we know too well, regulatory agencies are often staffed by past or future employees of the very industries they are supposed to regulate. The system is broken and the only way to fix it is to organize at the local level to pass prohibitory codes. In our example, a county would pass an ordinance forbidding the transit of a pipeline, further stating that the county does not recognize the corporation's power of eminent domain. Passing such an ordinance is an act of community civil disobedience and is, of course, illegal because the state has a "preemption clause" in its constitution and will void it. State law trumps local law and state and federal law recognize corporate rights as the rights of persons as defined in the 14th Amendment to the U.S. Constitution. On the basis of such clauses, either the state or the corporation will sue the county and win. However, this is a strategy. An act of community civil disobedience is the only means to get the question into court where the constitutional argument can be brought and light shed on the undemocratic nature of preemption.

This, they argue, is the only viable way forward—to create a fundamental change in the basic law of the land.

> Building economically and environmentally sustainable communities … cannot be done through a system that elevates the "rights" of corporate entities driven by unsustainability, above the rights of people and communities interested in establishing a different system. Nor can it be done through a corporate culture whose primary, overriding goal is—as historian Richard L. Grossman often said—the production of "endless more."[32]

Community Rights is not mere theory. Mora County, New Mexico, in the U.S., became the first county to ban oil and gas extraction through a new law, a "community bill of rights—which protects their right to clean air, clean water, and a sustainable energy future."[33] Over two hundred communities in the U.S. have begun to draft local sustainability ordinances with the assistance of the Community Environmental Legal Defense Fund.[34] They use community rights frameworks:

> ...to define sustainable farming, sustainable land development, and sustainable energy production within those municipalities. Those laws establish local bills of rights for communities, ecosystems and residents, and then prohibit unsustainable practices that would violate those rights. They protect those rights frameworks by redefining corporate rights and powers in those communities and by limiting other legal doctrines routinely used to overturn municipal lawmaking.[35]

The ultimate goal of this movement is to gather enough momentum to call for state constitutional conventions and eventually a federal convention which will overturn the long history of federal decisions awarding the rights of natural persons to corporations. This movement is about protecting people and the ecosystems on which they depend and love.

In Ecocivilization there will be a strong tendency to avoid privatization of the commons. National Parks may not be sold off, water may not be privatized. It will be illegal to patent genes or to sell them for profit. All scientific knowledge will be in the public domain. Genuine competitive trade will be encouraged. Monopolies will be discouraged in favor of small-scale, main street businesses. Moreover, the legal status of corporations will have to evolve. While the concept of a limited liability corporation can be beneficial for encouraging the pooling of resources for entrepreneurial innovation, two things must change. First, it must be understood that entrepreneurial enterprises are inherently risky and that the government is under no obligation to bail out those that "are too big to fail." Even more important, the original idea behind allowing such corporations was public benefit—for that reason alone were they chartered by the states. When corporations fail to secure public benefit, when they pollute nature's sinks or create unsafe products such as pesticides, their charters can and should be revoked. There is no constitutional right to a charter nor is there any such thing as a "taking," the complaint invented by corporations on the basis of previous court decisions that gave them 14th Amendment rights, that the government owes an individual or corporation compensation if it won't allow them to make money by polluting or by developing their property. Further, corporations will be encouraged by government incentives to adopt the triple bottom line, to re-form as S-corporations or cooperatives. Employee-owned corporations will be favored by the tax structure as will corporations using renewables and conservation technologies. Also privileged will be small-hold, organic farmers. Giant, corporate agriculture holdings will either be broken up or heavily discouraged by taxation to favor nature-restoring agriculture. Further, the practice adopted by many cities and countries of providing race-to-the-bottom subsidies to lure a business from one district to another will either become illegal or workers in the old location will have to be compensated for job loss. This will eliminate the problem of governments in the thrall of corporations serving their short-term interests rather than the long-term interests of the larger community.

## Governance

The ruling principle of government will be that adopted from ancient Roman law: "That which touches all is the concern of all," and not just of wealthy elites. Access to decision making will be truly democratic. Government will be in the hands of the governed because money will not be allowed to unduly influence elections. Campaign finance laws will prohibit large incursions of money into campaigns and campaigns will be limited in time. All candidates will have equal access to the airwaves. In the U.S., the *Citizen's United* decision of the U.S. Supreme Court which defined the unlimited use of money to influence elections as free speech, will be overturned.[36] Allowing big money in politics has led to thinly disguised political bribery in which government becomes the tool of the elites to maintain their privileged status. As James Bovard has pointed out: "Democracy must be something more than two wolves and a sheep voting on what to have for dinner."[37] High standards will be set at all levels of government for maintaining biodiversity, clean water, clean air and a stable climate. Regulators will not be able to move on to the industries they formerly regulated and vice versa.

Social policy will aim at rebalancing the urban—rural settlement patterns by making rural life economically feasible again. The current, drastic imbalance needs to be rebalanced insofar as possible so that one percent of the people are not growing food for 99 percent. A return of a significant portion of the population to the land and to rural villages will characterize Ecocivilization and stabilize the supply of wholesome food that is sustainably produced. Wendell Berry writes: "The people who do the land's work should own the land. It should not be owned in great monopolistic estates by absentee landlords, as in the latter days of the Roman Empire, and as increasingly now with us."[38]

When governments make decisions about the economy, a broad measure of national well-being that recognizes all costs and benefits will take the place of the crude measure of GDP and its sole reliance on gross economic growth. This implies a focus on the long-term well-being of communities. Berry says, "we need to stop thinking about the economic functions of individuals ... and try to learn to think of the economic functions of communities and households. We need to try to understand the long-term economies of places—places, that is, that are considered as dwelling places for humans and their fellow creatures, not as exploitable resources."[39] While the State may continue to provide old age insurance, in Ecocivilization we will recognize that the most effective social security comes from being embedded in a functioning, caring community of neighbors. This implies a revival of healthy rural communities and a complex of interrelated neighborhoods in our big cities—a revival of genuine civic life as it was before TV and other electronics isolated us in our houses and cut us off from one another. We will know each other's stories.

## Transition Towns, the Natural Step, Eco-cities and other Model Communities

Another hopeful sign is the rise of a community resilience movement under various names including the Transition Towns movement which began in England, and the Natural Step which started in Sweden, and many efforts going on around the world toward transforming our urban centers into "Ecocities." One of the pioneering experiments is Gaviotas, an ecovillage in the Llanos region of Colombia.

In *Gaviotas: A Village to Reinvent the World,* Alan Weisman has chronicled the remarkable story of a group of inventors, visionaries and ordinary folk who in 1971, on reflecting that the population explosion was going to require people to live and live lightly in some of the harshest areas of the world, set out to construct just such a place on the barren, rain-leached savannahs of eastern Colombia. It worked.

> For nearly three decades the scientists, artisans, rural peasants, ex-urban street kids, and Guahibo Indians ... have elevated phrases like *sustainable development* and *appropriate technology* from cliché to reality. Sixteen hours from the nearest major city, they invented windmills light enough to convert mild tropical breezes into energy, solar collectors that work in the rain, soil-free systems to raise edible and medicinal crops, solar "kettles" to sterilize drinking water, and ultra-efficient pumps to tap deep aquifers— pumps so easy to operate, they're hooked up to children's seesaws.[40]

They also planted 20,000 acres of trees.

When one of the founders of Gaviotas, Jorge Zapp, brought some of his graduate students out to look at the site, they were appalled. "It's fairly straight forward," he told them, "just figure out how to build the civilization of the future out of grass, sun and water."[41] They did. Weisman writes: "When a group from the Club of Rome visited Gaviotas in 1984,

Club founder Aurelio Peccei declared… 'This is what the world needs.'"[42] The United Nations agreed and provided a substantial development grant. Today the village is self-supporting.

The founders of Gaviotas did not set out to create the perfect eco-village that could be adapted and placed anywhere. To the contrary, their point is that solutions must be place-appropriate, growing out of the nature of the place itself. This is not to say that they eschew technology. In fact the founders were and are strong proponents of low impact technological innovation that grows out of local conditions and needs. Founder Paolo Lugari pointed out: "When we import solutions from the United States and Europe we also import their problems," and went on to enunciate an important principle for sustainable communities, that they live within "the economy of the near."[43] Zapp, said: "It's very romantic to build out of local natural materials, but it's dumb to be purists all the time. And impractical. The future will need nature and technology. We can't make solar panels out of whole wheat bread."[44] They have taken their appropriate, low impact technologies to some six hundred other villages in Colombia. Another key component of the success of Gaviotas is that while the scientists and technicians are important, the crucial component is helping people develop their own solutions. "In Zap's definition, development means renewing one's faith in the collective intelligence of humans" and he asserts: "What was spread in large part was that people learned to believe in their own abilities."[45] And, he concludes: "There's no such thing a sustainable technology or economic development without sustainable human development to match."[46] We see the same principles applied in a much bigger endeavor, the Transition Town movement.

In 2006, permaculture designer Rob Hopkins gave an assignment to his class at Kinsale College in County Cork, Ireland. They were to think about how communities could weather the coming changes. Out of that came the "Kinsale Energy Descent Action Plan."[47] The Transition model is a community organizing process to morph existing communities into resilient towns that can withstand the shocks that are coming with peak oil, climate deterioration and global financial crises. The model did not prescribe a solution but a rather a method to arrive at solutions by bringing together many constituencies in an imaginative visioning process. They presented the plan to the city council of Kinsale and the Council bought it, providing a little grant to get things going. The Transition movement had begun. Hopkins then took it on to his home town of Totness in Devon, England, and from there it began to spread world-wide. Eventually Hopkins wrote *The Transition Handbook* which is still the bible of the movement.[48] The number of such places continues to grow; there are Transition Initiatives in 1,196 places in over 50 countries including 159 in the U.S. in 37 different states, according to Transition U.S.[49] These include villages, urban neighborhoods, whole cities, and even regions. Transition U.S. states as its goal: "that every community in the United States will have engaged its collective creativity to unleash an extraordinary and historic transition to a future beyond fossil fuels; a future that is more vibrant, abundant and resilient; one that is ultimately preferable to the present."[50] These transition initiatives are "scalable microcosms of hope," stressing their bottom-up essence.[51]

While the current glut of oil from over-pumping may have put off the date of peak oil (or not—hurry up to run out is a possibility, too), the absolute necessity of leaving fossil fuels in the ground to prevent catastrophic global warming has made the movement even more urgent. Today our local economies are utterly dependent on the ability of much larger economies to function over long distances, all powered by fossil fuels. We are no longer self-reliant even to a small degree and thus are increasingly vulnerable. One of the trends that have made us vulnerable is the loss of skills that our grandparents had. Great changes are coming; the question is not if but when, and the when looks increasingly sooner. We can either wait for sudden collapse or engage each other in a creative devolution toward a world that is energy leaner but better off in many ways.

Resilience means more than just sustainability. Hopkins writes, "resilience refers to an ecosystem's ability to roll with the shocks.... In the context of communities and settlements it refers to their ability to not collapse at the first sight of oil or food shortages."[52] In practice this means two major changes: powering down, or what is called "energy descent," and localizing the economy insofar as possible to produce the necessities when necessary. The former is not a choice, and so we must ask ourselves how we can live within a realistic energy budget. Re-localizing does not mean walling a community off from the world. "What it does mean," he writes, "is being more prepared for a leaner future, more self-reliant, and prioritizing the local over the imported."[53] Nor does this have to mean deprivation. We will have more leisure time, better health, less haste, more wildlife, the satisfaction of knowing that much of our lives are produced by our own work, and we will have stronger communities—more neighborliness as well as cleaner air and water and less blight. Hopkins says: "I am not afraid of a world with less consumerism, less 'stuff' and no economic growth. Indeed, I am far more frightened of the opposite...."[54] We will be working together again. This is not to go backward in time to some imagined rosy past, nor is it to assume that the past was "all mud and incest and shoving young boys up chimneys and little else...."[55] Rather we need to be sensible about what we import and what we don't. Not every community is going to make computers and cars but neither should we be importing things we can produce locally, including, Hopkins suggests, "a wide range of seasonal fruits and vegetables, fresh fish, timber, mushrooms, dyes, many medicines, furniture, ceramics, insulation materials, soap, bread, glass, dairy products, wool and leather products, paper, building materials, perfumes and fresh flowers—to name but a few."[56] He stresses that "We aren't looking to create a 'nothing-in, nothing-out' economy, but to close economic loops where possible and to produce locally what we can."

How does resilience work? There are three principles: diversity, modularity, and tightness of feedbacks. Diversity relates to the number of parts of the system as well as the number of links between them and to a diversity of functions. Single income towns that rely on, for example tourism, or one big factory, are increasingly vulnerable. Diversity also refers to land use: no big monoculture but an array of mixed farms, woodlots, orchards, market gardens, aquaculture, and so forth. Building in resilience is "about working on small changes to lots of niches in the place, making lots of small interventions rather than a few large ones."[57] Modularity refers to insulating the local community from shocks from the outside by creating nested communities. "A more modular structure means that the parts of a system can more effectively self-organize in the event of shock."[58] So for example, local investment systems would be protected from housing bubbles thousands of miles away, local currencies from inflation or deflation of national currencies. Local food production protects against blight, drought, or unacceptable transport costs of food imported over long distances. Locally sourced jobs protect from layoffs by a national chain or a major corporation headquartered in a distant place. Tightness of feedbacks refers to "how quickly and strongly the consequences of change in one part of the system are felt and responded to in other parts."[59] The tighter the feedback the more likely a quick, in-time response before things get out of hand.

Transition United States lists seven guiding principles of transition. They are:

1. **Positive Visioning**. Transition Initiatives are based on a dedication to the creation of tangible, clearly expressed and practical visions of the community in question beyond its present-day dependence on fossil fuel. Our primary focus is … on creating positive, empowering possibilities and opportunities. The generation of new stories and myths are central to this visioning work.

2. **Help People Access Good Information and Trust Them to Make Good Decisions**. Transition initiatives dedicate themselves, through all aspects of their work, to raising awareness of peak oil and climate change and related issues such as critiquing economic growth. In doing so they recognize the responsibility to present this information in ways which are playful, articulate, accessible and engaging, and

which enable people to feel enthused and empowered rather than powerless.

3. **Inclusion and Openness.** Successful Transition Initiatives need an unprecedented coming together of the broad diversity of society. They dedicate themselves to ensuring that their decision-making processes and their working groups embody principles of openness and inclusion. This principle also refers to the principle of each initiative reaching the community in its entirety, and endeavoring, from an early stage, to engage their local business community, the diversity of community groups and local government authorities. It makes explicit the principle that there is no room for "them and us" thinking in the challenge of energy descent planning.

4. **Enable Sharing and Networking.** Transition Initiatives dedicate themselves to sharing their successes, failures, insights and connections at the various scales across the Transition network, so as to more widely build up a collective body of experience.

5. **Build Resilience.** This stresses the fundamental importance of building resilience, i.e., the capacity of our businesses, communities and settlements to withstand shock. Transition initiatives commit to building resilience across a wide range of areas (food, economics, energy, etc.) and also on a range of scales (from the local to the national) as seems appropriate—and to setting them within an overall context of the need to do everything we can to ensure environmental resilience.

6. **Inner and Outer Transition.** The challenges we face are not just caused by a mistake in our technologies but are a direct result of our world view and belief system. The impact of the information about the state of our planet can generate fear and grief—which may underlie the state of denial that many people are caught in. Psychological models can help us understand what is really happening and avoid unconscious processes sabotaging change. E.g., addictions models, models for behavioral change. This principle also honors the fact that Transition thrives because it enables and supports people to do what they are passionate about, what they feel called to do.

7. **Subsidiarity: Self-organization and Decision Making at the Appropriate Level.** This final principle embodies the idea that the intention of the Transition model is not to centralize or control decision making, but rather to work with everyone so that it is practiced at the most appropriate, practical and empowering level, and in such a way that it models the ability of natural systems to self-organize.[60]

They also suggest how to deal with negativism at the start including what they call the "& Buts"[61] as follows.

1. **But … we've got no funding.** This really is not an issue. Funding is a very poor substitute for enthusiasm and community involvement, both of which will take you through the first phases of your transition. Funders can also demand a measure of control, and may steer the initiative in directions that run counter to community interests and to your original vision.

2. **But … they won't let us.** "There is a fear among some green folks that somehow any initiative that actually succeeds in effecting any change will get shut down, suppressed, attacked by faceless bureaucrats or corporations. Transition Initiatives operate 'below the radar'; as such they don't incur the wrath of any existing institutions. On the contrary, with corporate awareness of sustainability and Climate Change building daily, you will be surprised at how many people in positions of power will be enthused and inspired by what you are doing, and will support, rather than hinder…."

3. **But … there are already green groups in this town**, and I don't want to step on their toes…. What your Transition Initiative will do is form a common goal and sense of purpose for the existing groups, some of which you might find are burnt out and will appreciate the new vigor you bring. Expect them to become your allies, crucial to the success of your Transition process.

4. **But … no one in this town cares about the environment anyway.** You'll find that people are already passionate about many aspects of what Transition Initiatives will focus on. The most surprising of people are keen advocates of key elements of a Transition Initiative—local food, local crafts, local history and culture. The key is to go to them, rather than expecting them to come to you.

5. **But … surely it's too late to do anything?** Don't let hopelessness sabotage your efforts. As Vandana Shiva says, "the uncertainty of our times is no reason to be certain about hopelessness." It is within your power to maximize the possibility that we can get through this— don't give that power away.

6. **But … I don't have the right qualifications.** If you don't do this, who else will? What's important is that you care about where you live, that you see the need to act, and that you are open to new ways of engaging people.

7. **But … I don't have the energy for doing that!** As the quote often ascribed to Goethe goes, "whatever you can do or dream you can, begin it. Boldness has genius, power and magic in it!" … there is

something about the energy unleashed by the Transition process that is unstoppable.

This all sounds good in theory, but does it work? Yes, it is working. Not all cases are towns; some are neighborhoods, valleys, even transition streets. In the latter case, just such a program was piloted in 2014 in the U.S. According to Transition U.S., a handful of neighbors got together for seven meetings, during which they follow clear and easy steps in a user-friendly workbook that covers five topics: food, energy, water, transportation, and waste. The focus was on low-cost (or no-cost) actions that result in lowering both expenses and carbon footprint. This popular program has been very successful. Households cut their bills by an average of $900/year and reduced their carbon emissions by 1.3 tons.[62]

A good example is the first transition effort, Transition Town Totnes in Devon, England where among other projects they plan a new low cost housing development that features twenty-seven new energy efficient homes using local natural materials, employing local trades people and suppliers; a community building with a wood-fired boiler and the whole site constructed with low carbon materials throughout to minimize impact. In addition the site will have food growing areas, a coppice, a community orchard, a village green, a community hub and laundry, training and education opportunities for locals, biodiversity improvement through hedgerow management and new hedgerows, wetlands, a wildflower meadow, and nesting for birds and bats. The project is managed as a non-profit and run not by the city government but by local volunteers.[63]

Also in Totnes the transition volunteers have planted nut trees and little food gardens all over the town and they plan another progressive co-housing project, the Baltic Wharf.

> Cohousing can be compared to a small village—there is a cluster of homes and a common house, which is a bit like a village hall. At Baltic Wharf, in common with most other cohousing schemes, we plan to share cars and keep them on the perimeter of the housing, leaving our streets a safe place for children to play. The layout of the housing will encourage social interaction and a real feeling of community. The houses will be designed so that they need minimum energy inputs in their use. There will also be a community garden and orchard for everyone to share and enjoy.[64]

They also helped get solar panels placed on the city hall.

Yet another project involves re-skilling workshops where people who have special skills teach others. The sessions are offered for free as part of the gift economy and "create a fundamental sense of 'can do' and feelings of positivity, creativity and empowerment [and] to establish and nurture links between old and young as skills are passed on."[65]

They also offer an annual entrepreneurs forum where people with

Passing on woodworking skills to the next generation. Hypercivilization has deprived most people of basic skills, from growing and preparing food to woodworking, sewing and mending clothes and many other skills our great grandparents practiced and which we may well need in the future (Monkey Business Images, Shutterstock).

ideas for local, low impact businesses can pitch their ideas to local funders including crowd funding. In addition, they operate a vibrant movement to bring together local food producers and the local market including institutions like schools as well as individuals. These are just a few of the kinds of projects that characterize the transition movement. More information and examples can be found in Hopkin's later book, *The Transition Companion*.

The Natural Step Program began in Sweden and is spreading around the world. The Natural Step process is laid out in detail in *The Natural Step for Communities: How Cities and Towns Can Change to Sustainable Practices*.[66] The authors write: "It is a how to book about changing the way we do things, moving away from practices that are harming the earth and its inhabitants, including us humans.... It gives examples of communities that have changed their practices and offers guiding principles and concrete steps to help other interested communities follow in their steps."[67] It is based on four "system conditions for sustainability" discussed previously. Scores of cities and towns in Sweden and the U.S. have gone thru the Natural Step process and are lowering their ecological footprint while creating a better quality of life for the citizens.

While Transition Towns and Natural Step programs can be

successfully adapted to urban neighborhoods, Richard Register is focused on a major transformation of our great cities. *Ecocities: Building Cities in Balance with Nature,* outlines the major environmental problems with our cities, their causes, and details solutions to retrofit and rebuild them in harmony with the natural world, making cities that are at the same time far more enjoyable to live in.[68] Cheap oil and the automobile have not only contributed to the growing disaster of climate deterioration, they have given us unworkable and environmentally destructive city designs, or more accurately lack of designs—just sprawl. Register writes: "As we build automobile sprawl infrastructures, we create a radically different social and ecological reality than if we build closely-knit communities for pedestrians … [and we will] demonstrate that cities can actually build soils, cultivate biodiversity, restore lands and waters, and make a net gain for the ecological health of the Earth."[69] It's not enough to tinker at the margins. He writes that "Cities need to be rebuilt from their roots in the soil, from their concrete and steel foundations on up. They need to be rebuilt upon ecological principles."[70] That means taking a systems approach. It means getting rid of cars and suburbs. The newly rebuilt cities will be oriented to pedestrians, streetcars and bicycles, green buildings, organic gardens, and dotted with green public spaces. The sensible thing to do is to build up rather than out. Suburbs that are now bedroom communities can become real towns with real commercial, light manufacturing and social centers. We will establish greenbelts and wildlife corridors between the city and the newly restructured towns. Both the city and the towns will be powered by renewables and will make recycling easy.

> …the anatomy of an ecologically healthy city will be characterized by walkable centers, transit villages, discontinuous boulevards, (rising over wildlife corridors or tunneling beneath them), and agricultural areas close to the center. Compact and very diverse town and city cores and neighborhood centers will be universal. The metropolis will have become several pedestrian cities of walkable centers, linked by bicycle, with support for longer distances from public transit.[71]

We would see skywalks and bridges connecting public terraces, verandahs, covered walkways, galleries and back alleys that become garden-like lanes, rooftop arboretums sporting tall trees that seem to float in the sky.

We would see greenhouses, fruit trees and berry bushes, window boxes that will attract birds and insects and everywhere water—brooks, cascades, and fountains. The miniaturized, three-dimensional pedestrian city he describes will be extraordinarily efficient with a far smaller ecological footprint than our current mega-sprawls. Getting from here to there will be exciting. Does it all seem hopelessly utopian? Register reminds us that a new idea "is initially derided as ridiculous, then opposed violently as dangerous, and finally embraced as self-evident."[72] If we are not already at the

Eco-cities. The cities of Ecocivilization will be characterized by green spaces that promote human health and biodiversity. This residential tower in Milan, Italy, sprouts trees and shrubs characteristic of the eco-cities envisioned by Richard Register in his book of that title (Pier Lugi. Palazzi, Shutterstock).

third stage we are, thanks to the crisis of climate deterioration, on the edge of it.

The Old Story is petering out. A new one is being born. The global economy is failing to serve the peoples of the globe while devastating the environmental basis of civilization. But we are developing new ideas and new forms of community, focusing more on local resilience, local self-reliance and local self-government in the interests of creating communities that are not only sustainable far into the future but also more just and more livable. The good news is that these trends are already underway.

# Conclusion
## We Can Make It—On Toward Ecocivilization

"It's the end of the world as we know it, but it's not the end of the world."[1]—Jack Nelson Pallmeyer, *Authentic Hope*

"Like it or not, we are not masters of this world but are simply part of a community. It is time to act like it."[2]—Randal Amster, *Peace Ecology*

Hypercivilization will come to an end, possibly in a spectacular failure involving nuclear weapons or a slow crash as resources run out and nature's sinks fill up and ecosystems that support organized human life fail. Or we can make a rational transition to Ecocivilization. We just might make it. I do know that the great anthropologist Margaret Mead, was right when she said: "We won't have a society if we destroy the environment."[3] What must we do when seemingly powerful forces are arrayed against us, when most of the corporate world currently controlling governments and their militaries are stuck in the Old Story of Hypercivilization and are even defending it, though ever more shrilly? When the governing elites benefit from the extraction-consumption model of human interaction with nature and they own most of the media which provides the world's story line on a daily basis.

Fundamental change takes time and we know time is short now. I have tried to lay out in detail the problems we need to solve lest we succumb to a false optimism that everything will turn out all right and we don't need to do much, don't need to change at a fundamental level of values and vision. Nevertheless, we must not simply focus our gaze on the problems and the difficulties, for that will defeat us. Paul Hawken suggests that we need to pay attention "to what is going *right* on this planet, narratives of imagination and conviction, not defeatist accounts about the limits. Wrong is an addictive, repetitive story. Right is where the movement

278

is."[4] And so much is going right and at an ever accelerating speed. Jack Pallmeyer writes:

> Authentic hope is rooted in the capacity to view or experience injustice without being overwhelmed because we also see existing evidence for and the possibility of greater goodness…. People who embody resilient hope, in the words of Meister Eckhart, a thirteenth century mystic, "don't walk sightless among miracles."[5]

And there are miracles aplenty, even if they are not reported in the conventional media. For example, a group called "Desert Wave," is "a community organization dedicated to exploration and creation of new models for sustainable food production and housing." They are gearing up to build what they call the "world's first agri-city" in the desert in Utah and they plan to expand their model globally.[6] Across the Atlantic an organization called "ReGen" is building "a new type of community designed to be fully self-sufficient, growing its own food, making its own energy, and handling its own waste in a closed loop."[7] They, too, plan to expand their model and build these self-sufficient communities world-wide. There are hundreds if not thousands of similar projects on the drawing boards or already underway. Our privilege is to learn about, collaborate and bolster them. We are at a hyper-critical turning point in human and earth history and as such it is the most exciting time to be alive in history. We can see the shore of the Promised Land from here and millions want to cross over to it. There is a tremendous amount to learn and a tremendous amount to do. There are wonderful opportunities for creative action, innovation and community building.

I am a realist; I know the how steep is the grade that lies before us, but as a realist I also know roads this steep have been climbed before. Rapid and unforeseen changes on a great scale do happen, have happened. In 1789 the French aristocracy thought their world would go on forever but it was swept away in a couple of years. No one predicted the sudden fall of the Berlin Wall or the sudden arrival of the internet age or the swift turn-around of attitudes regarding gays in the U.S., or any number of other historical revolutions one could mention. Note, I am not suggesting that these revolutions occurred spontaneously. Strong cultural forces were at work below the surface, preparing the way. Legalized human slavery, which had been around for thousands of years and was declared necessary for the economy and was sanctioned by philosophers and by the Church, fell quickly because of the initial work of a handful Methodists and Quakers at the end of the eighteenth century, men and women who had humane values, a vision, and authentic hope. The negative past is no guarantee of the future. Because some are still on the road to breakdown does not mean many others aren't already turning off toward breakthrough. John Schaar is right; human agency is very real.

The future is not a result of choices among alternative paths offered by the present, but a place that is created—created first in the mind and will, created next in activity. The future is not some place we are going to but one we are creating. The paths are not to be found, but made, and the activity of making them changes both the maker and the destination.[8]

Furthermore, the pressures on the Old Story of Hypercivilization are building every moment. Environmental writer David Orr points out: "The question is not whether we will change, but whether the transition will be done with more or less grace and whether the destination will be desirable or not."[9] And James Speth says there is much we can do in the meantime: "We need not wait for crises to deepen in order to change course; rather, our task is 'to develop alternatives to existing policies, to keep them alive and available until the politically impossible becomes the inevitable.'"[10]

Each of us faces a choice. We can start out defeated or we can push ahead with hope. In the words of Randall Amster, "there is no place for abdication—it is incumbent upon us to act well, and right now."[11] And we can't let that old argument defeat us, the one that goes: "I am just one person and what can one person do against big oil and against the Pentagon?" No one of us has to do it all. I love assertion attributed to the late Bill Mollison, the founder of permaculture, who said: "I can't save the world by myself. It's going to take at least three of us."[12] But even if there were only one of us, it would still be the right thing to do as the Rev. Everett Edward Hale said over a hundred years ago.

> I am only one,
> But still I am one.
> I cannot do everything,
> But still I can do something;
> And because I cannot do everything,
> I will not refuse to do the something that I can do.[13]

In fact, we are not just one; millions are already working to move us in the direction of Ecocivilization and they have been newly energized by the IPCC's warning that we only have twelve years to make significant reductions in emissions. As of this writing there is a new and growing enthusiasm in the U.S. for a Green New Deal, a comprehensive plan to replace Hypercivilization with Ecocivilization, leading with 100 percent renewable energy, green jobs, and a shutdown of the fossil fuel industry. A youth movement calling itself the Sunrise is mobilizing to stop climate change via education and protests. They recently occupied Congresswoman Nancy Pelosi's office, prompting arrests. In Great Britain, a nonviolent civil disobedience movement called "The Extinction Rebellion," focused on climate and on the extinction crisis, is mounting shutdown protests in London and attracting attention from around the world. Most exciting, on March 15,

2019, 1,400,000 children walked out of their schools in 128 countries and demonstrated, demanding decisive action on climate deterioration. It is too early to see where all this will lead, but it is encouraging to see the mounting fervor for change.

I choose to believe in a redemptive future and to work for it. We humans possess all we need to know in order to get under way. We know the terrible price of doing nothing, of letting Hypercivilization drift on. Will we make it? I choose to side with Lester Brown, the President of World Watch Institute, who observed: "First we need to decide what needs to be done. Then we do it. Then we ask if it is possible."[14]

There is no better advice.

# Chapter Notes

## Introduction

1. Peter Menzel, *Material World: A Global Family Portrait.* Text by Charles Mann (San Francisco: Sierra Club Books, 1994), p. 9.

2. Attributed to Thomas Gold by Timothy Ferris in *Coming of Age in the Milky Way* (Doubleday: New York, 1998), p. 338.

3. William K. Hartman and Ron Miller. *The History of the Earth: An Illustrated Chronicle of an Evolving Planet* (New York: Workman Publishing Co., 1991), p. 65.

4. Hartman and Miller, p. 23.

5. Loren Eiseley, *The Immense Journey: An Imaginative Naturalist Explores the Mysteries of Man and Nature* (New York: Time, Inc., 1957), p. 59.

6. Quoted in Paul Shepard, *Man in the Landscape* (New York: Ballantine, 1972), p. 43.

7. Brian Swimme, "How to Heal a Lobotomy," in Irene Diamond and Gloria Feman Orenstein, eds., *Reweaving the World: The Emergence of Ecofeminism* (San Francisco: Sierra Club Books, 2990), p. 20.

8. Invasive Species Advisory Committee, "Invasive Species Definition Clarification and Guidance White Paper," Submitted by the Definitions Subcommittee of the Invasive Species Advisory Committee (ISAC) Approved by ISAC April 27, 2006 Preamble: Executive Order 13112.

9. Waters Foundation, http://www.watersfoundation.org/index.cfm?fuseaction=content.display&id=93.

10. Susan Griffin, *The Eros of Everyday Life: Essays on Ecology, Gender and Society* (New York: Anchor Books, 1996), 22 & 36.

11. Brian Swimme, "How to Heal a Lobotomy" in Irene Diamond and Gloria

Feman Orenstein, eds. *Reweaving the World: The Emergence of Ecofeminism* (San Francisco: Sierra Club Books, 1990), p. 15.

12. Daniel Quinn, *Ishmael* (New York: Bantam/Turner, 1993), p. 26.

13. John Donne, *Meditation XVII*, The Literature Network, http://www.online-literature.com/donne/409/.

## Chapter 1

1. Lorraine Chow, *EcoWatch*, Nov. 15, 2017.

2. Gaylord Nelson, *Beyond Earth Day: Fulfilling the Promise* (Madison: University of Wisconsin Press, 2002).

3. Max Roser, "Economic Growth," Our World in Data, https://ourworldindata.org/economic-growth.

4. Victor Lebow, *Journal of Retailing*, Spring, 1955, http://www.gcafh.org/edlab/Lebow.pdf.

5. Muhammed Yunus, *A World of Three Zeros* (New York: Public Affairs, 2017), Pp.4–5.

6. https://d2gne97vdumgn3.cloudfront.net/api/file/QWz65NxtRXG58frH8BAU.

7. World population, https://en.wikipedia.org/wiki/World_population.

8. Donella Meadows, *The Limits to Growth* (New York: Mass Market Paperback, 1972), p. 19.

9. Charles H. Southwick, *Global Ecology in Human Perspective* (Oxford: Oxford University Press, 1996), p. 164.

10. Southwick, p. 161.

11. Max Roser and Hannah Ritchie, "Fertilizer and Pesticides," Our World in Data, https://ourworldindata.org/fertilizer-and-pesticides/.

12. Tom Philpott, "A Reflection on the Lasting Legacy of 1970s USDA Secretary Earl Butz," *Grist*, Feb 8, 2008, http://grist.org/article/the-butz-stops-here/.

13. Paul Gruchow, *Grass Roots: The Universe of Home* (Minneapolis: Milkweed Editions, 1995), pp., 87–88.

14. Frank Clifford, "Food or the Forest?" *Duluth News Tribune*, Aug. 7, 1997, p. 2a.

15. Norman Myers, *Gaia Atlas of Planet Management* (New York: Anchor, 1992), P. 64.

16. Clive Ponting, *A Green History of the World* (New York: Penguin, 1991), Pp. 309 & 310.

17. Archway Self-Publishing, "Cities with More Than One Million Inhabitants in 2005," http://data.mongabay.com/igapo/2005_world_city_populations/2005_city_population_01.html.

18. Ponting, p. 397.

19. Ponting, pp. 309, 310.

20. Frank Barnaby, ed., *Gaia Peace Atlas* (Berkeley: Pan Books, 1988), p. 111.

21. Robert Kaplan, *The Ends of the Earth: A Journey at the Dawn of the 21st Century* (New York: Random House, 1996), p. 192.

22. Lester Brown and Jodi Jacobson, "Assessing the Future of Urbanization," in Lester Brown, et al., *State of the World, 1987* (New York: Norton, 1987), p. 40.

23. Kaplan, p. 328.

24. Lester Brown and Jodi Jacobson, "Assessing the Future of Urbanization," in Lester Brown, et al., *State of the World, 1987* (New York: Norton, 1987), p. 40.

25. Wendell Berry, *The Unsettling of America,* available in *Counterpoint*; Reprint edition (September 1, 2015), p. 63.

26. Richard Louv, *Last Child in the Woods: Saving Our Children from Nature Deficit Disorder* (Chapel Hill, NC: Algonquin Books, 2006).

27. *Ibid.* p. 35.

28. *Ibid.*, p. 47.

29. Stewart Udall, *The Energy Balloon* (New York: McGraw Hill, 1974), p. 28–29.

30. Nicholas Hildyard, *Foxes in Charge of the Chickens*, in Wolfgang Sachs, ed., *Global Ecology: A New Arena of Political Conflict* (Halifax, Nova Scotia: Fernwood Publishing, 1993), p. 30.

31. Donella Meadows, et al., *Worldwide Growth in Selected Human Activities and Products, 1950–2000, Limits to Growth: The Thirty Year Update* (White River Junction, VT: Chelsea Green Publishing Co., 204) p. 8

32. *Ibid.*, p. 8.

33. Michael Bloch, "Green Living Tips," http://www.greenlivngtips.com/articles/185/1/Consumption-statistics.html.

34. Ewaste, "Electronic Waste by Numbers: Recycling & the World," March 2, 2016, http://www.ewaste.com.au/ewaste-articles/electronic-waste-by-numbers-recycling-the-world/.

35. Nicholas Freudenberg and Carol Steinsapir, "Not in Our Backyards; the Grassroots Environmental Movement," *American Environmentalism: The U.S. Environmental Movement: 1970–1990* (New York: Taylor and Francis, 1992), p. 28.

36. "Safe Chemicals Act Could Reverse Burden of Proof for Toxic Chemicals, Protect Children," Huffington Post, http://www.huffingtonpost.com/2012/07/24/safe-chemicals-act-flame-retardants_n_1699384.html Posted: 07/24/2012 7:30 pm Updated: 07/24/2012 7:57 pm.

37. Donella Meadows and Jorgen Randers, *Beyond the Limits: Confronting Global Collapse, Envisioning a Sustainable Future* (White River Junction, VT: Chelsea Green, 1992) figure 3–14, p. 81.

38. Pesticide Action Network, "At Long Last, EPA Releases Pesticide Use Statistics," http://www.panna.org/blog/long-last-epa-releases-pesticide-use-statistics.

39. Pesticide Action Network, "Roundup, Cancer and the Future of Food," http://www.panna.org/blog/roundup-cancer-future-food?gclid=Cj0KEQjw4pO7BRDl9ePazKzr1LYBEiQAHLJdR76WMGc0P-JIB777hzPqQ3Rh6Z4ra9Nv1reZf7C1ts4aAvwK8P8HAQ.

40. Pesticide Action Network, "Roundup, Cancer and the Future of Food," http://www.panna.org/blog/roundup-cancer-future-food?gclid=Cj0KEQjw4pO7BRDl9ePazKzr1LYBEiQAHLJdR76WMGc0P-JIB777hzPqQ3Rh6Z4ra9Nv1reZf7C1ts4aAvwK8P8HAQ.

41. Earthjustice, http://earthjustice.org/news/press/2011/as-embattled-ca-pesticide-chief-steps-down-feds-consider-petition-to-ban-cancer-causing-pesticide.

42. Earth Justice Action Alert, "Methyl Iodide Controversy: Warning About Strawberry Field Chemical Ignored." http://www.huffingtonpost.com/2010/06/07/methyl-iodide-controversy_n_602904.html, [and] "Tell EPA: Ban Cancer-Causing Strawberry Pesticide Before It's Too Late," April 19, 2011.

43. William Ashworth, *The Late Great Lakes: An Environmental History* (New York: Alfred A. Knopf, 1986), p. 163.

44. http://www.pvc.org/en/p/packaging.

45. Norman Meyers, *Gaia Atlas of Planet Management* (New York: Anchor, updated edition 1992), p. 122.

46. Reported in Joel Bleifus, "Sex and Toxics," *In These Times*, March 7, 1994, p. 12.

47. Adam Rogers, "Chemicals: The Great Impostors," *Newsweek*, March 18, 1996.

48. *Ibid.*, p. 48.

49. Wendy Koch, "Fda Officially Bans Bpa, or Bisphenol-A, from Baby Bottles," *USA Today*, 7/17/2012, http://usatoday30. usatoday.com/money/industries/food/story/ 2012-07-17/BPA-ban-baby-bottles-sippy-cups/56280074/1.

50. Mark B. Bush, *Ecology of a Changing Planet* (Upper Saddle River, NJ: Prentice Hall, 1997), p. 296.

51. Matt McDermot, "20% of World's Plant Species Threatened with Extinction—Yes, Human Activity Is Main Cause," *Treehugger*, http://www.treehugger. com/natural-sciences/20-of-worlds-plant-species-threatened-with-extinction-yes-human-activity-is-main-cause.html.

52. Richard Leakey and Roger Lewin, *The Sixth Extinction: Patterns of Life and the Future of Humankind* (New York: Anchor, 1996).

53. John C. Ryan, "Conserving Biological Diversity," *State of the World, 1992*, p. 9.

54. Ryan, p. 9.

55. World Lion Day, https://worldlionday. com/african-lion/.

56. Treehugger, "20% of World's Plant Species Threatened with Extinction—Yes, Human Activity Is Main Cause," http:// www.treehugger.com/natural-sciences/20-of-worlds-plant-species-threatened-with-extinction-yes-human-activity-is-main-cause.html.

57. Bird Life International, "Spotlight on Threatened Birds," http://www.birdlife.org/ datazone/sowb/spotthreatbirds.

58. Malaysiakini, "45 Bird Species in Malaysia Face Extinction: Report," https:// www.malaysiakini.com/news/73686.

59. "Malaysian Mammals Face Extinction," *The Star* http://www.thestar.com. my/news/nation/2015/04/14/malaysian-mammals-face-extinction-world-bank-fifth-of-species-threatened/.

60. The Center For Biological Diversity, "The Amphibian and Reptile Extinction Crisis," http://www.biologicaldiversity.org/ campaigns/amphibian_conservation/.

61. V.C. Chong, et al., "Diversity, Extinction Risk and Conservation of Malaysian Fishes," http://www.slideshare. net/PatrickLee37/diversity-extinction-conservation-of-malaysian-fishes-2010-jfb-v76.

62. "Introduced Species Summary Project, Nile Perch," Columbia University, http:// www.columbia.edu/itc/cerc/danoffburg/ invasion_bio/inv_spp_summ/Lates_ niloticus.htm.

63. James Tennent, "More than a Quarter of British Birds at Risk of Local Extinction," *International Business Times, Dec. 3, 2015*, http://www.ibtimes.co.uk/ more-quarter-british-birds-risk-local-extinction-1531645.

64. Global Fact Sheet #16: 1SC8, "Loss of Biodiversity," https://www.coursehero. com/file/7132306/16-1SC8-Loss-of-Biodiversity/.

65. Robert Gebelhoff, "Study of Land Snails Suggests Earth May Already Have Lost Seven Percent of Its Species," *Washington Post*, Aug. 8, 2015, https://www. washingtonpost.com/news/morning-mix/wp/2015/08/11/study-of-land-snails-suggests-earth-may-have-already-lost-7-percent-of-its-animal-species/.

66. Brooke Jarvis, "The Insect Apocalypse," *New York Times*, November 27, 2018.

67. Center for Biological Diversity, "The Southeast Freshwater Extinction Crisis," http://www.biologicaldiversity.org/ programs/biodiversity/1000_species/the_ southeast_freshwater_extinction_crisis/.

68. Peter Raven, "The Global Ecosystem in Crisis," a Macarthur Foundation Occasional Paper, Chicago, 1987, P. 15.

69. World Elephant Day, "Why Care?," http://Worldelephantday.Org/About/ Elephants.

70. Vital Ground, "The Grizzly Bear Predicament in North America," http:// www.Vitalground.Org/About-Bears/ North-American-Grizzlies/Why-Theyre-Threatened/.

71. Bush, P. 296.

72. *Sean Martin, World Rhino Day 2015: Rhinoceroses in Numbers, International Business Times, September 22, 2015*, http://www. Ibtimes.Co.Uk/World-Rhino-Day-2015-Rhinoceroses-Numbers-1520744.

73. Jaymi Heimbuch, "11 Critically Endangered Turtle Species," *Treehugger*, May 23, 2013. http://www.treehugger.com/ natural-sciences/11-critically-endangered-turtle-species.html.

74. Ryan, p. 92.

75. Cathryn Wellner, "Coming Soon: A World Without Penguins," Care2 E-Letter, June 24, 2012.

76. Ryan, p. 14.

## Chapter 2

1. Lester Brown, et al., *Vital Signs 1997*, p. 20.

2. Donella Meadows, et al., *Limits of Growth: The Thirty Year Update* (White River Junction, VT: Chelsea Green Publishing Co., 2004), p. 57.

3. Native Forest Council, *Forest Voice*, vol. 1, # 1, September, 1989, p. 1.

4. Richard Conniff, "Amid the Plunder of Forests, a Ray of Hope," *New York Times*, January 28, 2018.

5. Meadows, p. 58.

6. Nancy Roberts, "Top 3 Victims of Palm Oil: Wildlife, People and Planet," April 5, 2012, http://www.care2.com/causes/top-3-victims-of-palm-oil-wildlife-people-and-planet.html#ixzz1rGatLErK.

7. Robin Broad with John Cavanagh, *Plundering Paradise: The Struggle for the Environment in the Philippines* (Berkeley: University of California Press, 1993), p. 1.

8. Marcus Colchester and Larry Lohmann, eds., *The Struggle for the Land and the Fate of the Forests* (Sturminister Newton, England: World Rainforest Movement, Penang, Malaysia and *The Ecologist*, 1993), p. 21.

9. Meadows, p. 61.

10. Mark Bush, *Ecology of a Changing Planet* (Upper Saddle River, NJ: Prentice Hall, 1997), p. 260.

11. D.C. Everest, "Speech Delivered at Port Edwards, Wisconsin, October 4, 1949, at the Banquet Celebrating the 25th Anniversary of the Establishment of Nekoosa-Edwards Paper Company's Industrial Forestry Project," http://www.worldcat.org/identities/lccn-n85201504/.

12. Nancy Roberts, "Top 3 Victims of Palm Oil: Wildlife, People and Planet," April 5, 2012, http://www.care2.com/causes/top-3-victims-of-palm-oil-wildlife-people-and-planet.html#ixzz1rGatLErK.

13. Clive Ponting, *A Green History of the Earth* (New York: St. Martin's Press, 1991), p. 265.

14. Reported in Charles H. Southwick, *Global Ecology in Human Perspective* (Oxford: Oxford University Press, 1996), p. 104.

15. Charles H. Southwick, *Global Ecology in Human Perspective* (Oxford: Oxford University Press, 1996), p. 347.

16. *Ibid.*, p. 100.

17. *Ibid.*, p. 347.

18. *Ibid.*, p. 85.

19. "The Drought of 2012," Aug 28, 2012, http://www.theatlantic.com/infocus/2012/08/the-drought-of-2012/100360/.

20. *Christian Science Monitor*, Sept. 16, 1982, p. 12.

21. Southwick, p. 101.

22. Kodi, Yeager-Kozacek, "Droughts Hit World's Agricultural Regions: Without Water, U.S. Corn Crop Faces Setbacks," *Circle of Blue*, July 12, 2012.

23. Southwick, p. 105.

24. Mark B. Bush, *Ecology of a Changing Planet* (Upper Saddle River, NJ: Prentice Hall, 1997), pp. 234–235.

25. Ponting, p. 264.

26. George Gordon Lord Byron, "Apostrophe to the Ocean," from Canto IV of *Childe Harold's Pilgrimage* in Rewey Inglis and Josephine Spear, *Adventures in English Literature* (New York: Harcourt Brace, 1958), p. 412.

27. "The Effects of Bottled Water on the Environment" at All About Water. Org, http://www.Allaboutwater.Org/Environment.Html.

28. Center for Biological Diversity, "Ocean Plastics Pollution: A Global Tragedy for Our Oceans and Sea Life," http://www.biologicaldiversity.org/campaigns/ocean_plastics/.

29. Sarah Kaplan, "By 2050, There Will Be More Plastic Than Fish in the World's Oceans, Study Says," *Washington Post*, January 20, 2016 https://www.washingtonpost.com/news/morning-mix/wp/2016/01/20/by-2050-there-will-be-more-plastic-than-fish-in-the-worlds-oceans-study-says/?utm_term=.39632090c7.

30. "Demistifying the Great Pacific Garbage Patch," http://marinedebris.noaa.gov/info/patch.html#1.

31. Abigail Entwistle, 9th November 2017, https://www.fauna-flora.org/news/plastic-pollution-disposable-life?gclid=-Cj0KCQiA1afSBRD2ARIsAEvBsNkccJM-iTZFOxD3rWcsI7UrVWsK6sfdWQrBtprSBAZy1DT6qCvatQoaAhBtEALw_wcB.

32. Stiv Wilson, "Plastics in the Great Lakes: A Guest Post," *5gyres*, November 09, 2012, http://5gyres.org/posts/2012/11/09/plastics_in_the_great_lakes_a_guest_post.

33. Moms Clean Air Force, email alert, "Plastics in the Great Lakes," November 14, 2012.

34. Sierra Club, *Sierra* (January–February, 2018), p. 20.

35. Jennifer Mueller, "Are We Running Out of Water? World Water Crisis 101" (Part 1), April 7, 2012, http://www.care2.com/causes/world-water-crisis-101-part-1-quantity-and-quality.html#ixzz1rS5qRCxr.

36. National Oceanic and Atmospheric Agency, "Oil Pollution," 1995. http://seawifs.gsfc.nasa.gov/OCEAN_PLANET/HTML/peril_oil_pollution.html.

37. Southwick, p. 240.

38. Southwick, p. 240.

39. European Union, "Factory Farming Makes Baltic Sea One of the World's Most Polluted," http://www.arc2020.eu/2014/06/factory-farming-made-the-baltic-sea-one-of-the-worlds-most-polluted-seas/.

40. "A Clean Water Crisis," *National Geographic*, http://environment.nationalgeographic.com/environment/freshwater/freshwater-crisis/, Accessed June 7, 2016.

41. Bush, p. 229.

42. Dan Baum and Margaret L. Fox, "In Butte, Montana, A Is for Arsenic, Z Is for Zinc," *Smithsonian* Vol. 23, no. 8, Nov. 1992, p. 48.

43. Baum and Knox, p. 49.

44. Stewart Udall, former Secretary of the Interior, in a foreword to Harry Caudhill, *Night Comes to the Cumberlands* (Boston: Little, Brown and Company, 1962), pp. vii & viii.

45. Harry Caudhill, *Night Comes to the Cumberlands* (Boston: Little, Brown and Company, 1962), p. 150.

46. Caudhill, p. 74.

47. Caudhill, p. 93.

48. Earth Justice (http://earthjustice.org/features/campaigns/what-is-mountaintop-removal-mining).

49. *Ibid.*

50. Stewart Udall, *The Energy Balloon* (New York: McGraw-Hill, 1074), p. 66.

51. U.S. Energy Information Administration, Frequently Asked Questions, https://www.eia.gov/tools/faqs/faq.cfm?id=427&t=3.

52. Southwick, p. 198.

53. Weitz, quoted in Southwick, p. 211.

54. Southwick, p. 205.

55. "Air Pollution in China," Facts and Details http://factsanddetails.com/china.php?itemid=392&catid=10&subcatid=66.

56. *Ibid.*

57. *Ibid.*

58. Bush, p. 342.

59. Bush, p. 343.

60. Southwick, p. 204.

61. *Ibid.,* p. 202.

62. *Ibid.,* p. 203.

63. Norman Meyer, *Gaia Peace Atlas,* p. 118.

64. *Ibid.*

65. Bill McKibben, *The End of Nature* (New York: Random House, 1989), pp. 38–39.

66. Southwick, p. 215.

67. McKibben, pp. 39–40. The 100,000 figure is from Southwick, p. 215.

68. Southwick, p. 217.

69. McKibben, pp. 42–43.

70. Southwick, p. 216.

71. United Nations Environment Program, Ozone Secretariat, "The Montreal Protocol on Substances That Deplete the Ozone Layer," http://ozone.unep.org/en/treaties-and-decisions/montreal-protocol-substances-deplete-ozone-layer.

## Chapter 3

1. Some of the material in this section was published previously in my earlier book, *From War to Peace: A Guide to the Next Hundred Years* (Jefferson, NC: McFarland, 2011).

2. Tom Hastings, *Ecology of War and Peace: Counting the Costs of Conflict* (Lanham, MD: University Press of America, 2000), p. xvii.

3. Charles Southwick, *Global Ecology in Human Perspective* (Oxford: Oxford University Press, 1996), p. 315.

4. William Thomas, *Scorched Earth: The Military Assault on the Environment* (Philadelphia: New Society Publishers, 1995), pp. 110–111.

5. William Thomas, p. 112.

6. Tom Hastings, p. 47.

7. Southwick, p. 316.

8. Thomas, pp. 112,113 & 114.

9. Saul Bloom, John M. Miller, James Warner and Philippa Winkler, et al., eds., *Hidden Casualties: Environmental, Health and Political Consequences of the Gulf War* (Berkeley, CA: North Atlantic Books, 1993), p. 73.

10. Saul Bloom, et al., p. 82.

11. Saul Bloom, et al., p. 72.

12. Thomas, *Scorched Earth: The Military*

*Assault on the Environment* (Philadelphia: New Society Publishers, 1995), p. 16.

13. Thomas, p. 8.

14. Thomas, p. 67 and 68.

15. Tom Hastings, p. 40.

16. Thomas, p. 127.

17. Thomas, p. 80.

18. "List of Countries by Military Expenditures List of Countries by Military Expenditures," Stockholm Peace Research Institute, https://en.wikipedia.org/wiki/List_of_countries_by_military_expenditures.

19. National Priorities Project, "Military Spending in the United States," https://www.nationalpriorities.org/campaigns/military-spending-united-states/.

20. Arms Control Association, "U.S. Nuclear Modernization Programs," (August 2017), https://www.Armscontrol.Org/Factsheets/Usnuclearmodernization.

21. Kent Shifferd, *From War to Peace: A Guide to the Next Hundred Years* (Jefferson, NC: McFarland, 2011), pp. 25-26.

22. Tom Zoellnor, *Uranium: War, Energy and the Rock That Shaped the World* (New York: Penguin, 2009), p. 65.

23. Nuclear Regulatory Commission, Http://Www.Nrc.Gov/Waste/Spent-Fuel-Storage/Faqs.Html#20.

24. Paul Ehrlich and Carl Sagan, *In the Cold and the Dark: The World After Nuclear War* (New York: W.W. Norton, 1984).

25. Lester Brown, et al., *Vital Signs* (New York: Routledge, 1997), p. 48.

26. Peter Goin, *Nuclear Landscapes* (Baltimore: Johns Hopkins University Press, 1991), p. xxi.

27. Aubrey Wallace, *Eco-Heroes: Twelve Tales of Environmental Victory* (San Francisco: Mercury House, 1993), Pp. 96–97.

28. Thomas, P. 34.

29. Betty Eisendrath, "Military Ecocide: Man's Secret Assault on the Environment (World Milwaukee: Federalists, Milwaukee Chapter, 1992), p. 5.

30. Thomas, P. 133.

31. Peter Goin, P. Xx.

32. Goin, P. 25.

33. M. Stenehjem, "Indecent Exposure," *Natural History*, 9, 1990, pp. 6ff.

34. Thomas, P. 133

35. Debra Chasnoff, producer & director, *Deadly Deception: GE, Nuclear Weapons and Our Environment* (video), National INFACT, 256 Honover Street, Boston, MA 02113, 199.

36. Goin, p. 25.

37. Tom Hastings, p. 9.

38. John Gofman, "Control of the Atom," in *Progress as If Survival Mattered: A Handbook for a Conserver Society* (San Francisco: Friends of the Earth, 1981), p. 48.

39. *Nukewatch*, Fall; 2013, p. 5.

40. *Nukewatch*, "More Trouble for Vermont Yankee," *Nukewatch Quarterly*, Fall 2011, p. 7.

41. William Ashworth, *The Late Great Lakes: An Environmental History* (Detroit: Wayne State University Press, 1987), p. 169.

42. Christopher Flavin, "Reassessing Nuclear Power," *State of the World*, World Watch Institute, 1996, p. 58.

43. Wikipedia, "Fukushima Daiichi Nuclear Disaster," http://en.wikipedia.org/wiki/Fukushima_Daiichi_nuclear_disaster#cite_note-9 citing F. Tanabe, Journal of Nuclear Science and Technology, 2011, volume 48, issue 8, pages 1135 to 1139.

44. Wikipedia, "Fukushima Daiichi Nuclear Disaster," http://En.Wikipedia.Org/Wiki/Fukushima_Daiichi_Nuclear_Disaster#Cite_Note-9 Citing "3 Nuclear Reactors Melted Down After Quake, Japan Confirms". *Cnn*. 7 June 2011. And http://edition.cnn.com/2011/WORLD/asiapcf/06/06/japan.nuclear.meltdown/index.html?iref=NS1. And "'Melt-through' at Fukushima? / Govt report to IAEA suggests situation worse than meltdown," *Yomiuri*. 8 June 2011. http://www.Yomiuri.Co.Jp/Dy/National/T110607005367.Htm. *Retrieved 8 June 2011.*

45. David McNeill, "Why the Fukushima Disaster Is Worse Than Chernobyl," *Independent* (Uk) August 29, 2010.

46. Henry Fountain, "Cleanup Questions as Radiation Spreads," *New York Times*, April 1, 2011, on page A10.

47. Amy Goodman, "Hiroshima and Fukushima Censored," *Nukewatch Quarterly*, Fall 2011, p. 5. http://www.nukewatchinfo.org/japan/index.html.

48. "Radiation Gushes from Fukushima, Information Trickles," *Nukewatch Quarterly*, Fall 2011, p.5.

49. "Fukushima Disaster Produces World's Worst Nuclear Sea Pollution," The Maritime Executive, 2011–10–28 11:48:31.

50. E. F. Schumacher quoted in Gofman, "Control of the Atom," Hugh Nash, ed., *Progress as If Survival Mattered* (San Francisco: Friends of the Earth, 1982), p. 49.

51. Zoellnor, *Uranium: War, Energy and the Rock That Shaped the World* (New York: Penguin, 2009), p. 23.

## Chapter 4

1. Anne E. Platte, "Infecting Ourselves: How Environmental and Social Disruptions Trigger Disease," *World Watch Paper #129* (Washington, D.C: World Watch Institute, April, 1996), pp. 31–32.p. 5.

2. Jim Forsyth, "International Tourism Record: 1 Billion Travelers in 2012," Reuters, Posted: 12/06/2012, http://www.huffingtonpost.com/2012/12/06/international-tourism-to-_n_2253347.html.

3. Andrew Jacobs and Ian Johnson, "Pollution Killed 7 Million People Worldwide in 2012, Report Finds," *New York Times*, March 25, 2014.

4. John Collins Rudolf, "Polluted Air and Diabetes: A Link," *New York Times*, October 4, 2010.

5. David Swanson, *War No More: The Case for Abolition*, Charlottesville, VA, 2013, pp. 119–121.

6. Bruce Short, "War and Disease: War Epidemics in the Nineteenth and Twentieth Centuries," *ADF Health* Vol. 11, No. 1, 2010, p. 16.

7. Short, p. 16.

8. Platte, pp. 32–33.

9. latte, p. 46.

10. Laurie Garrett, *The Coming Plague: Newly Emerging Diseases in a World Out of Balance* (New York: Penguin, 1994), p. 52.

11. Garrett, pp. 53ff.

12. Platte, p. 33.

13. Garrett, p. 528.

14. The Harvard Working Group on New and Resurgent Diseases, "Globalization, Development and the Spread of Disease," in Jerry Mander and Edward Goldsmith, *The Case Against the Global Economy* (San Francisco: Sierra Club Books, 1996), p. 162.

15. Centers for Disease Control, World Map of Areas with Risk of Zika https://wwwnc.cdc.gov/travel/page/world-map-areas-with-zika.

16. Harvard Working Group, p. 161.

17. Harvard Working Group, p. 161.

18. Platt, p. 54.

19. Jeremy Rifkin, *The Biotech Century: Harnessing the Gene and Remaking the World* (New York: Penguin/Putnam, 1998), p. x.

20. Rifkin, p. 14.

21. Rifkin, p. 35.

22. Rifkin, p. 15.

23. Andrew Pollack, "A Dream of Trees Aglow at Night," *New York Times*, May 7, 2013.

24. "Genetic Engineering," Accessed Jan 7, 2014, http://en.wikipedia.org/wiki/Genetic_engineering.

25. Cited in Rifkin, p. 30.

26. "Synthetic Biology," Wikipedia, http://en.wikipedia.org/wiki/BioBrick.

27. *Ibid.*

28. *Ibid.*

29. "What Is Synthetic Biology?" http://sytheticbiology.org/FAQ.html.

30. "Biobrick," http://en.wikipedia.org/wiki/BioBrick.

31. *Ibid.*

32. *Ibid.*

33. "Genetic Engineering," Accessed Jan 7, 2014, http://en.wikipedia.org/wiki/Genetic_engineering.

34. N.I.H. website, https://ghr.nlm.nih.gov/primer/hgp/genome.

35. "Genome," *Biology Online Dictionary*, http://www.biology-online.org/dictionary/Genome.

36. "Tabula Rasa," *Smithsonian*, July/August, 2013, p. 16.

37. Adam Liptak, "Justices, 9–0, Bar Patenting Human Genes," *New York Times*, June 13, 2013.

38. *Ibid.*

39. Virginia Hughes, *Popular Science*, October, 2013, p. 40.

40. David Mclean in *Historical Events in the RDnA Debate*, 1997, http://www.ndsu.edu/pubweb/~mcclean/plsc431/debate/debate3.htm

41. Ed Yong, "Can We Save the World by Remixing Life?" *National Geographic*, http://phenomena.nationalgeographic.com/2013/04/11/can-we-save-the-world-by-remixing-life/.

42. *Ibid.*

43. "Books," *Smithsonian*, July/August 2013, p. 98.

44. Andrew Pollack, "A Dream of Trees Aglow at Night," *New York Times*, May 23, 2013.

45. Adele Peters, "Imagine a City Lit by Glowing Trees Instead of Streetlights," Fastcompany, May 14, 2918, https://www.fastcompany.com/40571215/imagine-a-city-lit-by-glowing-trees-instead-of-streetlights.

46. Andrew Pollack, "A Dream of Trees Aglow at Night," *New York Times*, May 23, 2013.

47. *Ibid.*

48. Biocurious, http://biocurious.org/about/.

49. S.A. Nickerson, "GMO Food, It's Worse Than We Thought," *Newsmax.Com*, Dec. 8, 2013.

50. Andrew Pollack, "Modified Wheat Is Discovered in Oregon," *New York Times*, May 29, 2013.

51. E-alert from the Center For Food Safety, 2018.

52. Kristina Hubbard, "Usda Tried to Hide Ge Bentgrass Contamination," *Seed Digest*, November 15, 2010, http://blog.seedalliance.org/2010/11/15/usda-tried-to-hide-ge-bentgrass-contamination/.

53. Andrew Pollack, "Exploring Algae as Fuel," *New York Times*, July 26, 2010.

54. Andrew Pollack, "Exploring Algae as Fuel," *New York Times*, July 26, 2010.

55. Jennifer Doudna and Samuel H. Sternberg, *A Crack in Creation: Gene Editing and the Unthinkable Power to Control Evolution* (New York: Houghton Mifflin Harcourt, 2017), p. 119.

56. Wikipedia, "CRISPR," https://en.wikipedia.org/wiki/CRISPR

57. Nathan J. Comp, "Tweaking Life," *Isthmus*, Vol. 43: Oc. 27–Nov. 2, 2016, p. 17.

58. Doudna, p. 239.

59. Michael Specter, "DNA Revolution," *National Geographic*, August 2016, *P. 36ff.*

60. Comp, p. 17.

61. Associated Press, "Chinese Researcher Claims First Edited Babies," *New York Times*, November 25, 2018, https://www.nytimes.com/aponline/2018/11/25/health/ap-us-med-genetic-frontiers-gene-edited-babies.html.

62. Jerry Adler, "Kill All the Mosquitoes?!" *Smithsonian*, June, 2016.

63. Gris Anik, "The Dangers of Crispr Technology," http://www.grisanik.com/blog/the-dangers-of-crispr-technology/.

64. Gris Anik, "The Dangers of Crispr Technology."

65. Doudna, p.113.

66. Doudna, p.174.

67. Esvelt quoted in Adler, "Kill All the Mosquitoes?!," *Smithsonian*, June 2016.

68. Michael Specter, "DNA Revolution," *National Geographic*, August 2016, P. 36ff.

69. Comp, p. 20.

70. Comp, p. 18.

71. Comp, p. 17.

72. Comp, p. 18.

73. "The Anthropocene," *Welcome to the Anthropocene: A Planet Transformed by Humanity*, http://www.anthropocene.info/en/anthropocene and "The Anthropocene," International Geosphere Biosphere Programme, http://www.igbp.net/5.d8b4c3c12bf3be638a8000578.html.

74. Michael Marshall, "Terraforming: Geoengineering Megaplan Starts Now," *New Scientist*, Oct. 9 2013, http://www.newscientist.com/article/mg22029382.500-terraforming-earth-geoengineering-megaplan-starts-now.html#.UthT0Tbnb4g.

75. Doudna and Samuel H. Sternberg, p. 189.

76. Doudna and Sternberg, pp. 86, 90, xiii, 96, 103, 117 and 243 resp.

77. Doudna and Sternberg, p. 152

78. Greely quoted in Adler, "Kill All the Mosquitoes?!," *Smithsonian* June 2016.

79. Michael Specter, Lander quoted in "DNA Revolution," *National Geographic*, August 2016, p. 36ff.

## Chapter 5

1. Quoted in Bill McKibben, "A Special Moment in History," *The Atlantic Online*, May 1998. http://www.theatlantic.com/past/docs/issues/98may/special3.htm. Originally in *The Atlantic Monthly: A Special Moment in History*, May 1998, Volume 281, No. 5, p. 70.

2. Bill McKibben, "A Special Moment in History," 55–78.

3. Michael Pollan, "Why Bother?" in *Drawdown: The Most Comprehensive Plan Ever Proposed to Reverse Global Warming*, ed. Paul Hawken (New York: Penguin Books 2017), p. 53.

4. Paul Erlich and Ann Ehrlich, quoted in John J. Ferber, *Climate Peril: The Intelligent Readers' Guide to the Climate Crisis* (Berkeley, CA: Northbrae Books, 2014), p. xxi.

5. Dr. Jerry Mahlman, "Climate Change," testimony before the U.S. Senate Committee on Commerce, Science and Transportation, March 3, 2004. http://www.commerce.senate.gov/public/index.cfm/hearings?Id=abed3dcd-e7e8-47fb-9506-93d703e4b634&Statement_id=CAA36353-8A4F-4719-AD7C-72860B4D41E8.

6. Natasha Geiling, "Methane Emissions Are Spiking, but It Might Be More Cow Than Car," https://thinkprogress.org/methane-emissions-are-spiking-but-it-might-be-more-cow-than-car-791e5233dc2a#.kgli9uw03.

7. James Hansen, *Storms of My Grandchildren: The Truth About the Coming Climate Catastrophe and the Last Chance to Save Humanity* (New York: Bloomsbury, 2009), p. 42.

8. John Berger, *Climate Peril: The Intelligent Readers' Guide to the Climate Crisis*

(Berkeley, CA: Northbrae Books, 2014), p, 24.

9. U.S. Environmental Protection Agency, "Atmospheric Concentrations of Greenhouse Gas," http://www.epa.gov/climatechange/images/indicator_downloads/ghg-concentrations-download2-2014.png.

10. Katia Moskvitch, "Mysterious Siberian Crater Attributed to Methane," *Nature: The International Weekly Journal of Science,* July 31, 2014, http://www.nature.com/news/mysterious-siberian-crater-attributed-to-methane-1.15649.

11. Bill McKibben, "A Special Moment in History."

12. *New York Times,* standing article, "Climate Change Is Complex: We've Got Answers to Your Questions," accessed Jan. 24, 2018.

13. Berger, pp. 55 & 61.

14. American Institute of Physics, "The Discovery of Global Warming," July, 2007, http://www.aiporg/history/climate/Revel.htm .

15. James Hansen, *The Observer,* Feb. 15, 2009.

16. Justin Gillis, "U.N. Panel Warns of Dire Effects from Lack of Action Over Global Warming," *New York Times,* November 2, 2014.

17. Justin Gillis, "U.N. Panel Warns of Dire Effects from Lack of Action Over Global Warming."

18. McKibben, "A Special Moment in History."

19. *Ibid.*

20. National Oceanic and Atmospheric Administration. Earth System Research Laboratoryhttp://www.esrl.noaa.gov/gmd/ccgg/trends/.

21. NOAA: /10187/NOAA-Carbon-dioxide-levels-reach-milestone-at-Arctic-sites.aspx http://research.noaa.gov/News/NewsArchive/LatestNews/TabId/684/ArtMID/1768/ArticleID.

22. $CO_2$ Earth, "Daily $Co_2$," March 19, 2019, https://www.co2.earth/daily-co2.

23. Berger, p. 64.

24. Mashable, "Carbon Dioxide Levels Climb into Uncharted Territory for Humans," http://mashable.com/2014/04/08/carbon-dioxide-highest-levels-global-warming/.

25. Berger, p. 66.

26. Berger, p. 65.

27. University of California/Riverside, "Global Climate Change: Evidence and Causes," https://globalclimate.ucr.edu/resources.html.

28. David G. Hallman, et al., *Climate Changes and the Quest for Sustainable Societies* (Geneva: World Council of Churches, 1998), p. 8.

29. Berger, p. 23.

30. Mark B. Bush, *Ecology of a Changing Planet* (Upper Saddle River, NJ: Prentice Hall, 1997), p. 204.

31. Donella Meadows, et al., *Beyond the Limits* (White River Junction, VT: Chelsea Green Publishing, 1992).

32. *Huffington Post,* "Heavy Snowfalls, Extreme Storms Linked to Climate Change, Scientists Claim," 3/2/2011, http://www.huffingtonpost.com/2011/03/02/snowfalls-storms-climate-change-link_n_830104.html?utm_source=DailyBrief&utm_campaign=030211&utm_medium=email&utm_content=NewsEntry&utm_term=Daily+Brief.

33. Union of Concerned Scientists, "Early Warning Signs of Global Warming: Spring Comes Earlier," http://www.ucsusa.org/global_warming/science_and_impacts/impacts/early-warning-signs-of-global-10.html#.VF0KtjbnbIU.

34. Bill McKibben, "A Special Moment in History."

35. American Institute of Physics, *The Discovery of Global Warming,* "Ice Sheets and Rising Seas," February, 2014. http://www.aip.org/history/climate/floods.htm.

36. Berger, pp. 108 and 31.

37. Berger, p. 3.

38. Agu 100, "Jgr Oceans," https://agupubs.onlinelibrary.wiley.com/doi/abs/10.1029/2018JC014388.

39. Justin Gillis, "Climate Model Predicts West Antarctic Ice Sheet Could Melt Rapidly," *New York Times,* March 30, 2016.

40. Jess Bidgood, "At a Cape Cod Landmark, a Strategic Retreat from the Ocean," *New York Times,* July 6, 2016.

41. Gail Collins, "Florida Goes Down the Drain: The Politics of Climate Change," *New York Times,* September 25, 2014.

42. Berger, p. 213.

43. Berger, pp. 211–212.

44. *Bloomberg Business Week,* "Drowning Kiribati," http://www.businessweek.com/articles/2013-11-21/kiribati-climate-change-destroys-pacific-island-nation.

45. Gardiner Harris, "Facing Rising Seas, Bangladesh Confronts the Consequences of Climate Change," *New York Times,* March 28, 2014.

46. Harris, *New York Times*, March 28, 2014.

47. Center for Biological Diversity, "Ocean Acidification," http://www.biologicaldiversity.org/campaigns/ocean_acidification/.

48. Berger, p. 215.

49. *Ibid.*, p. 216.

50. "Coral Reefs May Be Gone by 2050," *Huffington Post* Feb. 25, 2011, http://www.huffingtonpost.com/2011/02/25/coral-reefs-may-be-gone-b_n_827709.html?utm_source=DailyBrief&utm_campaign=022511&utm_medium=email&utm_content=NewsEntry&utm_term=Daily+Brief.

51. Berger, p, 184.

52. *Ibid.*, p. 195.

53. *Ibid.*, p. 174.

54. *Ibid.*

55. *Ibid.*

56. Kate Galbraith, "Amid Texas Drought, High-Stakes Battle Over Water," *New York Times*, June 18, 2011.

57. Manny Fernandez, "Hay Shortage Compounds Woe in Drought-Stricken Texas," *New York Times*, Published: October 31, 2011, http://www.nytimes.com/2011/11/01/us/hay-shortage-compounds-woe-in-drought-stricken-texas.html?_r=0.

58. U.S.D.A. "California Drought 2014: Farm and Food Impacts," http://ers.usda.gov/topics/in-the-news/california-drought-2014-farm-and-food-impacts.aspx.

59. Pacific Institute, "California Drought: Impacts and Solutions," http://www.californiadrought.org/drought/current-conditions/.

60. Julien Goldstein, "Searching for Crumbs in Syria's Breadbasket," New York Times, October 13, 2010.

61. Julien Goldstein, "Searching for Crumbs in Syria's Breadbasket."

62. Marine Stewardship Council, "How Will Climate Change Affect Fish and Fisheries?" http://www.msc.org/.

63. "Fisheries and Climate Change," http://en.wikipedia.org/wiki/Fisheries_and_climate_change.

64. Mike Pearl. "Phoenix Will Be Almost Unlivable by 2050 Thanks to Climate Change," *Vice*, Sep 18 2017, 1:03pm, https://www.vice.com/en_us/article/vb7mqa/phoenix-will-be-almost-unlivable-by-2050-thanks-to-climate-change.

65. Shaoni Bhattacharya "European Heatwave Caused 35,000 Deaths," *NewScientist.Com*, Oct. 10, 2003, http://www.heatisonline.org/contentserver/objecthandlers/index.cfm?id=4485&method=full.

66. Huizhong Wu, "India Facing Another Summer of Deadly Heat," CNN, April 24th, 2017, https://www.cnn.com/2017/04/24/asia/india-heat-wave-deaths/index.html.

67. National Wildlife Federation, "Global Warming and Heat Waves," http://www.nwf.org/Wildlife/Threats-to-Wildlife/Global-Warming/Global-Warming-is-Causing-Extreme-Weather/Heat-Waves.aspx.

68. Jerry Adler, "Hot Enough for You?" *Smithsonian*, May 2014, p. 54.

69. James Hansen, *Storms of My Grandchildren*, p., xv.

70. Bill McKibben, Radio Interview, *Democracy Now*, May 26, 2011, http://www.democracynow.org/2011/5/26/bill_mckibben_from_storms_to_droughts.

71. Hansen, p. 255.

72. "Hurricane Katrina," http://www.history.com/topics/hurricane-katrina.

73. Berger, p. 168.

74. *Ibid.*, p. 166–167.

75. *Ibid.*, p. 167.

76. *Ibid.*, 167.

77. Government of the Philippines, "Ndrrmc Updates Re Effects of Ty Yolanda (Haiyan)," National Disaster Risk Reduction and Management Council. April 17, 2014. http://reliefweb.int/report/philippines/ndrrmc-updates-re-effects-ty-yolanda-haiyan-17-apr-2014.

78. Scott Neuman, "2 Feet of Rain Causes Massive Flooding in Florida, Alabama," *National Public Radio*, May 01, 2014, http://www.npr.org/blogs/thetwoway/2014/05/01/308587006/2-feet-of-rain-causes-massive-flooding-in-fla-alabama.

79. *Democracy Now*, May 26, 2011, http://www.democracynow.org/2011/5/26/bill_mckibben_from_storms_to_droughts.

80. Johan Engle Bromwich, "Flooding in the South Looks a Lot Like Climate Change," *New York Times*, Aug. 16, 2016.

81. Elizabeth Chuck, "Hurricane Harvey: How Many Billions of Dollars in Damage Will Historic Hurricane Cost?" https://www.Nbcnews.Com/Storyline/Hurricane-Harvey/How-Many-Billions-Damage-Will-Harvey-Cost-S-Anyone-S-N797521.

82. Justine Gillis, "Bipartisan Report Tallies High Toll on Economy from Global Warming," *New York Times Now*, June 24, 2014.

83. Henry Paulson, Jr., "The Coming

Climate Crash: Lessons for Climate Change in the 2008 Recession," *New York Times,* June 21, 2014.

84. Union of Concerned Scientists, "Is Global Warming Fueling Increased Wildfire Risks?" http://www.ucsusa.org/global_warming/science_and_impacts/impacts/global-warming-and-wildfire.html#.VJRNL2AABg.

85. Tom Kenworthy, "A Nation on Fire: Climate Change and the Burning of America," Posted on July 31, 2013 at 8:35 a.m. Updated: July 31, 2013 at http://thinkprogress.org/climate/2013/07/31/2312591/climate-change-wildfires/.

86. Christopher Meleaug, "Tens of Thousands Evacuated as Fire Rages in Southern California," *New York Times,* August 17, 2016.

87. Jeremy Berke and Bryan Logan, "5 Major Fires Are Raging in Southern California—Here's the Latest Size, Location, and Damage of Each," *Business Insider,* Dec. 11, 2017, 2:13 p.m. http://www.businessinsider.com/california-ventura-county-wildfires-locations-news-updates-damage-2017-12.

88. John H. Berger, *Climate Myths: The Campaign Against Climate Science* (Berkeley: Northbrae Books, 2013), p. 58.

89. Berger, *Climate Peril,* p. 120.

90. Berger, *Climate Peril,* p. 120.

91. Berger, *Climate Peril,* p. 122.

92. McKibben, "Special Moment," accessed at https://www.theatlantic.com/search/?q=McKibben+Earth2.

93. Tim Radford, "Fossil Fuel Must Fall Twice as Fast as Thought to Contain Global Warming—Study," *The Guardian,* http://www.theguardian.com/environment/2016/feb/25/fossil-fuel-use-must-fall-twice-fast-thought-contain-global-warming.

94. Robert Pollin, *Greening the Global Economy* (Cambridge: MIT Press, 2015), pp. 1–17.

95. Coral Davenport, "Pentagon Signals Security Risks of Climate Change," *New York Times,* Oct. 13, 2014.

96. Jerry Adler, "Hot Enough for You?" *Smithsonian,* May 2014, p. 56.

97. Nicholas D. Kristof, "Our Beaker Is Starting to Boil," *New York Times,* July 16, 2010.

98. Peace And Justice Studies Association, "Connecting Militarism and Climate Change," Peace & Justice Studies Association, February 14, 2011, http://www.peacejusticestudies.org.

99. Berger, *Climate Myths,* p, 9.

100. *Ibid.,* p. 10.

101. *Ibid.,* p. 13.

102. *Ibid.,* pp. 14–15.

103. "Exxon Funding Climate Denial. Yes, Again," Left Foot Forward, 19 July, 2010, http://leftfootforward.org/2010/07/exposed-exxon-funding-climate-denial-yes-again/.

104. Shannon Hall, "Exxon Knew About Climate Change Almost 40 Years Ago," *Scientific American,* October 26, 2015, www.scientificamerican.com/article/exxon-knew-about-climate-change-almost-40-years-ago/.

105. Berger, *Climate Myths,* pp. 35–63.

106. *Ibid.,* p. 3.

107. *Ibid.*

108. "United Nations Convention on Climate Change," *Wikipedia,* http://en.wikipedia.org/wiki/United_Nations_Framework_Convention_on_Climate_Change.

109. Kent Shifferd, "In the Land of the Blind: Evaluating Obama's Environmental Record," in *Grading the 44th President,* eds. Luigi Esposito and Laura Finley (Santa Barbara: Praeger, 2012), pp. 99 ff.

110. Coral Davenport, "A Climate Deal, 6 Fateful Years in the Making," *New York Times,* December 13, 2015.

111. Coral Davenport, "Nations Approve Landmark Climate Accord in Paris" *New York Times,* December 12, 2015.

112. Eduardo Porter, "Fighting Climate Change? We're Not Even Landing a Punch," *New York Times,* Jan. 23, 2018, https://www.nytimes.com/2018/01/23/business/economy/fighting-climate-change.html.

113. Coral Davenport, "Nations Approve Landmark Climate Accord in Paris" *New York Times,* December 12, 2015.

114. Coral Davenport, "Major Climate Report Describes a Strong Risk of Crisis as Early as 2040" New York Times, October 7, 2018.

115. Bill McKibben, "A Special Moment in History."

116. "Ipcc Keeps Feeding the Addiction," *Arctic News Blogspot,* October 2018, http://arctic-news.blogspot.com/2018/10/ipcc-keeps-feeding-the-addiction.html.

117. David Wallace Wells, *The Uninhabitable Earth: Life After Warming* (New York: Tim Duggan Books, 2019), p. 3.

## Chapter 6

1. *The Guardian,* https://www.theguardian.com/environment/earth-

insight/2014/mar/14/nasa-civilisation-irreversible-collapse-study-scientists.

2. Ernest Callenbach, "Epistle to the Ecotopians," 2012, published by Tom Engelhardt in the blog Tomdispatch under the title "Tomgram: Ernest Callenbach, Last Words to an America in Decline," May 6, 2012.

3. This is actually the popular transmutation of their longer slogan, "Better Things for Better Living ... Through Chemistry." Wikipedia, "Better Living Through Chemistry," accessed August 29th, 2016.

4. Jared Diamond, *Collapse: How Societies Choose to Fail or Succeed* (New York: Penguin Books, revised edition, 2011), p. 6.

5. Diamond, p. 421.

6. Arthur Demarest in Edward Fischer, December 22, 2014, "The Indiana Jones of Collapsed Cultures: Our Western Civilization Itself Is a Bubble," *PBS Newshour,* http://www.pbs.org/newshour/making-sense/indiana-jones-collapsed-cultures-western-civilization-bubble/.

7. Demarest in Edward Fischer.

8. Demerest in Edward Fischer.

9. Diamond, p. 275.

10. *The Guardian,* https://www.theguardian.com/environment/earth-insight/2014/mar/14/nasa-civilisation-irreversible-collapse-study-scientists.

11. Jaymi Heimbuch, "*Interactive Map Shows Worldwide Water and Energy Tug-O-War,*" http://www.treehugger.com/feeds/authors/jaymi.xml.

12. *Ibid.*

13. Demarest in Fischer.

14. Steven Radelet, *The Great Surge: The Ascent of the Developing World* (New York: Simon & Schuster, 2015), kindle edition.

15. Thomas Friedman, "Stuff Happens to the Environment, Like Climate Change," *New York Times,* Oct. 7, 2015.

16. Peter Goodchild, "The Imminent Collapse of Industrial Society," Countercurrents.org, 09 May, 2010.

17. Demarest in Fischer.

18. Dimitri Orlov, *The Five Stages of Collapse* (New Society Publishers: Gabriola Island, BC, 2013), p. 9.

19. Christopher Ingraham, "Mapping America's Most Dangerous Bridges," *Washington Post,* http://www.washingtonpost.com/news/wonkblog/wp/2015/02/04/mapping-Americas-Most-Dangerous-Bridges/.

20. Orlov, p. 9.

21. *Ibid.*, p. 13.

22. *Ibid.*, pp. 14–15.

23. Demarest in Fischer.

24. *Ibid.*

25. Jack Nelson-Pallmeyer, *Authentic Hope* (Maryknoll, NY: Orbis Books, 2012), introduction and pp. 5–7.

## Chapter 7

1. The term belongs to Elizabeth Marshall Thomas, *The Old Way: A Story of the First People* (Sarah Crichton Books; 2003).

2. Elizabeth Marshall Thomas.

3. Thomas Hobbes' *Leviathan, or the Matter, Forme, and Power of a Commonwealth, Ecclesiasticall and Civill,* 1651. The Phrase Finder, http://www.phrases.org.uk/meanings/254050.html. This description sounds more like life in Europe during the Hundred Years War or in the slums of industrial revolution London.

4. Jack Harlan, *Crops and Man* (Madison, WI: American Society of Agronomy, Crop Science Society of America, 1975).

5. Calvin Martin, *In the Spirit of the Earth: Rethinking History and Time* (Baltimore: The Johns Hopkins University Press, 1992) p. 10.

6. Martin, p. 80.

7. *Ibid.*, p. 20.

8. *Ibid.*, p. 17.

9. *Ibid.*, p. 19.

10. Martin, p. 20.

11. Mircea Eliade, *Cosmos and History: The Myth of the Eternal Return* (New York: Harper, 1959), p. vii.

12. Henri Frankfort, Mrs. H.A. Frankfort, John A. Wilson, and Thorkild Jackobson, *Before Philosophy: The Intellectual Adventure of Ancient Man* (Baltimore: Pelican, 1949)p. 12.

13. Riane Eisler, The Chalice and the Blade: Our History, Our Future (New York: HarperOne, 2011).

14. "Latin America's Agricultural Exports to China, Recent Trends," http://google.com/url?sa=t&rct=j&q=&esrc=s&source=web&cd=1&ved=0CB4QFjAA&url=http%3A%2F%2Fwww.iamo.de%2Ffile admin%2Fuploads%2Fforum2011%2FPa pers%2FCaballero_IAMO_Forum_2011. pdf&ei=ev7cVI_iLYWYPJypgNgE&usg=-AFQjCNFQanDzYQRIh-AeDcDmD2EJMY kiGQ&bvm=bv.85970519,d.ZWU.

15. J. Donald Hughes, *Ecology in Ancient Civilizations* (Albuquerque: University of New Mexico Press, 1975), p. 29.

16. Eisler, *The Chalice and the Blade* (New York: Harper Collins, 1987).

17. *Ibid.*, p. 49.
18. Hughes, p.32.
19. *Ibid.*, p. 35.
20. *Ibid.*, p. 33.
21. Lewis Mumford, *The City in History* as excerpted in L.G. Schaefer, et al., *The Shaping of Western Civilization* (New York: Holt, Rinehart and Winston, 1970), p. 36.
22. Plato quoted in Dale and Carter, p. 31.
23. Rolf Eldberg, "Earth, Time and Man," in Robert Disch, *The Dying Generations* (New York: Dell Publishing Company, 1971), p. 23.
24. Edward Goldsmith, "The Fall of the Roman Empire: A Social and Ecological Interpretation," *Ecologist*, July 1977, p. 203.
25. Fritz M. Heichelheim, "Effects of Classical Antiquity on the Land," in William L. Thomas, ed., *Man's Role in Changing the Face of the Earth* (Chicago: University of Chicago Press, 1956), pp. 165–181, quote p. 171.
26. Frederick Cartwright, *Disease and History* (New York: Crowell, 1972), p. 10.
27. Riane Eisler, pp. 94, 95 & 99.
28. *Random House Dictionary of the English Language*, College Edition, New York: Random House, 1968 ( "Pagan").
29. Lynn White, Jr., in Robert Detweiller, John N. Sutherland and Michael S. Werthman, eds., *Environmental Decay in Its Historical Context* (Glenville, IL: Scott, Foresman and Co., 1973) p. 15.
30. White in Detweiller, pp. 25 and 24 respectively.
31. Victor Ferkiss, *Nature, Technology and Society: Cultural Roots of the Current Environmental Crisis* (New York: New York University Press, 1993), p. 22.
32. Peter Bowler, *The Norton History of the Environmental Sciences* (New York: W.W. Norton, 1992), p.58.
33. Url Lanham, *Origins of Modern Biology* (New York: Columbia University Press, 1968), p. 79.

## Chapter 8

1. Frederick Nussbaum, *The Triumph of Science and Reason: 1660–1685* (New York: Harper, 1953), p. 1.
2. Morris Berman, *The Reenchantment of the World* (Toronto: Bantam, 1984), pp. 83 and 65 resp.
3. Book blurb for *The Golden Chain of Homer,* https://www.iuniverse.com/ Bookstore/BookDetail.aspx?BookId= SKU-000507629.
4. Carolyn Merchant, *The Death of Nature: Women, Ecology and the Scientific Revolution* (San Francisco: Harper and Row, 1980), p. 2.
5. Berman, p.108.
6. Merchant, pp. 38–39.
7. Berman, p. 57.
8. *Ibid.*, p. 21.
9. Berman, p. 34.
10. Bacon in Will Durant, *The Story of Civilization*, vol. VII, *The Age of Reason Begins* (New York: Simon and Schuster, 1961), p. 172.
11. *Ibid.*, p. 179.
12. *Ibid.*, p. 176.
13. *Ibid.*, p. 173.
14. R.R. Palmer and Joel Colton, *A History of the Modern World*, 5th ed. (New York: Knopf, 1978), p. 273.
15. Berman, p. 15.
16. Durant, p. 183.
17. Berman, p. 14.
18. *Ibid.*, p. 17.
19. Descartes in Berman, p. 13, italics added.
20. Floyd W. Matson, *The Broken Image: Man, Science and Society* (New York: Anchor Books, 1966), p. 6.
21. Durant, p. 639.
22. Berman, pp. 20–21
23. *Ibid.*, p. 48.
24. *Ibid.*, p. 49.
25. *Ibid.*, p.30.
26. E.A. Burtt, *The Metaphysical Foundations of Modern Science* as quoted in Matson, pp. 4–5.
27. K. Sale in Alwyn Jones, "From Fragmentation to Wholeness: A Green Approach to Science and Society," *The Ecologist*, Vol. 17, No. 6, 1987, pp. 238–239.
28. Keynes in Berman, p. 108.
29. Berman, p. 121 (Cornell University Press edition).
30. Matson, p. 9.
31. Merchant, pp. 21–22.
32. Anne Barstow, *Witchcraze: A New History of the European Witch Hunts* (San Francisco: HarperCollins, 1994), p. 88.
33. Barstow, p. 159.
34. Herman Daly, "Economics Unmasked," *The Daly News, Casse*, Posted: 11 Sep 2011 11:19 p.m. Pdt).
35. Herman Daly, "Krugman's Growthism," *The Daly News, Casse,* Posted: 29 Apr 2014 12:26 p.m. Pdt.
36. David Korten, *When Corporations*

*Rule the World* (Oakland, CA: Barrett-Koehler, 3rd ed., 2015)

37. http://www.websterworld.com/websterworld/prose/s/sirthomas moreinutopia2504.html.

38. Anup Shah, "A Primer on Neoliberalism," http://www.globalissues.org/article/39/a-primer-on-neoliberalism#-globalissues-org.

39. Jacob Burckhardt, *The Civilization of the Renaissance in Italy*, quoted in James Bruce Ross (ed.) *The Portable Renaissance Reader* (New York: Viking, 1968), p. 3.

40. Quoted in Clive Ponting, *A Green History of the World* (New York: Penguin Books, new edition, 1992), p. 146.

41. *Protagoras*, Plato in the *Theaetetus*, 160d.

42. Pico Della Mirandola, *Oration on the Dignity of Man* in Franklin Le Van Baumer, *Main Currents of Western Thought* (New York: Alfred A. Knopf, 1970), p.126.

43. Daniel Defoe, *The Complete English Tradesman*, vol. 2. (London: J. Rivington, St. Paul's Courtyard, 1745), p. 73.

44. Daniel Defoe, *An Essay on Projects*, in Henry Morely (ed.), *The Earlier Life and Chief Earlier Works of Daniel Defoe* (London: George Routledge and Sons, 1889), p. 61.

45. Defoe, *Caledonia, A Poem in Honour of Scotland, and the Scots Nation* (London: Printed by J. Matthews and Sold by John Morphew, Near Stationers Hall, 1707), p.55.

46. Defoe, *Caledonia*, pp. 195–196.

47. Defoe, *The Complete English Tradesman*, p. 153.

48. Defoe, *A Tour Through the Whole Island of Great Britain*, 4 vols. (London: Printed for S. T. Birt, et al., 1748), vol. 1, p. 4.

49. Defoe, Daniel, A Plan of the English Commerce, Being a Compleat Prospect of the Trade of This Nation, as Well the Home Trade as the Foreign. In Three Parts (London: Charles Rivington, 1728), p. 19.

50. Defoe, *The Complete English Tradesman*, p. 151.

51. Defoe, *Caledonia*, p. 11.

52. Defoe, *Caledonia*, p. 10.

53. Defoe, *A Plan of the English Commerce*, p. 241.

54. Defoe, *The Complete English Tradesman*, vol. 2, p.95.

55. William Derham, *Physico-Theology*, 1713, quoted in Naomi Klein, *This Changes Everything: Capitalism Vs. the Climate* (New York: Simon and Schuster, 2014), p. 171.

56. Daniel Defoe, *An Essay Upon the South Sea Trade* (London: Printed for F.

Baker at the Black Boy in Pater Noster Row, 1712), p. 39.

57. W. Fred Cottrell, *Energy and Society: The Relation Between Energy, Social Change, and Economic Development* (Westport, CT: Greenwood Press, 1970), p.46.

58. Cottrell, p. 61.

59. Vaclav Smil, *Energy in World History* (Boulder: Westview Press, 1994), pp. 137–8.

60. Carl Cipolla, *Before the Industrial Revolution: European Society and Economy, 1000–1700* (New York: Norton, 1976), p. 209.

61. Roxanne Dunbar-Ortiz, *An Indigenous Peoples' History of the United States* (Boston: Beacon Press, 2014), xi.

62. Daniel Defoe, *The Trueborn Englishman* in James Southerland, ed., *Robinson Crusoe and Other Writings* (Boston: Houghton Mifflin, 1967) p. 266.

63. Ian Watt, *The Rise of the English Novel* (Berkeley: University of California Press, 1957), p. 66.

64. Maximilian Novak, "Crusoe's Fear and the Search for Natural Man," *Modern Philology*, 58 (May, 1961), pp. 241–244.

65. Defoe, Daniel *Robinson Crusoe and the Farther Adventures of Robinson Crusoe* (New York: Washington Square Press, 1957), pp. 256 and 260.

66. Defoe, *Robinson Crusoe and the Farther Adventures*, p. 65.

67. *Ibid.*, pp. 378–383.

68. *Ibid.*, p. 180.

69. *Encyclopedia Britannica*, 15th edition, 1981, p. 861.

70. Defoe, *Robinson Crusoe and the Farther Adventures of Robinson Crusoe*, p. 27.

71. *Ibid.*, pp. 338–9.

## Chapter 9

1. F. Roy Willis, *Western Civilization: An Urban Perspective* (Lexington, MA: D.C. Heath, 1973), p. 659.

2. Carl Cipolla, *Before the Industrial Revolution: European Society and Economy, 1000–1700* (New York: Norton, 1976), p. 275.

3. Jean-Claude Debeir, Jean-Paul Deleage and Daniel Hemery, *In the Servitude of Power: Energy and Civilization Through the Ages* (London: Zed Books, 1987), p. 87.

4. Kevin Reilly, *The West and the World: A Topical History of Civilization* (New York: Harper and Row, 1980), p. 385.

5. Andrew Ure, *The Philosophy of Manufactures* cited in Alisdair Clayre, *Nature and*

*Industrialization* (Oxford: Oxford University Press, 1977), p. 70.

6. Clive Ponting, *A Green History of the World* (New York: St. Martin's Press, 1991), p. 284.

7. *Ibid.*, p. 695.

8. R. R. Palmer and Joel Colton, *A History of the Modern World* (New York: Knopf, 1978), p. 420.

9. Karl Marx, *Capital*, vol. 1 (New York: International Publishers, 1967), p. 696.

10. Mrs. Burrows, "Work in the Fields," from *A Childhood in the Fens* (1931), in Alisdair Clayre, *Nature and Industrialization* (Oxford: Oxford University Press, 1977), p. 18.

11. Eric Williams, *Capitalism and Slavery*, cited in J. M. Blaut, *The Colonizer's Model of the World* (New York: The Guilford Press, 1993), p. 55.

12. J.M. Blaut, *The Colonizer's Model of the World: Geographical Diffusionism and Eurocentric* History (New York: Guilford Press, 1993), pp. 200–201.

13. Adam Smith in Alasdir Clayre, *Nature and Industrialization* (Oxford: Oxford University Press, 1977), p. 183.

14. Debeir, p. 98.

15. Kevin Reilly, *The West and the World: A Topical History of Civilization* (New York: Harper and Row, 1980), p. 390.

16. Debeir, p. 87.

17. Vaclav Smil, *Energy in World History* (Boulder: Westview Press, 1994), p. 185.

18. "World Coal Facts 2013," World Coal Association, www.worldcoal.org/bin/pdf/... pdf.../coal_facts_2013(11_09_2013).pdf.

19. Willis, p. 746.

20. Reilly, p. 390.

21. *Ibid.*, p. 392.

22. Smil, p. 165

23. Debeir, p. 103.

24. *Ibid.*, p. 104.

25. Reilly, p. 392–392.

26. Cited in Willis, pp. 660–661.

27. de Tocqueville in Clayre, pp. 117–118.

28. Karl Marx, *Economic and Philosophic Manuscripts of 1844* (New York: International Publishers, 1964), pp. 148–149.

29. In Alisdair Clayre, *Nature and Industrialization* (Oxford: Oxford University Press, 1977), p. 118.

30. In Clayre, p. 194.

31. Engels in Willis, p. 669.

32. Debeir, p. 100.

33. *Ibid.*, p. 92.

34. Debeir, p. 110. In J. M. Blaut, *The Colonizer's Model of the World: Geographical* *Diffusionism and Eurocentric* History (New York: Guilford Press, 1993),

35. Victor Ferkis, *Nature, Technology and Society: Cultural Roots of the Current Environmental* Crisis (New York: New York University Press, 1993), p. 79.

36. Madhav Gadgil and Ramachandra Gutkind, *This Fissured Land: An Ecological History of India* (Berkeley: University of California Press, 1992), pp. 116–117.

37. Willis. p. 668.

38. *Ibid.*, p. 668.

39. J.M. Blaut, *The Colonizer's Model of the World: Geographical Diffusionism and Eurocentric History* (New York: Guilford Press, 1993), p. 153.

40. I.G. Simmons, *Environmental History: A Concise Introduction* (Oxford: Blackwell, 1993), p. 39.

41. Told in a speech by Winona LaDuke, Northland College, March 16, 1998.

42. Debeir, p. 101.

43. Peter N. Stearns, *The Industrial Revolution in World History* (Boulder: Westview Press, 1993), p. 30.

44. *Ibid.*, p. 46.

45. *Ibid.*, p. 47.

46. Willis, p. 697.

47. Cited in Willis, p. 700.

48. Cited in Willis, pp. 696–697.

49. Cited in Willis, p. 701.

50. Blaut, p. 16.

51. Rudyard Kipling, *Rudyard Kipling's Verse*, definitive edition (Garden City, NY: Doubleday, 1940). pp. 321ff.

52. F. Roy Willis, *Western Civilization: An Urban Perspective* (Lexington, MA: D.C. Heath, 1973), p. 718.

53. Aron J. Ihde, *The Development of Modern Chemistry* (New York: Dover, 1984), pp. 443–44.

54. Smil, p. 170.

55. *Ibid.*, p. 168.

56. *Ibid.*, p. 168.

57. Clive Ponting, *A Green History of the World* (New York: St. Martin's Press, 1991), p. 26.

58. Joseph Petulla, *American Environmental History* (San Francisco: Boyd and Fraser, 1977), pp. 170–171.

59. Richard Bartlett, *The New Country* (London: Oxford University Press, 1974), p. 263.

60. Roxanne Dunbar-Ortiz, *An Indigenous Peoples' History of the United States*, (Boston: Beacon Press, 2014).

61. Alan Trachtenberg, *The Incorporation of America* (Culture And Society In

The Gilded Age) (New York: Hill and Wang, 1982), p. 41.

62. Bartlett, p. 263

63. *Ibid.*, p. 266.

64. *Ibid.*, p. 269.

65. *Ibid.*, p. 271.

66. *Ibid.*, p. 272.

67. Trachtenberg, pp. 26–27.

## Chapter 10

1. Chellis Glendinning, *My Name Is Chellis and I Am in Recovery from Western Civilization* (Boston: Shambhala, 1994), p. 17.

2. David Pepper, *The Roots of Modern Environmentalism* (London: Routledge, 1984), p. 91.

3. See Garett Hardin, "The Tragedy of the Commons," *Science* 13 December 1968: Vol. 162 no. 3859 pp. 1243–1248, DOI: 10.1126/science. 162.3859.1243 and "Lifeboat Ethics: The Case Against Helping the Poor," *Psychology Today*, September 1974 and Paul Ehrlich, *The Population Bomb* (New York: Ballantine, 1971).

4. Alexander Pope, *An Essay on Man*, Epistle I, l 25 in John Butt, ed., *The Poems of Alexander Pope* (New Haven: Yale University Press, 1963), p. 505.

5. Pope, *An Essay on Man*, Epistle I, ll.267, 268, in Butt, p. 514.

6. Pope, "Solitude" in Oscar Williams, *Immortal Poems of the English Language* (New York: Washington Square Press, 1960), p. 161.

7. Yi Fu Tuan, *Topophilia: A Study of Environmental Perception, Attitudes, and Values* (New York: Columbia University Press, 1990 reprint).

8. E.O. Wilson, "Biophilia" (Address to the President and Fellows of Harvard College, 1984*).*

9. Arne Naess, *The Ecology of Wisdom: Writings by Arne Naess* (Berkeley, CA: Counterpoint Press, 2010).

10. Alan Badiner, ed., *Dharma Gaia: A Harvest of Essays in Buddhism and Ecology* (Berkeley: Parallax Press, 1990).

11. Susan Griffin, *The Eros of Everyday Life: Essays on Ecology, Gender and Society* (New York: Doubleday, 1995).

12. Oliver Goldsmith, "The Deserted Village" in Rewey Belle Inglis and Josephine Spear, *Adventures in English Literature* (New York: Harcourt Brace, 1958), p. 317.

13. Thomas Gray, "Elegy Written in a Country Churchyard," The Poetry

Foundation, https://www.poetryfoundation.org/poems/44299/elegy-written-in-a-country-churchyard.

14. Denis Diderot, *Supplement to Bougainville's Voyage* in Jean Stewart and Jonathan Kemp, *Diderot: Interpreter of Nature* (New York: International Publishers, 1963), pp. 188–189.

15. Will Durant and Ariel Durant, *Rousseau and Revolution* (New York: Simon and Schuster, 1967), p. 11.

16. *Ibid.*, p. 14.

17. *Rousseau and Revolution* (New York: Simon and Schuster, 1967), p. 6.

18. Rousseau, "From the Social Contract" in *The World's Great Thinkers, Man and the State: The Political Philosophers*, eds. Saxe Commins and Robert N. Linscott (New York: Random House, 1947), p. 268.

19. Gilbert F. LaFreniere, "Rousseau and the European Roots of Environmentalism," *Environmental History Review*, Winter 1990, p. 57.

20. Rousseau quoted in LaFreniere, p. 48.

21. Quoted in LaFreniere, p. 51.

22. LaFreniere, p. 63.

23. John Morley, ed. *The Complete Poetical Works of William Wordsworth* (London: MacMillan and Co., 1891), p. 85.

24. Wordsworth in Morely, p. 94.

25. Wordsworth in Morely, pp. 353 & 4.

26. Jonathan Bates, *Romantic Ecology: Wordsworth and the Environmental* Tradition (London: Routledge, 1991), p. 57.

27. *Ibid.*, p. 40.

28. Quoted in Bates, p. 102.

29. Wordsworth in his *Guide* quoted in Bates, p. 47.

30. James A. W. Heffernan, *The Recreation of Landscape: A Study of Wordsworth, Coleridge, Constable and Turner* (London: University Press of New England, 1985), p. xvii.

31. *Ibid.*, p. 2.

32. Wolfgang Born, *American Landscape Painting: An Interpretation* (New Haven: Yale University Press, 1948), p. 38.

33. *Ibid.*, p. 46.

34. La Freniere, p. 63.

35. *Ibid.*, p. 64.

36. Ralph Waldo Emerson, *Essays* (Boston: Houghton Mifflin & Co., 1897), p. 161.

37. Emerson, *Nature*, p. 185.

38. Emerson, *Nature*, p. 186.

39. Arthur Ekirch, *Man and Nature in America* (New York: Columbia University Press, 1963), p. 49.

40. Emerson in Ekirch, p. 50.
41. Emerson in Ekirch, p. 53.
42. Emerson, *Nature*, p. 187.
43. Emerson, *Nature*, p. 165.
44. Emerson quoted in Joseph Petulla, *American Environmentalism: Values, Tactics, Priorities* (College Station: Texas A&M University Press, 1980), p. 27.
45. Quoted in Sam Inkinen, ed., *Mediopolis: Aspects of Texts, Hypertexts, and Multimedial Communications* (Berlin: De Gruyter, 2011), p. 248.
46. Leo Marx, *The Machine in the Garden: Technology and the Pastoral Ideal in America* (London: Oxford University Press, 1964), pp. 230–231.
47. Emerson, quoted in Marx, p. 231.
48. Emerson quoted in Marx, p. 234.
49. Emerson, *Poems*, 4th ed. (Boston: 1847), p.119.
50. Ekirch, p. 60.
51. Roderick Nash, *Wilderness and the American Mind* (Hartford, CT: Yale University Press, 1967), p. 84.
52. Benjamin Kline, *First Along the River: A Brief History of the U.S. Environmental Movement* (San Francisco: Acada Books, 1997), pp. 35 & 34 resp.
53. Joseph M. Petulla, *American Environmental History* (College Station: Texas A&M University Press, 1980), p. 27.
54. Henry David Thoreau, *Walden* (Secaucus, NJ: Long River Press, 1976), p. 78.
55. *Ibid.*, pp. 78 and 79.
56. *Ibid.*, p. 70.
57. *Ibid.*, p. 27.
58. *Ibid.*, p. 79.
59. *Ibid.*, p. 56.
60. Quoted in Ekirch, p. 61.
61. Thoreau, *Walden*, p. 57.
62. *Ibid.*, p. 45.
63. *Ibid.*, p. 32.
64. Quoted in Ekirch. p. 62.
65. Thoreau, *Walden*, p. 13.
66. *Ibid.*, p. 77.
67. *Ibid.*, p. 77.
68. *Ibid.*, p. 5.
69. *Ibid.*, p. 175.
70. *Ibid.*, p. 59.
71. Thoreau, quoted in Douglas H. Strong, *Dreamers and Defenders: American Conservationists* (Lincoln, Neb: University of Nebraska Press, 1997), p. 16.
72. Thoreau, his April 23, 1851 lecture before the Concord Lyceum, printed in vol. 9 of *The Writings of Henry David Thoreau* as "Walking" (Riverside Edition: Boston, 1893) and quoted here in Rod Nash, p. 84.

73. Thoreau in Ekirch, p. 66.
74. Thoreau, *Walden*, p. 23.
75. Quoted in Strong, p. 11.
76. Quoted in Ekirch, p. 59.
77. Thoreau, *Walden*, p. 113.
78. Thoreau, quoted in Strong, p. 12.
79. Thoreau, *Walden*, p. 180.
80. George Perkins Marsh, *Man and Nature*, quoted in Roderick Nash, *The American Environment: Readings in the History of Conservation* (Reading, MA: Addison-Wesley, 1968), p. 14.
81. Marsh in Nash, *Readings*, p. 14.
82. Petulla, *American Environmentalism*, p. 32.
83. David Pepper, *The Roots of Modern Environmentalism* (London: Routledge, 1984), p. 103.
84. Alan Axelrod and Charles Phillips, *The Environmentalists: A Biographical Dictionary from the 17th Century to the Present* (New York: Zenda, 1993), p. 105.
85. Peter Bowler, *The Norton History of the Environmental Sciences* (New York: Norton, 1992), p. 309.

## Chapter 11

1. Quoted in Marybeth Lorbiecki, *Aldo Leopold: A Fierce Green Fire* (Helena, MT: Falcon Publishing Co., 1996), p. 175.
2. Garrett Harden, "The Tragedy of the Commons," *Science*, Dec. 13, 1968. See also Hardin's *Living Within Limits*, Oxford University Press (April 6, 1995).
3. Hal K. Rothman, *The Greening of America: Environmentalism in the United States Since 1945* (Fort Worth: Harcourt Brace, 1998), p. 5.
4. Naomi Klein, *This Changes Everything: Capitalism Vs. the Climate* (New York: Simon and Schuster, 2014), title of chapter 9.
5. Leopold quoted in Lorbiecki, p. 43.
6. Lorbiecki, pp. ix and xi.
7. Leopold in Lorbiecki, pp. 80–81.
8. Douglas H. Strong, *Dreamers and Defenders: American Conservationists* (Lincoln: University of Nebraska Press, 1988), p. 140.
9. Lorbiecki, p. 135.
10. Quoted in Lorbiecki, p. 157.
11. Lorbiecki, p. 158.
12. *Ibid.*, p. 116.
13. *Ibid.*, p. 160.
14. Aldo Leopold quoted in Strong, p. 150–151.
15. Lorbiecki, p. 121.

16. Quoted in Lorbiecki, p. 142.

17. Quoted in Lorbiecki, p. 143.

18. Strong, p. 135.

19. Rachel Carson, *Silent Spring* (New York: Mariner Books Anniversary Edition, 2002).

20. J.E. de Steiguer, *Age of Environmentalism* (New York: McGraw-Hill, 1997), p. 29.

21. de Steiguer, p. 30.

22. Rachel Carson, *The Sea Around Us* (New York: Oxford University Press, 1989).

23. Strong, p. 189.

24. Strong, p. 191.

25. Klein, p. 78.

26. Archibald MacLeish, "Riders on the Earth Together: Brothers in the Eternal Cold," http://cecelia.physics.indiana.edu/life/moon/Apollo8/122568sci-nasa-macleish.html.

27. Donella H. Meadows, et al., *Limits to Growth* (New York: Signet, 1972). Subsequent updates include *The Limits to Growth: The 30 Year Update* (White River Junction, VT: Chelsea Green, 2012).

28. E.F. Schumacher, *Small Is Beautiful: Economics as If People Mattered* (New York: Harper Perennial Reprint edition, 2010).

29. Gro Harmon Bruntland, *Report of the World Commission on Environment and Development: Our Common Future*: United Nations, www.un-documents.net/our-common-future.pdf.

30. World Watch Institute: *Annual State of the World* reports can be found here. http://www.worldwatch.org/bookstore/state-of-the-world.

31. Snopes.com. http://www.snopes.com/quotes/reagan/redwoods.asp.

32. The American Presidency Project/George Bush, "The President's News Conference in Rio De Janeiro," June 13, 1992 http://www.presidency.ucsb.edu/ws/?pid=21079.

33. "Tx Sen. Candidate Ted Cruz Spouts Paranoid Fantasy About United Nations/George Soros Conspiracy to Eliminate Golf," *Think Progress*, thinkprogress.org/justice/2012/03/16/446352/tx-sen-candidate-ted....

34. Rebecca Leber, "Republican Platform Says Coal Is a Clean Energy Resource," *Newsweek*, 7/12/16. http://www.newsweek.com/republican-party-coal-clean-energy-479745.

35. John Gibler, "Under the Gun," *Sierra*, July/August 2017, pp. 28ff.

36. "Almost Four Environmental Defenders a Week Killed in 2017," *The Guardian*, https://www.theguardian.com/environment/2018/feb/02/almost-four-environmental-defenders-a-week-killed-in-2017.

37. Klein, chapter 9, beginning p. 294.

38. Klein, p. 205.

39. EduGreen, "The Chipko Movement—EduGreen," http://edugreen.teri.res.in/*explore*/forestry/chipko.htm.

40. Klein, p. 310.

41. Klein, p. 311.

42. Sharon Delgado, *Shaking the Gates of Hell: Faith Led Resistance to Corporate Globalization* (Minneapolis: Fortress Press, 2007), p. 156–157.

43. Delgado, p. 157.

44. Klein, p. 313.

45. Klein, p 296–297.

46. See Paul Engler and Mark Engler, *This Is an Uprising: How Nonviolent Revolt Is Shaping the Twenty-First Century* (New York: Nation Books, 2016) and Paul Cienfuegos and Matt Guynn, *The Community Rights Movement and the Arc of Nonviolent Social Change*, 5/23/2012, http://paulcienfuegos.com/community-rights-movement-and-arc-nonviolent-social-change.

47. Klein, p. 304.

48. Greta Thunberg tweet, https://twitter.com/gretathunberg/status/11247238911239 61856?lang=en

49. Greta Thunberg, "Almost Everything Is Black and White," in *No One Is Too Small to Make a Difference* (New York: Penguin Books, 2019), p. 11.

50. quoted in "Greta Thunberg Biography," Biography.com editors, https://www.biography.com/activist/greta-thunberg, updated Dec. 12, 2019.

51. Elizabeth Weise, "'How Dare You?' Read Greta Thunberg's Emotional Climate Change Speech to UN and World Leaders," *USA Today*, Sept. 23, 2019, https://www.usatoday.com/story/news/2019/09/23/greta-thunberg-tells-un-summit-youth-not-forgive-climate-inaction/2421335001/

52. Greta Thunberg, "Wherever I Go I Seem to Be Surrounded by Fairy Tales," in *No One Is Too Small to Make a Difference*. p. 87.

53. The Sunrise Movement, https://www.sunrisemovement.org/calls/2019/3/15/prepare-for-the-us-climate-strike, March 15 2019.

54. Greta Thunberg, "Cathedral Thinking," in *No One Is Too Small to Make a Difference*, p. 43.

55. From a collection of tweets from an unofficial #COP25 account to share news & information.

56. Quoted in "Greta Thunberg Biography," Biography.com editors, https://www.biography.com/activist/greta-thunberg, updated Dec. 12, 2019.

## Chapter 12

1. Willis Harmon, quoted in Rob Hopkins, *The Transition Handbook: from Oil Dependency to Local Resilience* (White River Junction, VT: Chelsea Green Publishing Co., 2008), p. 93.
2. Rob Hopkins, *The Transition Handbook*, p. 94.
3. Northwest Earth Institute, *Seeing Systems: Peace, Justice and Sustainability* (Portland, OR: Northwest Earth Institute, 2014), p. 11.
4. Wendell Berry, *What Matters? Economics for a New Commonwealth* (Berkeley: Counterpoint, 2010), p. 66.
5. Stephen Jay Gould, "Unenchanted Evening," in *Eight Little Piggies: Reflections in Natural History* (New York: Norton, 1993 and reprinted 2010), p. 40.
6. Edward O. Wilson, quoted in "Biophilia Hypothesis," Wikipedia, https://www.google.com/?gws_rd=ssl#q=Biophilia.
7. Ursula K. Le Guin, *The Dispossessed* http://www.goodreads.com/quotes/tag/environment.
8. Richard Schiffman, "Bigger Than Science, Bigger than Religion," *Yes!Magazine*, Spring 2015, p. 19.
9. David C. Korten, The Great Turning: from Empire to Earth Community (West Hartford, CT: Kumarian Press, 2007).
10. Joanna Macy, "The Great Turning," *Joanna Macy and Her Work*, http://www.joannamacy.net/thegreatturning.html.
11. Russell Schweikert, http://www.context.org/ICLIB/IC03/Schweik.htm.
12. Quoted in Richard Schiffman, "Bigger than Science, Bigger than Religion," *Yes!Magazine!*, Spring 2015, p. 20.
13. Pope Francis, *On Care for Our Common Home* (Washington, DC: Conference of United States Catholic Bishops), 2015, p. 25.
14. Pope Francis, pp. 17 and 27 respectively.
15. Pope Francis, p. 6.
16. Jim Wallis, *On God's Side: What Religion Forgets and Politics Hasn't Learned About Serving the Common Good* (Grand Rapids, MI: Baker Publishing Group, Brazos Press, 2013), p. 12.
17. Rabbi Lawrence Troster, *Ten Jewish Teachings on Judaism and the Environment*, http://www.greenfaith.org/religious-teachings/jewish-statements-on-the-environment/ten-jewish-teachings-on-judaism-and-the-environment.
18. Troster, *Ten Jewish Teachings*.
19. Dr. Abdullah Omar Naseef, secretary general of the Muslim World League. "Islamic Statements on the Environment, Green Faith," http://www.greenfaith.org/religious-teachings/islamic-statements-on-the-environment.
20. Hyder Ihsan Mahasneh, "Islamic Declaration on Nature," ARC Alliance Of Religion And Conservation, "Faiths and Ecology," quoting John Seymour, http://www.arcworld.org/faiths.asp?pageID=75.
21. Hyder Ihsan Mahasneh, "Islamic Declaration on Nature," ARC Alliance Of Religion And Conservation.
22. United Nations, "Islamic Declaration on Climate Change Calls for 1.6 Billion Muslims to Support Strong Paris Agreement," Statement, UN Climate Change Newsroom, 18. Aug, 2015, Http://Newsroom.Unfccc.Int/Unfccc-Newsroom/Islamic-Declaration-On-Climate-Change/.
23. Allen Hunt Badiner, Ed., *Dharma Gaia: A Harvest of Recent Essays in Buddhism and Ecology* (Berkeley, Ca: Parallax Press, 1990).
24. Ken Jones, "Getting Out of Our Own Light," in Badiner, *Dharma Gaia*, p. 183.
25. Dalai Lama, *Ethics for the New Millennium* (New York: Riverhead Books, 1999), pp. 42–43.
26. Aldo Leopold, "The Land Ethic," here from a reprint in http://www.waterculture.org/uploads/Leopold_TheLandEthic.pdf.
27. Northwest Earth Institute, "Enjoying a Sense of Place," http://www.nwei.org/enjoying-sense-place/.
28. Gloria Flora, "Remapping Relationships," in Richard Heinberg and Daniel Lerch, eds., *The Post Carbon Reader* (Healdsburg, CA: Watershed Media, 2010), pp. 184–193, p. 189 & 190.
29. Northwest Earth Institute.
30. Wendell Berry, *What Matters? Economics for a Renewed Commonwealth* (Washington, DC: Counterpoint Books, 2010), pp. 34–35.
31. David Toolan, *At Home in the Cosmos* (Maryknoll, NY: Orbis Books, 2001), p. 10.
32. Llewellyn Vaughan-Lee, "The Call of the Earth," in Joanna Macy and Thich Nhat Hanh, *Spiritual Ecology: The Cry of the Earth* (Point Reyes, CA: The Golden Sufi Center,

2013) and quoted in Kimerer LaMothe, "Our Sacred Earth: What Does It Mean to Call the Earth Divine?" *Psychology Today*, Posted Jun 30, 2014. https://www.psychologytoday.com/blog/what-body-knows/201406/our-sacred-earth.

33. Quoted in Geoffrey Shugen Arnold, Sensei, "Just Enough," in Alice Peck, Ed., *Bread, Body, Spirit: Finding the Sacred in Food* (Woodstock, VT: Skylight Paths Pub., 2008), p. 102.

34. Rachel Carson, "Excerpts from Silent Spring," Diane Ravitch, ed., *The American Reader: Words That Moved a Nation* (New York: HarperCollins, 1990), 323 http://www.uky.edu/Classes/NRC/381/carson_spring.pdf.

35. Quoted in Geoffrey Shugen Arnold, Sensei, "Just Enough," in Alice Peck, ed., p. 102.

36. Wendell Berry, *Life Is a Miracle: An Essay Against Modern Superstition* (Washington, DC: Counterpoint, 2011), p. 11.

37. Janine M. Benyus, *Biomimicry: Innovation Inspired by Nature* (New York: William Morrow and Co., 1997).

38. Benyus, p. 1.

39. Benyus, fronticepiece.

40. Benyus, p. 5.

41. Rob Dietz and Dan O'Neil, *Enough Is Enough: Building a Sustainable Economy in a World of Finite Resources* (San Francisco: Barrett-Koehler Pub., 2013), p. 18.

42. *Ibid.*, p. 33.

43. *Ibid.*, p. 35.

44. Allen Weisman, *Gaviotas: A Village to Reinvent the World* (White River Junction, VT: Chelsea Green, 1995), pp. 5–6.

45. Wendell Berry, *The Unsettling of America: Culture and Agriculture* (San Francisco: Sierra Club Books, 1977), p. 11.

46. Sulak Sivaraska, *The Wisdom of Sustainability: Buddhist Economics for the 21st Century* (Kihei, HI: Koa Books, 2009).

47. Patrick H.T. Doyle, http://www.goodreads.com/quotes/tag/needs.

48. Thomas Merton, "Silence," in *No Man Is an Island* (New York: Fall River Press, 2003 edition, originally 1955), p. 257.

49. Sivaraska, p.72.

50. Dietz and O'Neil, pp. 25–26.

51. Sivaraska, p. 66.

52. Dietz and O'Neil, p. 116.

53. Dietz and O'Neil, p. 119.

54. Andras Patrilla and Daniel Tarr, "Herding the Ox'" [A Commentary on the Ten Ox Herding Pictures] *Buddhism and Oriental Thought*. Picture Number 7. "The Ox Forgotten, Leaving the Man Alone." http://www.tarrdaniel.com/documents/ZenBuddhizmus/oxherding.html.

55. Cf. Kent Shifferd, *From War to Peace: A Guide to the Next Hundred Years* (McFarland Pubs., 2011) for a full analysis of the war system and the road to peace.

56. Randall Amster, *Peace Ecology* (Boulder: Paradigm Publishers, 2015), p. 23.

57. David Suzuki, quoted in Randal Amster, *Peace Ecology* (Boulder, CO: Paradigm Publishers, 2015), p. 3.

58. Randal Amster, p 21.

59. Anita Wenden, *Educating for a Culture of Social and Ecological Peace*, quoted in Randal Amster, p. 17.

60. Kent D. Shifferd, *From War to Peace: A Guide to the Next Hundred Years* (Jefferson, NC: McFarland, 2011) and Kent Shifferd, Patrick Hiller and David Swanson, *A Global Security System: An Alternative to War*, published by World Beyond War, 2018 edition and updated annually (worldbeyondwar.org).

61. Sarah van Gelder, *The Revolution Where You Live: Stories from a 12,000-Mile Journey Through a New America* (Oakland, CA: Barrett-Koehler Publishers, 2017).

62. Sarah van Gelder, pp. 60 & 106.

63. Earth Charter, http://www.earthcharter.

## Chapter 13

1. Sarah James and Torbörn Lahti, *The Natural Step: How Cities and Towns Can Change to Sustainable Practices* (Gabriola Island, BC: New Society Publishers, 2004), pp. 7 & 8.

2. Martin Luther King, Jr., "Remaining Awake Through a Great Revolution," Commencement Address, Oberlin College, 1965. http://www.oberlin.edu/external/EOG/BlackHistoryMonth/MLK/CommAddress.html.

3. Both quoted in Richard Louv, *Last Child in the Woods: Saving Our Children from Nature Deficit Disorder* (Chapel Hill, NC: Algonquin Books, 2008), pp. 224 & 223 respectively.

4. Richard Louv, *Last Child in the Woods*. The quote is from an interview in *Orien Magazine*, https://orionmagazine.org/article/leave-no-child-inside/, n.d.

5. Lilian Mongeaude, "Preschool Without Walls," *New York Times*, Dec. 29th, 2015.

6. Alan Weisman, *Gaviotas: A Village to*

*Reinvent the World* (White River Junction, VT: Chelsea Green Publishing, 1998), p. 142.

7. Louv, *Last Child in the Woods,* p. 135.

8. *Ibid.,* p. 141.

9. Philip Ackerman-Leist, *Rebuilding the Foodshed: How to Create Local, Sustainable, and Secure Food Systems* (White River Junction, VT: Chelsea Green Publishing, 2013, p. xxiii.

10. *Ibid.,* p. xxiii.

11. *Ibid.,* pp. 171–172.

12. Aubrey Wallace, *Green Means: Living Gently on the Planet* (San Francisco: KQED Books, 1994), p. 146.

13. Ackerman-Leist, p. xxx.

14. Wendell Berry, *The Unsettling of America: Culture & Agriculture* (San Francisco: Sierra Club Books, 1996), p. 62.

15. Bill Mollison, *Introduction to Permaculture* (Sisters Creek, Tasmania, AU: Tagari, 1991), np.

16. Ackerman-Leist, p. xxvii.

17. *Ibid.,* p. xxix.

18. Evan George, "The Urban Agriculture Movement: History and Current Trends," Michigan State University College of Law, 3/14/2013, p. 1. https://www.law.msu.edu/clinics/food/UrbanAgMvmnt.pdf. Citation for the Internal Quote Is: http://ase.tufts.edu/polsci/faculty/portney/gittlemanThesisFinal.pdf.

19. Rohit Kumar, "Five Reasons Why Urban Farming Is the Most Important Movement of Our Time," *The Good Life,* https://magazine.good.is/articles/five-reasons-why-urban-farming-is-the-most-important-movement-of-our-time.

20. D-Town Farm, http://www.d-townfarm.com/.

21. Wisconsin Public Radio NPR News, "Rooftop Farming Is Getting Off the Ground," Sept. 27, 2013, http://www.npr.org/sections/thesalt/2013/09/24/225745012/why-aren-t-there-more-rooftop-farms.

22. Evan George, "The Urban Agriculture Movement: History and Current Trends," Michigan State University College of Law, 3/14/2013, p. 2.

23. Ackerman-Leist, p. 54.

24. "Triple Bottom Line," *The Economist,* Nov. 17, 2009 http://www.economist.com/node/14301663.

25. http://www.sustainabilitydictionary.com/cradle-to-cradle/.

26. "Cradle to Cradle Design," Wikipedia, https://en.wikipedia.org/wiki/Cradle-to-cradle_design.

27. Green Living Ideas, http://green livingideas.com/2015/08/31/cradle-to-cradle-manufacturing/.

28. "Cradle to Cradle Design," Wikipedia, https://en.wikipedia.org/wiki/Cradle-to-cradle_design.

## Chapter 14

1. Robert Pollin, *Greening the Global Economy* (Boston: MIT Press, 2015), p. 1.

2. See David Wallace-Wells, *The Uninhabitable Earth: Life After Warming* (New York: Tim Dugan Books, 2019).

3. Pollin, p. 26.

4. Coral Davenport, "In Climate Move, Obama to Halt New Coal Mining Leases on Public Lands," *New York Times,* January 14, 2016.

5. Lester Brown, et al., *The Great Transition: Shifting from Fossil Fuels to Solar and Wind Energy* (New York: W.W. Norton: 2015), p. 12.

6. Brown, et al., pp. 23–26.

7. Pollin, p. 24.

8. Pollin, p. 25.

9. Brown, et al., p. 9.

10. Kari Lydersen, "Amid Climate Concerns, Nuclear Plants Feel the Heat of Warming Water," *Energy News Network,* September 9, 2016, https://energynews.us/2016/09/09/midwest/nuclear-plants-feel-the-heat-of-warming-water/.

11. Pollin, p. 48.

12. Brown, et al., p. xiii.

13. Tom Zeller, Jr., "Net Benefits of Biomass Power Under Scrutiny, Green: A Blog About Energy and the Environment," *New York Times,* June 18, 2010.

14. Andrew Pollack, "Exploring Algae as Fuel," in "Green: A Blog About Energy and the Environment," *New York Times,* July 26, 2010, on page B1 of the New York edition.

15. Andrew Pollack, "Exploring Algae as Fuel."

16. Daniel Clery, "Exclusive: Secretive Fusion Company Claims Reactor Breakthrough," *Science Magazine,* 24 August 2015 8:15 pm, http://news.sciencemag.org/physics/2015/08/secretive-fusion-company-makes-reactor-breakthrough.

17. Scott Simonson, "Will These Massive Geoengineering Projects Fix the Earth, or Break It?" *Singularity Hub,* Feb. 3, 2019, https://singularityhub.com/2019/02/03/will-these-massive-geoengineering-projects-fix-the-earth-or-break-it-more/.

18. *Ibid.*

19. *Ibid.*

20. Several sources attribute this to Carter's February 2, 1972 television address to the nation but it does not appear in the transcript.

21. ODYSSEE—MURE 2012, *Trends and Policies for Energy Savings and Emissions in Transport,* http://www.odyssee-mure.eu/publications/br/energy-efficiency-trends-policies-transport.pdf.

22. Sarah James and Torböjrn Lahti, *The Natural Step: How Cities and Towns Can Change to Sustainable Practices* (Gabriola Island, BC: New Society Publishers, 2004), p. 84.

23. Henry Fountain, "Rethinking the Airplane, for Climate's Sake," *New York Times* (e-edition), January 11, 2016.

24. U.S. Department of Energy, "Energy Efficiency Trends in Residential and Commercial Buildings," http://apps1.eere.energy.gov/buildings/publications/pdfs/corporate/bt_stateindustry.pdf.

25. Lauren Urbanek and Noah Horowitz, "The Trump Administration's Assault on Energy Efficiency Standards," Natural Resources Defense Council, March 05, 2019, https://www.nrdc.org/resources/trump-administrations-assault-energy-efficiency-standards.

26. Justin Gillis, "Oil Industry's New Threat? the Global Growth of Electric Cars," *New York Times,* November 7, 2016.

27. http://www.cnbc.com/2016/04/07/tesla-motors-reports-325k-deposits-For-Model-3.Html.

28. Justin Gillis, "Oil Industry's New Threat? the Global Growth of Electric Cars," *New York Times,* November 7, 2016.

29. *Ibid.*

30. *Ibid.*

31. "History Was Made in Sacramento, Https://Www.Youtube.Com/Watch?V=D4uktvg17ae.

32. Brown, et al., p. 104.

33. *Ibid.,* p. 105.

34. *Ibid.,* p. 130–131.

35. *Ibid.,* p. 69.

36. *Ibid.,* p. 3–4.

37. *Ibid.,* p. 7.

38. *Ibid.,* p. 68.

39. Googled Dictionary Definition.

40. United States Department of Energy, "Grid Modernization and the Smart Grid," https://www.Energy.Gov/Oe/Activities/Technology-Development/Grid-Modernization-And-Smart-Grid.

41. The Editorial Board, "A Renewable Energy Boom," *New York Times*, April 4, 2016, p. A18.

42. Chris Mooney, "Why Storing Solar Energy and Using It at Night Is Closer Than You Think," *Washington Post,* September 16, 2015, https://www.washingtonpost.com/news/energy-environment/wp/2015/09/16/why-using-solar-energy-at-night-is-closer-than-you-think/.

43. Steve Hanley, "Tesla-Solarcity Project in Connecticut Underlines Need for Merger," November 7, 2016, http://www.teslarati.com/tesla-solarcity-project-connecticut-underlines-need-merger/.

44. Shannon Hall, "Exxon Knew About Climate Change Almost 40 Years Ago," *Scientific American,* October 26, 2015, http://www.scientificamerican.com/article/exxon-knew-about-climate-change-almost-40-years-ago/.

45. Ted Trainer, "Renewable Energy: No Solution for Consumer Society," Counter Currents.org, http://www.countercurrents.org/trainer240411.htmre.

46. Pollin, p. 4.

47. President Carter, "Report to the American People on Energy," February 2, 1977, http://millercenter.org/president/speeches/speech-3396.

48. Pollin, p. 4.

49. Pollin, p. 7.

50. Stanley Reed, "Business Leaders Support Steps to Rescue Climate," *New York Times,* November 7, 2016.

51. Mark Scott, "Pollution Accord Is Set for Global Flights, but Tasks Remain," *New York Times,* November 3rd, 2016.

52. *Ibid.*

53. Stanley Reed, "Business Leaders Support Steps to Rescue Climate," *New York Times,* November 7, 2016.

54. Draft Text for Proposed Addendum to House Rules for 116th Congress of the United States, https://docs.google.com/document/d/1jxUzp9SZ6-VB-4wSm8sselVMsqWZrSrYpYC9slHKLzo/edit#.

55. Richard Heinberg, "There's No App for That: Technology and Morality in the Age of Climate Change, Overpopulation and Biodiversity Loss," Post Carbon Institute, 2017, noappforthat.org.

## Chapter 15

1. Rob Hopkins, *The Transition Handbook: from Oil Dependency to Local Resilience*

(White River Junction, VT: Chelsea Green Publishing Co., 2008.) P. 49.

2. Rob Hopkins, Transition Culture, http://www.transitionnetwork.org/blogs/rob-hopkins/2016-03/budget-2016-what-we-are-facing-isnt-financial-crisis-crisis-imagination.

3. Sarah van Gelder, *The Revolution Where You Live: Stories from a 12,000—Mile Journey Through a New America* (Berkeley, CA: Barrett-Koehler, 2017), p. 16.

4. van Gelder, p. 60 and elsewhere.

5. Wendell Berry, *The Unsettling of America* (San Francisco: Sierra Club Books, 1992), p. 22.

6. David Morris, "Free Trade: The Great Destroyer," In Jerry Mander and Edward Goldsmith, eds., *The Case Against the Global Economy and for a Turn Toward the Local* (San Francisco: Sierra Club Books, 1996), p. 220.

7. David Korton, *Agenda for a New Economy: from Phantom Wealth to Real Wealth* (San Francisco: Berrett-Koehler, 2010) p. 14.

8. John Nichols and Bob McChesney, *People Get Ready: The Fight Against the Jobless Economy and a Citizenless Democracy* (New York: Nation Books, 2016).

9. David Korton, *The Post Corporate World: Life After Capitalism* (San Francisco: Berrett-Koehler Publishers and Kumarian Press: West Hartford, CT, 1999), p. 197.

10. Aldo Leopold, Foreword to A Sand County Almanac (1949), ASCA viii. Cited at "Aldo Leopold Quotes, Http://Www.Aldoleopold.Org/Greenfire/Quotes.Shtml.

11. Wendell Berry, *What Matters? Economics for a Renewed Commonwealth* (Berkeley CA: Counterpoint, 2010), P. 48.

12. Berry, P. 91.

13. Jerry Mander and Edward Goldsmith, Eds., the *Case Against the Global Economy and for a Turn Toward the Local* (San Francisco: Sierra Club Books, 1996), P. 5.

14. *Ibid.*, p. 5.

15. *Ibid.*, p. 6.

16. Kalle Lasn, Http://Www.Goodreads.Com/Quotes/Tag/Environment?Page=5. See Also "Kalle Lasn: The Man Who Inspired the Occupy Movement," http://www.theguardian.com/world/2012/nov/05/kalle-lasn-man-inspired-occupy.

17. van Gelder, pp. 60 & 106.

18. Schumacher Center For A New Economics, "Local Currencies," http://www.centerforneweconomics.org/content/local-currencies?gclid=CjwKEAjw55K4BRC53L6

x9pyDzl4SJAD_21V1XyfzTloAQi3ZU7NrJoOIlDInirQb2AJ8uJky1fw09BoCQi_w_wcB.

19. Berkshares, Inc., http://www.berkshares.org/what_are_berkshares.

20. Korton, *Agenda*, p. 196.

21. Berry, p. 52.

22. Herman E. Daly, *For the Common Good: Redirecting the Economy Toward Community, the Environment, and a Sustainable Future*, http://www.goodreads.com/quotes/tag/environment?page=5.

23. Vandana Shiva, *Soil Not Oil: Environmental Justice in an Age of Climate Crisis* http://www.goodreads.com/quotes/tag/environment?page=2.

24. Eric Zencey, "Two Schools and the Path to the Steady State," http://steadystate.org/learn/blog/ CASSE website, Center for the Advancement of a Steady State Economy, May 15, 2011.

25. Kenneth Boulding, http://www.goodreads.com/quotes/627148-anyone-who-believes-that-exponential-growth-can-go-on-forever.

26. Eric Zencey, "Adjusting the Fifth to a Finite Planet, Part II," *The Daily News*, Posted: 18 Mar 2015 09:00 a.m. PDT.

27. Berry, p. 80.

28. Berry, p. 4.

29. Van Gelder, p. 188.

30. Community Environmental Legal Defense Fund, *On Community Civil Disobedience in the Name of Sustainability: The Community Rights Movement in the United States* (Oakland, CA: PM Press, Series No, 0013, 2015), p.2.

31. Community Environmental Legal Defense Fund, p. 1.

32. Community Environmental Legal Defense Fund, p. 3.

33. Community Environmental Legal Defense Fund, *Common Sense: Community Rights Organizing, the Spirit of '73 and the Right to Local Self-Government*, June 2015, vol. 2, www.celdf, p 24.

34. *Ibid.*, p. 26. The Fund is at PO Box 360, Mercer, PA, 17236, http://www.cldf.org.

35. *Ibid.*, p. 26.

36. In 2010, the Supreme Court case *Citizens United vs. Federal Election Commission* overturned the ban on certain types of corporate expenditures for political candidates. http://uscommonsense.org/research/citizens-united/.

37. James Bovard, "Lost Rights: The Destruction of American Liberty," http://www.goodreads.com/quotes/tag/government.

38. Berry, p. 64.
39. Berry, p. 74.
40. Alan Weisman, *Gaviotas: A Village to Reinvent the World* (White River Junction, VT: Chelsea Green Publishing Co., 1998) from the jacket cover summary of the book.
41. Weisman, p. 47.
42. Weisman, p. 18.
43. Seth Biderman and Christian Casillas, "Gaviotas: Village of Hope," *Yes Magazine*, April 2010.
44. Weisman, p. 61.
45. Biderman and Casillas.
46. Weisman, p. 13.
47. Hopkins, p. 134.
48. Rob Hopkins, *The Transition Handbook: from Oil Dependency to Local Resilience* (White River Junction, Vt: Chelsea Green Publishing Co., 2008.)
49. Transition United States, http://www.transitionus.org/.
50. "Transition Towns," Wikipedia. https://en.wikipedia.org/wiki/Transition_town#From_Kinsale_to_Totnes.
51. Hopkins, p. 134.
52. *Ibid.*, p. 54.
53. *Ibid.*, p. 55.
54. *Ibid.*, p. 15.
55. *Ibid.*, p. 57.
56. *Ibid.*, p. 68.
57. *Ibid.*, p. 55.
58. *Ibid.*, p. 56.
59. *Ibid.*, p. 56.
60. Transition United States, http://transitionus.org/transition-town-movement.
61. Transition United States, http://transitionus.org/transition-town-movement.
62. Transition United States, "About Transition Streets," http://www.transitionus.org/about-transition-streets.
63. Transition Town Totnes, www.transitiontowntotnes.org.
64. www.transitiontowntotnes.org.
65. Transition Town Totnes, www.skillshare.transitiontowntotnes.org.
66. Sarah James and Torabjorn Lahti, *The Natural Step for Communities* (Gabriola Island, BC: New Society Publishers, 2004).
67. *Ibid.*, p. xv.
68. Richard Register, *Ecocities: Building*

*Cities in Balance with Nature* (Berkeley, CA: Berkeley Hills Books, 2002).
69. *Ibid.*, p. 15.
70. *Ibid.*, p. 19.
71. *Ibid.*, p. 34.
72. *Ibid.*, p. 116.

## Conclusion

1. Jack Nelson-Pallmeyer, *Authentic Hope* (Maryknoll, NY: Orbis Books, 2012), p.171.
2. Randal Amster, *Peace Ecology* (Boulder, CO: Paradigm Publishers, 2015), *P. 198.*
3. https://www.goodreads.com/author/quotes/61107.Margaret_Mead.
4. Paul Hawken quoted in Jack Nelson-Pallmeyer, *Authentic Hope* (Maryknoll, NY: Orbis Books, 2012), p. 164.
5. Nelson-Pallmeyer, p. 152.
6. Desert Wave, http://www.desertwave.org/about-us/.
7. http://www.fastcoexist.com/3060167/this-new-neighborhood-will-grow-its-own-food-power-itself-and-handle-its-own-waste/7.
8. http://www.quotationspage.com/quote/4633.html.
9. David Orr quoted in Amster, p, 197.
10. Quoting Speth in Amster, p. 197 (James Speth, *A New American Environmentalism and the New Economy,* Tenth annual John H. Chafee Memorial Lecture on Science and the Environment, Washington, DC: National Council for Science and the Environment.)
11. Amster, p. 203.
12. Cited without reference in many places including, http://victorygardeninitiative.org/Blog/907260.
13. Jeanie Ashley Bates Greenough, *A Year of Beautiful Thoughts* (1902), p. 172.
14. Lester Brown, *Plan B 4.0 Saving Civilization We Need to Think Like a Life-Giving Planet: Part 6: Overpopulation In America,* http://rense.com/general95/thinklike6.html. Brown says this is a paraphrase of something Paul Hawken said.

# Bibliography

## Books and Periodicals

Ackerman-Leist, Philip. *Rebuilding the Foodshed: How to Create Local, Sustainable, and Secure Food Systems*. White River Junction, VT: Chelsea Green Publishing, 2013.

Adler, Jerry. "Hot Enough for You?" *Smithsonian*, May 2014.

_____. "Kill All the Mosquitoes?!" *Smithsonian*, June, 2016.

Amster, Randall. *Peace Ecology*. Boulder: Paradigm Publishers, 2015.

Ashworth, William. *The Late Great Lakes: An Environmental* History. New York: Alfred A. Knopf, 1986.

Axelrod, Alan, and Charles Phillips. *The Environmentalists: A Biographical Dictionary from the 17th Century to the Present*. New York: Zenda, 1993.

Badiner, Alan, ed. *Dharma Gaia: A Harvest of Essays in Buddhism and Ecology*. Berkeley, CA: Parallax Press, 1990.

Barnaby, Frank, ed. *Gaia Peace Atlas*. London: Pan Books, 1988.

Barstow, Anne. *Witchcraze, a New History of the European Witch Hunts*. New York: HarperCollins, 1994.

Bartlett, Richard. *The New Country*. London: Oxford University Press, 1974.

Bates, Jonathan. *Romantic Ecology: Wordsworth and the Environmental* Tradition. London: Routledge, 1991.

Baum, Dan, and Margaret L. Fox. "In Butte, Montana, A Is for Arsenic, Z Is for Zinc." *Smithsonian* 23, 38, Nov. 1992.

Benyus, Janine M. *Biomimicry: Innovation Inspired by Nature*. New York: William Morrow, 1997.

Berger, John J. *Climate Peril: The Intelligent Readers' Guide to the Climate Crisis*. Berkeley, CA: Northbrae Books, 2014.

Berman, Morris. *The Reenchantment of the World*. Toronto: Bantam, 1984.

Berry, Wendell. *Life Is a Miracle: An Essay Against Modern Superstition*. Washington, D.C.: Counterpoint, 2011.

_____. *The Unsettling of America*. Berkeley: Counterpoint (reprint edition), 2015.

_____. *What Matters? Economics for a New Commonwealth*. Berkeley: Counterpoint, 2010.

Biderman, Seth, and Christian Casillas. "Gaviotas: Village of Hope." *Yes! Magazine*, April 2010.

Bidgood, Jess. "At a Cape Cod Landmark, a Strategic Retreat from the Ocean." *New York Times*. July 6, 2016.

Blaut, J.M. *The Colonizer's Model of the World: Geographical Diffusionism and Eurocentric History*. New York: Guilford Press, 1993.

Bleifus, Joel. "Sex and Toxics." *In These Times*, March 7, 1994.

Bloom, Saul, John M. Miller, James Warner, and Philippa Winkler, eds. *Hidden Casualties: Environmental, Health and Political Consequences of the Gulf War*. Berkeley, CA: North Atlantic Books, 1993.

Born, Wolfgang. *American Landscape Painting: An Interpretation*. New Haven: Yale University Press, 1948.

Bowler, Peter. *The Norton History of the Environmental Sciences*. New York: W.W. Norton, 1992.

Broad, Robin, with John Cavanagh. *Plundering Paradise: The Struggle for the Environment in the Philippines.* Berkeley: University of California Press, 1993.

Bromwich, John Engle. "Flooding in the South Looks a Lot Like Climate Change." *New York Times.* August 16, 2016.

Brown, Lester, and Jodi Jacobson. "Assessing the Future of Urbanization," in Lester Brown, et al. *State of the World, 1987.* New York: Norton, 1987.

Brown, Lester, et al. *The Great Transition: Shifting from Fossil Fuels to Solar and Wind Energy.* New York: W.W. Norton, 2015.

_____. *State of the World, 1987.* New York: W.W. Norton, 1987.

_____. *Vital Signs.* New York: W.W. Norton, 1997.

Burckhardt, Jacob. *The Civilization of the Renaissance in Italy.* Quoted in James Bruce Ross, ed. *The Portable Renaissance Reader.* New York: Viking, 1968.

Burrows, Mrs. "Work in the Fields," from *A Childhood in the Fens* (1931) in Alistair Clayre, *Nature and Industrialization.* New York: Oxford University Press, 1977.

Burtt, E.A. *The Metaphysical Foundations of Modern Science,* as quoted in Matson, *The Broken Image: Man, Science and Society.* New York: Anchor Books, 1966.

Bush, Mark B. *Ecology of a Changing Planet.* Upper Saddle River, NJ: Prentice Hall, 1997.

Byron, George Gordon Lord. "Apostrophe to the Ocean," from Canto IV of *Childe Harold's Pilgrimage.* In Rewey Inglis and Josephine Spear, eds. *Adventures in English Literature.* New York: Harcourt Brace, 1958.

Callenbach, Ernest. "Epistle to the Ecotopians." Published by Tom Engelhardt in the blog *Tomdispatch* under the title "Tomgram: Ernest Callenbach, Last Words to an America in Decline," May 6, 2012.

Carson, Rachel. *The Sea Around Us.* New York: Oxford University Press, 1989.

_____. *Silent Spring.* New York: Mariner Books Anniversary Edition, 2002.

Cartwright, Frederick. *Disease and History.* New York: Crowell, 1972.

Caudhill, Harry. *Night Comes to the Cumberlands.* Boston: Little, Brown and Company, 1962.

Chasnoff, Debra, producer & director. *Deadly Deception: Ge, Nuclear Weapons and Our Environment.* National INFACT, 1991.

Cipolla, Carl. *Before the Industrial Revolution: European Society and Economy, 1000–1700.* New York: Norton, 1976.

Clifford, Frank. "Food or the Forest?" *Duluth News Tribune,* Aug. 7, 1997.

Colchester, Marcus, and Larry Lohmann, eds. *The Struggle for the Land and the Fate of the Forests. World Rainforest Movement, Penang, Malaysia and the Ecologist.* Sturminister Newton, England: 1993.

Collins, Gail. "Florida Goes Down the Drain: The Politics of Climate Change." *New York Times,* September 25, 2014.

Collins, John Rudolf. "Polluted Air and Diabetes: A Link." *New York Times,* October 4, 2010.

Community Environmental Legal Defense Fund. *On Community Civil Disobedience in the Name of Sustainability: The Community Rights Movement in the United States.* Oakland, CA: PM Press, Series No. 0013, 2015.

Comp, Nathan J. "Tweaking Life." *Isthmus* Vol. 43: Oct. 27–2 Nov. 2, 2016, p. 17.

Conniff, Ricard. "Amid the Plunder of Forests, a Ray of Hope." *New York Times.* January 28, 2018.

_____. "California: Desert in Disguise." *Water. National Geographic* 184 (Special Issue), 1993.

Cottrell, Fred. *Energy and Society: The Relation Between Energy, Social Change, and Economic Development.* Westport, CT: Greenwood Press, 1970.

Dalai Lama. *Ethics for the New Millennium.* New York: Riverhead Books, 1999.

Davenport, Coral. "A Climate Deal, 6 Fateful Years in the Making." *New York Times,* December 13, 2015.

_____. "In Climate Move, Obama to Halt New Coal Mining Leases on Public Lands." *New York Times.* January 14, 2016.

_____. "Major Climate Report Describes a Strong Risk of Crisis as Early as 2040." *New York Times,* October 7, 2018.

_____. "Nations Approve Landmark Climate Accord in Paris." *New York Times,* December 12, 2015.

_____. "Pentagon Signals Security Risks of Climate Change." *New York Times,* Oct. 13, 2014.

Debeir, Jean-Claude, Jean-Paul Deleage, and Daniel Hemery. *In the Servitude of Power: Energy and Civilization Through the Ages.* London: Zed Books, 1987.

Defoe. *An Essay Upon the South Sea Trade*. London: Printed for F. Baker at the Black Boy in Pater Noster Row, 1712.

Defoe, Daniel. *Caledonia, a Poem in Honour of Scotland, and the Scots Nation*. London: Printed by J. Matthews and sold by John Morphew, near Stationers Hall, 1707.

_____. *An Essay on Projects*. In Henry Morely [ed.] *The Earlier Life and Chief Earlier Works of Daniel Defoe*. London: George Routledtge and Sons, 1889.

_____. *A Plan of the English Commerce: Being a Compleat Prospect of the Trade of This Nation, as Well the Home Trade as the Foreign. In Three Parts*. Charles Rivington, 1728. ......00

_____. *Robinson Crusoe and the Farther Adventures of Robinson Crusoe*. New York: Washington Square Press, 1957.

_____. *A Tour Through the Whole Island of Great Britain*. 4 vols. London: Printed for S. T. Birt, et al., 1748.

_____. *The Trueborn Englishman* in James Southerland, ed. *Robinson Crusoe and Other Writings*. Boston: Houghton Mifflin, 1967.

Delgado, Sharon. *Shaking the Gates of Hell: Faith Led Resistance to Corporate Globalization*. Minneapolis: Fortress Press, 2007.

Derham, William. *Physico-Theology*, 1713, quoted in Naomi Klein. *This Changes Everything: Capitalism Vs. the Climate*. New York: Simon & Schuster, 2014.

de Steiguer, J.E. *Age of Environmentalism*. New York: McGraw-Hill, 1997.

Diamond, Jared. *Collapse: How Societies Choose to Fail or Succeed*. New York: Penguin Books; Revised edition. January 4, 2011.

Diderot, Denis. *Supplement to Bougainville's Voyage* in Jean Stewart and Jonathan Kemp. *Diderot: Interpreter of Nature*. New York: International Publishers, 1963.

Dietz, Rob, and Dan O'Neil. *Enough Is Enough: Building a Sustainable Economy in a World of Finite Resources*. San Francisco: Barrett-Koehler, 2013.

Disch Robert, ed. *The Dying Generations*. New York: Dell Publishing Company, 1971.

Doudna, Jennifer, and Samuel H. Sternberg. *A Crack in Creation: Gene Editing and the Unthinkable Power to Control Evolution*. San Francisco: Houghton Mifflin Harcourt, 2017.

Dunbar-Ortiz, Roxanne. *An Indigenous Peoples' History of the United States*. Boston: Beacon Press, 2014.

Durant, Will. *The Story of Civilization*. Vol. VII. *The Age of Reason Begins*. New York: Simon & Schuster, 1961.

Ehrlich, Paul. *The Population Bomb*. New York: Ballantine, 1971.

Eiseley, Loren. *The Immense Journey: An Imaginative Naturalist Explores the Mysteries of Man and Nature*. New York: Time, Inc., 1957.

Eisendrath, Bettie. *Military Ecocide: Man's Secret Assault on the Environment*. Milwaukee: World Federalist Association, 1992.

Eisler, Riane. *The Chalice and the Blade: Our History, Our Future*. New York: HarperOne, 2011.

Ekirch, Arthur. *Man and Nature in America*. New York: Columbia University Press, 1963.

Eldberg, Rolf. "Earth, Time and Man," in Robert Disch, *The Dying Generations*. New York: Dell Publishing Company, 1971.

Eliade, Mircea. *Cosmos and History: The Myth of the Eternal Return*. New York: Harper, 1959.

Emerson, Ralph Waldo. *Essays*. Boston: Houghton Mifflin & Co., 1897.

Engler, Paul, and Mark Engler. *This Is an Uprising: How Nonviolent Revolt Is Shaping the Twenty-First Century*. New York: Nation Books, 2016.

Ferkiss, Victor. *Nature, Technology and Society: Cultural Roots of the Current Environmental Crisis*. New York: New York University Press, 1993.

Flavin, Christopher. "Reassessing Nuclear Power." *State of the World*. N.p.: World Watch Institute, 1996.

Fountain, Henry. "Cleanup Questions as Radiation Spreads." *New York Times*. April 1, 2011.

_____. "Rethinking the Airplane, for Climate's Sake." *New York Times*, January 11, 2016.

Frankfort, Henri, H.A. Frankfort, John Wilson, and Thorkild Jacobson. *Before Philosophy: The Intellectual Adventure of Ancient Man*. Baltimore: Pelican, 1949.

Freudenberg, Nicholas, and Carol Steinsapir. "Not in Our Backyards; the Grassroots Environmental Movement." *American Environmentalism: The U.S. Environmental Movement: 1970–1990*. New York: Taylor and Francis, 1992.

Friedman, Thomas. "Stuff Happens to the Environment, Like Climate Change." *New York Times*, Oct. 7, 2015.

Gadgil, Madhav, and Ramachandra Gutkind. *This Fissured Land: An Ecological History of India.* Berkeley: University of California Press, 1992.

Garrett, Laurie. *The Coming Plague: Newly Emerging Diseases in a World Out of Balance.* New York: Penguin, 1994.

Geoffrey Shugen Arnold, Sensei. "Just Enough," in Alice Peck, ed. *Bread, Body, Spirit: Finding the Sacred in Food.* Woodstock, VT: Skylight Paths Pub., 2008.

Gibler, John. "Under the Gun." *Sierra.* July/August 2017, pp. 28ff.

Gillis, Justin. "Climate Model Predicts West Antarctic Ice Sheet Could Melt Rapidly." *New York Times,* March 30, 2016.

_____. "Oil Industry's New Threat? The Global Growth of Electric Cars." *New York Times,* November 7, 2016.

_____. "U.N. Panel Warns of Dire Effects from Lack of Action Over Global Warming." *New York Times,* November 2, 2014.

Glendinning, Chellis. *My Name Is Chellis and I Am in Recovery from Western Civilization.* Boston: Shambhala, 1994.

Gloria Flora. "Remapping Relationships," in Richard Heinberg and Daniel Lerch, eds. *The Post Carbon Reader.* Healdsburg, CA: Post Carbon Institute and Watershed Media, 2010.

Gofman, John. "Control of the Atom," in *Progress as If Survival Mattered: A Handbook for a Conserver Society.* San Francisco: Friends of the Earth, 1981.

Goin, Peter. *Nuclear Landscapes.* Baltimore: Johns Hopkins University Press, 1991.

Gold, Thomas, attributed to by Timothy Ferris. *Coming of Age in the Milky Way.* New York: Doubleday, 1998.

Goldsmith, Edward. "The Fall of the Roman Empire: A Social and Ecological Interpretation." *Ecologist,* July 1977, p. 203.

Goldsmith, Oliver. "The Deserted Village" in Rewey Belle Inglis and Josephine Spear. *Adventures in English Literature.* New York: Harcourt Brace, 1958.

Goodchild, Peter. "The Imminent Collapse of Industrial Society." *Countercurrents.Org,* 09 May, 2010.

Gould, Stephen J. "Unenchanted Evening." *Eight Little Piggies: Reflections in Natural History.* New York: Norton, 1993; reprinted 2010.

Greenough, Jeanie Ashley Bates *A Year of Beautiful Thoughts.* Philadelphia: G.W. Jacobs, 1902.

Griffin, Susan. *The Eros of Everyday Life: Essays on Ecology, Gender and Society.* New York: Anchor Books, 1996.

Gruchow, Paul. *Grass Roots: The Universe of Home.* Minneapolis: Milkweed Editions, 1995.

Hallman, David G., et al. *Climate Changes and the Quest for Sustainable Societies.* Geneva: World Council of Churches, 1998.

Hansen, James. "Coal-Fired Power Stations Are Death Factories. Close Them." *The Guardian,* Feb. 15, 2009.

_____. *Storms of My Grandchildren: The Truth About the Coming Climate Catastrophe and the Last Chance to Save Humanity.* New York: Bloomsbury, 2009.

Hardin, Garret. "Lifeboat Ethics: The Case Against Helping the Poor." *Psychology Today.* September 1974.

_____. "The Tragedy of the Commons." *Science* 13 December 1968, vol. 162 no. 3859: pp. 1243–1248.

Harlan, Jack. *Crops and Man.* Madison, WI: American Society of Agronomy, Crop Science Society of America, 1975.

Harris, Gardiner. "Facing Rising Seas, Bangladesh Confronts the Consequences of Climate Change." *New York Times,* March 28, 2014.

Hartman, William K., and Ron Miller. *The History of the Earth: An Illustrated Chronicle of an Evolving Planet.* New York: Workman Publishing Co., 1991.

Harvard Working Group On New And Resurgent Diseases. "Globalization, Development and the Spread of Disease," in Jerry Mander and Edward Goldsmith, eds. *The Case Against the Global Economy.* San Francisco: Sierra Club Books, 1996.

Hastings, Tom. *Ecology of War and Peace: Counting the Costs of Conflict.* Lanham, MD: University Press of America, 2000.

Hawkin, Paul, ed. *DRAWDOWN: The Most Comprehensive Plan Ever Proposed To Reverse Global Warming.* New York: Penguin Books, 2017.

Heffernan, James. *Recreation of Landscape: A Study of Wordsworth, Coleridge, Constable and Turner.* London: University Press of New England, 1985.

Heichelheim, Fritz. "Effects of Classical Antiquity on the Land," in William L. Thomas, ed. *Man's Role in Changing the Face of the Earth.* Chicago: University of Chicago Press, 1956.

Hopkins, Rob. *The Transition Companion: Making Your Community More Resilient in Uncertain Times.* White River Junction, VT: Chelsea Green, 2011.

_____. *The Transition Handbook: from Oil Dependency to Local Resilience.* White River Junction, VT: Chelsea Green, 2008.

Hughes, J. Donald. *Ecology in Ancient Civilizations.* Albuquerque, NM: University of New Mexico Press, 1975.

Hughes, Virginia. "How Having Three Parents Leads to Disease Free Kids." *Popular Science.* October, 2013, p. 40.

Ihde, Aron J. *The Development of Modern Chemistry.* New York: Dover, 1984.

Inkinen, Sam, ed. *Mediopolis: Aspects of Texts, Hypertexts, and Multimedial Communications.* Berlin: De Gruyter, 2011.

Invasive Species Advisory Committee. "Invasive Species Definition Clarification and Guidance White Paper." Submitted by the Definitions Subcommittee of the Invasive Species Advisory Committee (ISAC). Approved by ISAC April 27, 2006.

Jacobs, Andrew, and Ian Johnson. "Pollution Killed 7 Million People Worldwide in 2012, Report Finds." *New York Times,* March 25, 2014.

Jarvis, Brooke. "The Insect Apocalypse." *New York Times.* November 27, 2018.

Kaplan, Robert. *The Ends of the Earth: A Journey at the Dawn of the 21st Century.* New York: Random House, 1996.

Kipling, Rudyard. *Rudyard Kipling's Verse,* definitive edition. Garden City, NY: Doubleday, 1940.

Klein, Benjamin. *First Along the River: A Brief History of the U.S. Environmental Movement.* San Francisco: Acada Books, 1997.

Klein, Naomi. *This Changes Everything: Capitalism Vs. the Climate.* New York: Simon & Schuster, 2014.

Kodi, Yeager-Kozacek. "Droughts Hit World's Agricultural Regions: Without Water, U.S. Corn Crop Faces Setbacks." *Circle of Blue,* July 12, 2012.

Korten, David C. *The Great Turning: from Empire to Earth Community.* West Hartford, CT: Kumarian Press, 2007.

Korton, David. *Agenda for a New Economy: from Phantom Wealth to Real Wealth.* San Francisco: Berrett-Koehler, 2010.

_____. *The Post Corporate World: Life After Capitalism.* San Francisco: Berrett-Koehler, 1999.

Kristof, Nicholas. "Our Beaker Is Starting to Boil." *New York Times,* July 16, 2010.

LaFreniere, Gilbert. "Rousseau and the European Roots of Environmentalism." *Environmental History Review.* Winter 1990, p. 57.

Lahti, James, and Torbörn Lahti. *The Natural Step: How Cities and Towns Can Change to Sustainable Practices.* Gabriola Island, BC: New Society Publishers, 2004.

Lanham, Url. *Origins of Modern Biology.* New York: Columbia University Press, 1968), p. 79.

Leakey, Richard and Roger Lewin. *The Sixth Extinction: Patterns of Life and the Future of Humankind.* New York: Anchor, 1996.

Liptak, Adam. "Justices, 9–0, Bar Patenting Human Genes." *New York Times, June 13, 2013.*

Lorbiecki, Marybeth. *Aldo Leopold: A Fierce Green Fire.* Helena, MT: Falcon Publishing Co., Inc., 1996.

Louv, Richard. *Last Child in the Woods: Saving Our Children from Nature Deficit Disorder.* Chapel Hill, NC: Algonquin Books, 2006.

Mander, Jerry, and Edward Goldsmith, eds. *The Case Against the Global Economy and for a Turn Toward the Local.* San Francisco: Sierra Club Books, 1996.

MAREX. "Fukushima Disaster Produces World's Worst Nuclear Sea Pollution." *The Maritime Executive,* 2011–10–28 11:48:31.

Marsh, George Perkins. *Man and Nature.* Quoted in Roderick Nash. the *American Environment: Readings in the History of Conservation.* Reading, MA: Addison-Wesley, 1968.

Martin, Calvin. *In the Spirit of the Earth: Rethinking History and Time.* Baltimore: The Johns Hopkins University Press, 1992.

Marx, Karl. *Capital.* Vol. 1. New York: International Publishers, 1967.

_____. *Economic and Philosophic Manuscripts of 1844*. New York: International Publishers, 1964.

Marx, Leo. the *Machine in the Garden: Technology and the Pastoral Ideal in America*. London: Oxford University Press, 1964.

Matson, Floyd W. *The Broken Image: Man, Science and Society*. New York: Anchor Books, 1966.

McKibben, Bill. *The End of Nature*. New York: Random House, 1989.

McNeill, David. "Why the Fukushima Disaster Is Worse than Chernobyl." *The Independent* (UK) August 29, 2010.

Meadows, Donella, and Jorgen Randers. *Beyond the Limits: Confronting Global Collapse, Envisioning a Sustainable Future*. White River Junction, VT: Chelsea Green, 1992.

Meadows, Donella, et al. *Limits of Growth: The Thirty Year Update*. White River Junction, VT: Chelsea Green Publishing Co., 2004.

_____. *The Limits to Growth*. New York: Universe Books, 1972.

_____. *Worldwide Growth in Selected Human Activities and Products, 1950–2000, Limits to Growth: The Thirty Year Update*. White River Junction, VT: Chelsea Green Publishing Co., 2004.

Meleaug, Christopher. "Tens of Thousands Evacuated as Fire Rages in Southern California." *New York Times*, August 17, 2016.

Menzel, Peter. *Material World: A Global Family Portrait*. Text by Charles Mann. San Francisco: Sierra Club Books, 1994.

Merchant, Carolyn. *The Death of Nature: Women, Ecology and the Scientific Revolution*. San Francisco: Harper and Row, 1980.

Merton, Thomas. "Silence," in *No Man Is an Island*. New York: Fall River Press, 2003 edition, originally 1955.

Mollison, Bill. *Introduction to Permaculture*. Tasmania, Australia: Tagari, 1991.

Mongeaude, Lilian. "Preschool Without Walls." *New York Times*, Dec. 29th, 2015.

"More Trouble for Vermont Yankee." *Nukewatch Quarterly*. Fall 2011, P. 7.

Morley, John, ed. *The Complete Poetical Works of William Wordsworth*. London: Macmillan and Co., 1891.

Morris, David. "Free Trade: The Great Destroyer," in Jerry Mander and Edward Goldsmith, eds. *The Case Against the Global Economy and for a Turn Toward the Local*. San Francisco: Sierra Club Books, 1996.

Mumford, Lewis. *The City in History*, as excerpted in L.G. Schaefer, et al. *The Shaping of Western Civilization*. New York: Holt, Rinehart and Winston, 1970.

Myers, Norman. *Gaia Atlas of Planet Management*. London: Anchor, 1992.

Naess, Arne. *The Ecology of Wisdom: Writings by Arne Naess*. Berkeley, CA: Counterpoint Press, 2010.

Nash, Roderick. *Wilderness and the American Mind*. Hartford, CT: Yale University Press, 1967.

Native Forest Council. *Forest Voice*. Vol. 1, # 1, September, 1989.

Nelson, Gaylord, *Beyond Earth Day: Fulfilling the Promise*. University of Wisconsin Press, 2002.

Nelson-Pallmeyer, Jack. *Authentic Hope*. Maryknoll, NY: Orbis Books, 2012.

*New York Times*, Editorial Board. "A Renewable Energy Boom." *New York Times*, April 4, 2016, p. A18.

*New York Times*, standing article. "Climate Change Is Complex: We've Got Answers to Your Questions." www.nytimes.com, accessed Jan. 24, 2018.

Nichols, John, and Bob McChesney. *People Get Ready: The Fight Against the Jobless Economy and a Citizenless Democracy*. New York: Nation Books, 2016.

Northwest Earth Institute. *Seeing Systems: Peace, Justice and Sustainability*. Portland, OR: Northwest Earth Institute, 2014.

Novak, Maximillian. "Crusoe's Fear and the Search for Natural Man." *Modern Philology* 58, May, 1961.

Nussbaum. Frederick. *The Triumph of Science and Reason: 1660–1685*. New York: Harper, 1953.

Orlov, Dimitri. *The Five Stages of Collapse*. New Society Publishers: Gabriola Island, BC, 2013.

Palmer, R.R., and Joel Colton. *A History of the Modern World*. 5th ed. New York: Knopf, 1978.

Alice Peck, ed. *Bread, Body, Spirit: Finding the Sacred in Food*. Woodstock, VT: Skylight Paths Pub., 2008.

Pepper, David. *The Roots of Modern Environmentalism*. London: Routledge, 1984.

Petulla, Joseph. *American Environmental* History. San Francisco: Boyd and Fraser, 1977.

Pico della Mirandola, Giovanni. *Oration on the Dignity of Man,* in Franklin Le Van Baumer, ed. *Main Currents of Western Thought.* New York: Alfred A. Knopf, 1970.

Platte, Anne E. "Infecting Ourselves: How Environmental and Social Disruptions Trigger Disease." *World Watch Paper* #129. Washington, D.C: World Watch Institute, April, 1996.

Pollack, Andrew. "A Dream of Trees Aglow at Night." *New York Times,* May 7th, 2013.

_____. "Exploring Algae as Fuel." *Green: A Blog About Energy and the Environment. New York Times,* July 26, 2010, on page B1 of the New York edition.

_____. "Modified Wheat Is Discovered in Oregon." *New York Times,* May 29, 2013.

Pollan, Michael. "Why Bother?" in *DRAWDOWN: The Most Comprehensive Plan Ever Proposed To Reverse Global Warming,* ed. Paul Hawken, New York: Penguin Books 2017.

Pollin, Robert. *Greening the Global Economy.* Cambridge, MA: MIT Press, 2015.

Ponting, Clive. *A Green History of the World.* New York: Penguin, 1991.

Pope, Alexander. *An Essay on Man.* Epistle I, l 25 in John Butt, ed., *The Poems of Alexander* Pope. New Haven: Yale University Press, 1963.

_____. "Solitude," in Oscar Williams, ed. *Immortal Poems of the English Language.* New York: Washington Square Press, 1960.

Pope Francis. *On Care for Our Common Home.* Washington, DC: Conference of United States Catholic Bishops, 2015.

Quinn, Daniel. *Ishmael.* New York: Bantam/Turner, 1993.

Radelet, Steven. *The Great Surge: The Ascent of the Developing World.* New York: Simon & Schuster, 2015 (kindle edition).

"Radiation Gushes from Fukushima, Information Trickles." *Nukewatch Quarterly.* Fall 2011, p. 5.

Raven, Peter. "The Global Ecosystem in Crisis." A MacArthur Foundation Occasional Paper, Chicago, 1987.

Reed, Stanley. "Business Leaders Support Steps to Rescue Climate." *New York Times,* November 7, 2016.

Register, Richard. *Ecocities: Building Cities in Balance with Nature.* Berkeley, CA: Berkeley Hills Books, 2002.

Reilly, Kevin. *The West and the World: A Topical History of Civilization.* New York: Harper and Row, 1980.

Rifkin, Jeremy. *The Biotech Century: Harnessing the Gene and Remaking the World.* New York: Penguin/Putnam, 1998.

Rogers, Adam. "Chemicals: The Great Impostors." *Newsweek.* March 18, 1996.

Rothman, Hal K. *The Greening of America: Environmentalism in the United States Since 1945.* Fort Worth: Harcourt Brace, 1998.

Rousseau, Jean Jaques. "From the Social Contract." In *The World's Great Thinkers, Man and the State: The Political Philosophers*, eds. Saxe Commins and Robert N. Linscott. New York: Random House, 1947.

Ryan, John C. "Conserving Biological Diversity." *State of the World 1992.* Washington, DC: Worldwatch Institute, 1992.

Sale, K. in Alwyn Jones. "From Fragmentation to Wholeness: A Green Approach to Science and Society." *The Ecologist,* Vol. 17, No. 6, 1987.

Schiffman, Richard. "Bigger than Science, Bigger than Religion." *Yes! Magazine.* Spring 2015, p. 19.

Schumacher, E.F. *Small Is Beautiful: Economics as If People Mattered.* New York: Harper Perennial Reprint edition, 2010.

Scott, Mark. "Pollution Accord Is Set for Global Flights, but Tasks Remain." *New York Times,* November 3rd, 2016.

Shepard, Paul. *Man in the Landscape.* New York: Ballantine, 1972.

Shifferd, Kent. "In the Land of the Blind: Evaluating Obama's Environmental Record," in *Grading the 44th President.* Luigi Esposito and Laura Finley, eds. Santa Barbara: Praeger, 2012.

Shifferd, Kent D. *From War to Peace: A Guide to the Next Hundred Years.* Jefferson, NC: McFarland,, 2011.

Short, Bruce. "War and Disease: War Epidemics in the Nineteenth and Twentieth Centuries." *ADF Health,* Vol. 11 No. 1 2010, p. 16.

Sierra Club. *Sierra*. January-February, 2018.

Simmons, I.G. *Environmental History: A Concise Introduction*. Oxford: Blackwell, 1993.

Sivaraska, Sulak. *The Wisdom of Sustainability: Buddhist Economics for the 21st Century*. Kihei, HI: Koa Books, 2009.

Smil, Vaclav. *Energy in World History*. Boulder: Westview Press, 1994.

Smith, Adam, in Alasdir Clayre. *Nature and Industrialization*. Oxford: Oxford University Press, 1977.

Southwick, Charles H. *Global Ecology in Human Perspective*. Oxford: Oxford University Press, 1996.

Specter, Michael. "DNA Revolution." *National Geographic*, August 2016, p. 36ff.

Stearns, Peter. *The Industrial Revolution in World History*. Boulder: Westview Press, 1993.

Stenehjem, M. "Indecent Exposure." *Natural History* 9, 1990, pp. 6ff.

Strong, Douglas. *Dreamers and Defenders: American Conservationists*. Lincoln: University of Nebraska Press, 1997.

Swanson, David. *War No More: The Case for Abolition*. Charlottesville, VA: N.p., 2013.

Swimme, Brian. "How to Heal a Lobotomy," in Irene Diamond and Gloria Feman Orenstein, eds. *Reweaving the World: The Emergence Of Ecofeminism*. San Francisco: Sierra Club Books, 1990.

Thomas, Elizabeth Marshall. *The Old Way: A Story of the First People*. New York: Sarah Crichton Books, 2003.

Thomas, William. *Scorched Earth: The Military Assault on the Environment*. Philadelphia: New Society Publishers, 1995.

Thoreau, Henry David. *Walden*. Princeton: Princeton University Press, 1973.

Thunberg, Greta. *No One Is Too Small to Make a Difference*. London: Penguin Books, 2019.

Toolan, David. *At Home in the Cosmos*. Maryknoll, NY: Orbis Books, 2001.

Trachtenberg, Alan. *The Incorporation of America. Culture and Society In the Gilded Age*. New York: Hill and Wang, 1982.

Tuan, Yi Fu. *Topophilia: A Study of Environmental Perception, Attitudes, and Values*. New York: Columbia University Press, 1990 reprint.

Udall, Stewart. *The Energy Balloon*. New York: McGraw-Hill, 1974.

_____, foreword to Harry Caudhill. *Night Comes to the Cumberlands*. Boston: Little, Brown and Company, 1962.

Ure, Andrew. *The Philosophy of Manufactures*, cited in Alisdair Clayre. *Nature and Industrialization*. Oxford: Oxford University Press, 1977.

Van Gelder, Sarah. *The Revolution Where You Live: Stories from a 12,000-Mile Journey Through a New America*. Oakland, CA: Barrett-Koehler, 2017.

Vaughan-Lee, Llewellyn. "The Call of the Earth," in Joanna Macy and Thich Nhat Hanh. *Spiritual Ecology: The Cry of the Earth*. Point Reyes, CA: The Golden Sufi Center, 2013.

Wallace, Aubrey. *Eco-Heroes: Twelve Tales of Environmental Victory*. San Francisco: Mercury House, 1993.

_____. *Green Means: Living Gently on the Planet*. San Francisco: Kqed Books, 1994.

Wallace Wells, David. *The Uninhabitable Earth: Life After Warming*. New York: Tim Duggan Books, 2019.

Wallis, Jim. *On God's Side: What Religion Forgets and Politics Hasn't Learned About Serving the Common Good*. Grand Rapids, MI: Baker Publishing Group, Brazos Press, 2013.

Watt, Ian. *The Rise of the English Novel*. Berkeley: University of California Press, 1957.

Weisman, Allen. *Gaviotas: A Village to Reinvent the World*. White River Junction, VT: Chelsea Green, 1995.

Williams, Eric. *Capitalism and Slavery*, cited in J. M. Blaut. *The Colonizer's Model of the World*. New York: The Guilford Press, 1993.

Willis, F. Roy. *Western Civilization: An Urban Perspective*. Lexington, MA: D.C. Heath, 1973.

Wilson, E. O. "Biophilia." Address to the President and Fellows of Harvard College, 1984.

Yong, Ed. "Can We Save the World by Remixing Life?" *National Geographic*. April 11, 2013.

Yunus, Muhammad. *A World of Three Zeros*. New York: Public Affairs, 2017.

Zeller, Tom, Jr. "Net Benefits of Biomass Power Under Scrutiny." *Green: A Blog About Energy and the Environment. New York Times*, June 18, 2010.

Zoellnor, Tom. *Uranium: War, Energy and the Rock That Shaped the World*. New York: Penguin, 2009.

# Websites

Alliance of Religions and Conservation. (ARC) www.arcworld.org.
American Geophysical Union. www.agu.org.
American Institute of Physics. www.aip.org.
Anthropocene. info "Welcome to the Anthropocene: A Planet Transformed by Humanity." http://www.anthropocene.info/en/anthropocene.
Arms Control Association. "U.S. Nuclear Modernization Programs." August 2017, Https://Www.Armscontrol.Org/Factsheets/Usnuclearmodernization.
Arctic News Blogspot. Arctic-news.blogspot.com.
Berkshares, Inc. http://www.berkshares.org/what_are_berkshares.
Biocurious. http://biocurious.org/about/.
Biology Online Dictionary. http://www.Biology-Online.Org/Dictionary.
Bird Life International. "Spotlight on Threatened Birds." datazone/birdlife.org /sowb/ spotthreatbirds.
Bloch, Michael. "Green Living Tips." http://www.greenlivingtips.com.
Bruntland, Gro Harman. *Report of the World Commission on Environment and Development: Our Common Future*: United Nations, www.un-documents.net/our-common-future.pdf.
Center for Biological Diversity. www.biologicaldiversity.org.
Centers for Disease Control. www.cdc.gov.
Clery, Daniel. "Exclusive: Secretive Fusion Company Claims Reactor Breakthrough." Science Magazine, 24 August 2015. http://news.sciencemag.org/physics/2015/08/secretive-fusion-company-makes-reactor-breakthrough.
Cloudfront.net. https://d2gne97vdumgn3.cloudfront.net/api/file/QWz65NxtRXG58frH8BAU
CO₂ Earth. "Daily Co₂" https://www.co2.earth/daily-co2.
Colombia University. "Introduced Species Summary Project." http://www.columbia.edu/itc/cerc/danoff-burg/invasion_bio/inv_spp_summ/invbio_plan_report_home.html.
Community Environmental Legal Defense Fund. www.celdf.org.
D-Town Farm. http://www.d-townfarm.com/.
Desert Wave. http://www.desertwave.org/about-us/.
Earth Charter. http://www.earthcharter.org.
Earthjustice. http://earthjustice.org.
EduGreen. "The Chipko Movement." http://edugreen.teri.res.in/explore/forestry/chipko.htm.
European Union. "Factory Farming Makes Baltic Sea One of the World's Most Polluted." http://www.arc2020.eu/2014/06/factory-farming-made-the-baltic-sea-one-of-the-worlds-most-polluted-seas/.
Ewaste.com "Electronic Waste by Numbers: Recycling & the World." March 2, 2016, http://www.ewaste.com.au/ewaste-articles/electronic-waste-by-numbers-recycling-the-world/.
Facts and Details (factsanddetails.com). "Air Pollution in China," http://factsanddetails.com/china.php?itemid=392&catid=10&subcatid=66.
George, Evan. *The Urban Agriculture Movement: History and Current Trends*. Michigan State University College of Law. 3/14/2013. https://www.law.msu.edu/clinics/food/UrbanAgMvmnt.pdf.
Gebelhoff, Robert. "Study of Land Snails Suggests Earth May Already Have Lost Seven Percent of Its Species." *Washington Post*. Aug. 8, 2015. https://www.washingtonpost.com/news/morning-mix/wp/2015/08/11/study-of-land-snails-suggests-earth-may-have-already-lost-7-percent-of-its-animal-species/.
Goldberg, Jeffrey. "Drowning Kiribati." *Bloomberg Business Week*. http://www.businessweek.com/articles/2013-11-21/kiribati-climate-change-destroys-pacific-island-nation.
Green Faith. www.greenfaith.org.
Green Living Ideas. http://greenlivingideas.com/2015/08/31/cradle-to-cradle-manufacturing/.
Hall, Shannon. "Exxon Knew About Climate Change Almost 40 Years Ago." *Scientific American*, October 26, 2015. https://www.scientificamerican.com/article/exxon-knew-about-climate-change-almost-40-years-ago/.
Heimbuch, Jaymie. "11 Critically Endangered Turtle Species," *Treehugger*. May 23, 2013. http://www.treehugger.com/natural-sciences/11-critically-endangered-turtle-species.html.
Heinberg, Richard. "There's No App for That: Technology and Morality in the Age of Climate

Change, Overpopulation and Biodiversity Loss." Post Carbon Institute, 2017. Noappforthat. org.

International Rivers.org.

Kaplan, Sarah. "By 2050, There Will Be More Plastic than Fish in the World's Oceans, Study Says." *Washington Post.* January 20, 2016 https://www.washingtonpost.com/news/morning-mix/wp/2016/01/20/by-2050-there-will-be-more-plastic-than-fish-in-the-worlds-oceans-study-says/?utm_term=.39632090c7.

King, Jr., Martin Luther. "Remaining Awake Through a Great Revolution." Commencement Address, Oberlin College, 1965. http://www.oberlin.edu/external/EOG/BlackHistoryMonth/MLK/CommAddress.html.

Koch, Wendy. "FDA Officially Bans BPA, or Bisphenol-A, from Baby Bottles." *USA Today,* July 17, 2012. http://usatoday30.usatoday.com/money/industries/food/story/2012-07-17/BPA-ban-baby-bottles-sippy-cups/56280074/1.

Lamothe, Kimerer. "Our Sacred Earth: What Does It Mean to Call the Earth Divine?" *Psychology Today.* Posted Jun 30, 2014. https://www.psychologytoday.com/blog/what-body-knows/201406/our-sacred-earth.

Lydersen, Kari. "Amid Climate Concerns, Nuclear Plants Feel the Heat of Warming Water." *Energy News Network.* September 9, 2016, https://energynews.us/2016/09/09/midwest/nuclear-plants-feel-the-heat-of-warming-water/.

Macy, Joanna. "Joanna Macy and Her Work." http://www.joannamacy.net.

Mahlman, Jerry. "Climate Change," testimony before the U.S. Senate Committee on Commerce, Science and Transportation, March 3, 2004. http://www.commerce.senate.gov/public/index. cfm/hearings?Id=abed3dcd-e7e8-47fb-9506-93d703e4b634&Statement_id=CAA36353-8A4F-4719-AD7C-72860B4D41E8.

Marshall, Michael. "Terraforming: Geoengineering Megaplan Starts Now." *New Scientist.* Oct. 9 2013, http://www.newscientist.com/article/mg22029382.500-terraforming-earth-geoengineering-megaplan-starts-now.html#.UthT0Tbnb4g.

McKibben, Bill. "A Special Moment in History." *The Atlantic Online,* May 1998. http://www. theatlantic.com/past/docs/issues/98may/special3.htm.

_____. Radio Interview. *Democracy Now.* May 26, 2011. http://www.democracynow. org/2011/5/26/bill_mckibben_from_storms_to_droughts.

Mooney, Chris. "Why Storing Solar Energy and Using It at Night Is Closer than You Think." *Washington Post.* September 16, 2015. https://www.washingtonpost.com/news/energy-environment/wp/2015/09/16/why-using-solar-energy-at-night-is-closer-than-you-think/.

Moskvitch, Katia. "Mysterious Siberian Crater Attributed to Methane." *Nature: The International Weekly Journal of Science.* July 31, 2014. http://www.nature.com/news/mysterious-siberian-crater-attributed-to-methane-1.15649.

National Institutes of Health. "What Is a Genome?" https://ghr.nlm.nih.gov/primer/hgp/genome.

National Priorities Project. "Military Spending in the United States." https://www. nationalpriorities.org/campaigns/military-spending-united-states/.

Natural Resources Defense Council. https://www.nrdc.org/.

National Wildlife Federation. "Global Warming and Heat Waves." http://www.nwf.org/Wildlife/Threats-to-Wildlife/Global-Warming/Global-Warming-is-Causing-Extreme-Weather/Heat-Waves.aspx.

Noaa. "Demistifying the Great Pacific Garbage Patch." http://marinedebris.noaa.gov/info/patch.html#1.

NOAA, Earth System Research Laboratory. http://www.esrl.noaa.gov.

Northwest Earth Institute. "Enjoying a Sense of Place." http://www.nwei.org/enjoying-sense-place/.

Nuclear Regulatory Commission. http://www.nrc.gov.

ODYSSEE—MURE 2012. *Trends and Policies for Energy Savings and Emissions in Transport.* http://www.odyssee-mure.eu/publications/br/energy-efficiency-trends-policies-transport. pdf.

Pacific Institute. "California Drought: Impacts and Solutions." http://www.californiadrought. org/drought/current-conditions/.

Peace and Justice Studies Association. "Connecting Militarism and Climate Change." February 14, 2011. http://www.peacejusticestudies.org.

Pesticide Action Network (panna.org). "At Long Last, EPA Releases Pesticide Use Statistics." http://www.panna.org/blog/long-last-epa-releases-pesticide-use-statistics.

_____. "Roundup, Cancer and the Future of Food." http://www.panna.org/blog/roundup-cancer-future-food?gclid=Cj0KEQjw4pO7BRDl9ePazKzrlLYBEiQAHLJdR76WMGc0P-JIB 777hzPqQ3Rh6Z4ra9NvlreZf7Clts4aAvwK8P8HAQ.

Peters, Adele. "Imagine a City Lit by Glowing Trees Instead of Streetlights." *Fast Company.* May 14, 2018. https: //www.fastcompany.com/40571215/imagine-a-city-lit-by-glowing-trees-instead-of-streetlights.

Philpott, Tom. "A Reflection on the Lasting Legacy of 1970s USDA Secretary Earl Butz." *Grist.* Feb 8, 2008. http://grist.org/article/the-butz-stops-here/.

Porter, Eduardo. "Fighting Climate Change? We're Not Even Landing a Punch." *New York Times,* Jan. 23, 2018. https://www.nytimes.com/2018/01/23/business/economy/fighting-climate-change.html.

Pvc.Org. http://www.pvc.org/en/p/packaging.

Roser, Max. "Economic Growth." Our World in Data. https://ourworldindata.org/economic-growth.

Roser, Max, and Hannah Ritchie. "Fertilizer and Pesticides." Our World in Data. https://ourworldindata.org/fertilizer-and-pesticides/.

Schumacher Center for a New Economics. "Local Currencies Program." http://www.centerforneweconomics.org/content/local-currencies?gclid=CjwKEAjw55K4BRC53L6x9py Dzl4SJAD_21V1XyfzTloAQi3ZU7NrJoOIlDInirQb2AJ8uJkylfw09BoCQi_w_wcB.

Shah, Anup. "A Primer on Neoliberalism." http://www.globalissues.org/article/39/a-primer-on-neoliberalism#globalissues-org.

Simonson, Scott. "Will These Massive Geoengineering Projects Fix the Earth, or Break It?" Singularity Hub, Feb. 3, 2019. https://singularityhub.com/2019/02/03/will-these-massive-geoengineering-projects-fix-the-earth-or-break-it-more/.

The Sunrise Movement. https://www.sunrisemovement.org/calls/2019/3/15/prepare-for-the-us-climate-strike, March 15 2019.

Sustainability Dictionary http://www.sustainabilitydictionary.com/.

Tennent. James. "More than a Quarter of British Birds at Risk of Local Extinction." *International Business Times,* Dec. 3, 2015. http://www.ibtimes.co.uk/more-quarter-british-birds-risk-local-extinction-1531645.

Transition Network. www.transitionnetwork.org.

Transition Town Totnes. www.transitiontowntotnes.org.

Transition United States. "About Transition Streets." http://www.transitionus.org/about-transition-streets.

Transition United States. http://transitionus.org/transition-town-movement.

Transition United States. http://www.transitionus.org/.

Tree Hugger. www.treehugger.com.

Troster, Rabbi Lawrence. "Ten Jewish Teachings on Judaism and the Environment." http://www.greenfaith.org/religious-teachings/jewish-statements-on-the-environment/ten-jewish-teachings-on-judaism-and-the-environment.

Union of Concerned Scientists. "Early Warning Signs of Global Warming: Spring Comes Earlier." http://www.ucsusa.org/global_warming/science_and_impacts/impacts/early-warning-signs-of-global-10.html#.VF0KtjbnbIU.

_____. www.ucsusa.org/.

United Nations Environment Program, Ozone Secretariat. "The Montreal Protocol on Substances That Deplete the Ozone Layer." http://ozone.unep.org/en/treaties-and-decisions/montreal-protocol-substances-deplete-ozone-layer.

United States Department of Agriculture. www.usda.gov.

United States Department of Energy. "Energy Efficiency Trends in Residential and Commercial Buildings." http://apps1.eere.energy.gov/buildings/publications/pdfs/corporate/bt_stateindustry.pdf.

_____. "Grid Modernization and the Smart Grid." https://www.energy.gov/oe/activities/technology-development/grid-modernization-and-smart-grid.

United States Energy Information Administration. "Frequently Asked Questions." https://www.eia.gov/tools/faqs/faq.cfm?id=427&t=3.

United States Environmental Protection Agency. "Atmospheric Concentrations of Greenhouse

Gas." http://www.epa.gov/climatechange/images/indicator_downloads/ghg-concentrations-download2-2014.png.

University of California–Riverside. "Global Climate Change: Evidence and Causes." https://globalclimate.ucr.edu/resources.html.

Urbanek, Lauren, and Noah Horowitz. "The Trump Administration's Assault on Energy Efficiency Standards." Natural Resources Defense Council, March 05, 2019, https://www.nrdc.org/resources/trump-administrations-assault-energy-efficiency-standards.

Waters Foundation. http://www.watersfoundation.org/index.cfm?fuseaction=content.display&id=93.

Weise, Elizabeth, '*How Dare You?*' *Read Greta Thunberg's Emotional Climate Change Speech to UN and World Leaders,*" *USA Today*. Sept. 23, 2019, https://www.usatoday.com/story/news/2019/09/23/greta-thunberg-tells-un-summit-youth-not-forgive-climate-inaction/2421335001/.

Wikipedia.org.

Wilson, Stiv. "Plastics in the Great Lakes: A Guest Post." *5Gyres*, November 09, 2012, http://5gyres.org/posts/2012/11/09/plastics_in_the_great_lakes_a_guest_post.

Wisconsin Public Radio NPR News. "Rooftop Farming Is Getting Off the Ground." Sept. 27, 2013. http://www.npr.org/sectons/the Salt/2013/09/24/225745012/Why-Aren-T-There-More-Rooftop-Farms.

World Coal Association. "World Coal Facts 2013." www.worldcoal.org/bin/pdf/...pdf.../coal_facts_2013(11_09_2013).pdf.

World Elephant Day. *Why Care?* http://worldelephantday.org/about/elephants.

World Lion Day. https://worldlionday.com/african-lion/.

Yong, Ed. "Can We Save the World by Remixing Life?" *National Geographic.* http://phenomena.nationalgeographic.com/2013/04/11/can-we-save-the'world-by-remixing-life/.

Zelman, Joanna. "Coral Reefs May Be Gone by 2050." *Huffington Post*, Feb. 25, 2011, http://www.huffingtonpost.com/2011/02/25/coral-reefs-may-be-gone-b_n_827709.html?utm_source=DailyBrief&utm_campaign=022511&utm_medium=email&utm_content=NewsEntry&utm_term=Daily+Brief.

Zencey, Eric. "Two Schools and the Path to the Steady State." http://steadystate.org/learn/blog/ CASSE website, Center for the Advancement of a Steady State Economy, May 15, 2011.

# Index

Numbers in **bold italics** indicate pages with illustrations

**319**